Minimax Theory and Applications

Nonconvex Optimization and Its Applications

Volume 26

The titles published in this series are listed at the end of this volume.

Minimax Theory and Applications

Edited by

Biagio Ricceri

Department of Mathematics,
University of Catania
Catania, Italy

and

Stephen Simons

Department of Mathematics,
University of California at Santa Barbara,
Santa Barbara, California, U.S.A.

KLUWER ACADEMIC PUBLISHERS

DORDRECHT / BOSTON / LONDON

A C.I.P. Catalogue record for this book is available from the Library of Congress.

ISBN 0-7923-5064-2

Published by Kluwer Academic Publishers,
P.O. Box 17, 3300 AA Dordrecht, The Netherlands.

Sold and distributed in North, Central and South America
by Kluwer Academic Publishers,
101 Philip Drive, Norwell, MA 02061, U.S.A.

In all other countries, sold and distributed
by Kluwer Academic Publishers,
P.O. Box 322, 3300 AH Dordrecht, The Netherlands.

Printed on acid-free paper

Printed in the Netherlands

Contents

Preface

The present volume contains the proceedings of the workshop on "Minimax Theory and Applications" that was held during the week 30 September – 6 October 1996 at the "G. Stampacchia" International School of Mathematics of the "E. Majorana" Centre for Scientific Culture in Erice (Italy) .

The main theme of the workshop was minimax theory in its most classical meaning. That is to say, given a real-valued function f on a product space $X \times Y$, one tries to find conditions that ensure the validity of the equality

$$\sup_{y\in Y}\inf_{x\in X} f(x,y) = \inf_{x\in X}\sup_{y\in Y} f(x,y).$$

This is not an appropriate place to enter into the technical details of the proofs of minimax theorems, or into the history of the contributions to the solution of this basic problem in the last 7 decades. But we do want to stress its intrinsic interest and point out that, in spite of its extremely simple formulation, it conceals a great wealth of ideas. This is clearly shown by the large variety of methods and tools that have been used to study it.

The applications of minimax theory are also extremely interesting. In fact, the need for the ability to "switch quantifiers" arises in a seemingly boundless range of different situations. So, the good quality of a minimax theorem can also be judged by its applicability.

We hope that this volume will offer a rather complete account of the state of the art of the subject.

We first thank the "E. Majorana" Centre for the perfect organization of the workshop. We also thank the participants to the workshop and the contributors to this volume. Thanks are also due to Kluwer Academic Publishers for having offered us the opportunity to publish this volume. Finally, we wish to express our deepest gratitude to our colleague Professor Michele Frasca of the University of Catania who, with great patience and expertise, has taken upon himself the task of doing the technical work needed to prepare the volume.

Biagio Ricceri, University of Catania

Stephen Simons, University of California at Santa Barbara

NONLINEAR TWO FUNCTIONS MINIMAX THEOREMS

CAO-ZONG CHENG
Department of Mathematics
Computer Institute, Beijing Polytechnic
Beijing 100044, China

AND

BOR-LUH LIN
Department of Mathematics
University of Iowa
Iowa City, IA 52242, U.S.A.
e-mail: bllin@ math.uiowa.edu

1. Introduction

Let X and Y be nonempty sets. Let $f, g : X \times Y \to \mathbb{R}$ be functions. Throughout this paper, we assume that $f(x, y) \leq g(x, y)$ for all $(x, y) \in X \times Y$. A minimax theorem for f implies that, under certain conditions, the following equation holds:

$$(*) \qquad \inf_Y \sup_X f(x, y) = \sup_X \inf_Y f(x, y) .$$

A two functions minimax theorem for f and g implies that, under certain conditions, the following inequality holds:

$$(**) \qquad \inf_Y \sup_X f(x, y) \leq \sup_X \inf_Y g(x, y) ,$$

It is clear that (**) generalizes (*) by letting $f = g$. For $h : X \times Y \to$

B. Ricceri and S. Simons (eds.), Minimax Theory and Applications, 1–20.

$\mathbb{R}$, $A \subset X$, $B \subset Y$ and $\lambda \in \mathbb{R}$, let

$$h^* = \inf_Y \sup_X \ h(x,y) \ ,$$

$$h_* = \sup_X \inf_Y \ h(x,y) \ ,$$

$$Y(A,h,\leq,\lambda) = \{y : y \in Y, h(x,y) \leq \lambda \text{ for all } x \in A\} \ ;$$

$$Y(A,h,<,\lambda) = \{y : y \in Y, h(x,y) < \lambda \text{ for all } x \in A\} \ ;$$

$$X(B,h \leq,\lambda) = \{x : x \in X, h(x,y) \leq \lambda \text{ for all } y \in B\} \ ;$$

$$X(B,h <,\lambda) = \{x : x \in X, h(x,y) < \lambda \text{ for all } y \in B\} \ .$$

In the statements of theorems in the paper, unless specifically pointed out, we shall always assume that f, g are real-valued functions on $X \times Y$ satisfying the condition that $f \leqq g$ on $X \times Y$. The first two functions minimax theorem was obtained by K. Fan [6] in generalizing Sion's minimax theorem [38].

Theorem 1.1 [Sion] *Assume that*

(1) X is compact convex set in topological vector space and Y is a convex set in a topological vector space not necessarily the same space as X;

(2) For all y in Y, $f(\cdot,y)$ is quasiconcave on X, i.e., for all $\lambda \in \mathbb{R}$,

$$\{x : x \in X, \ f(x,y) \geq \lambda\}$$

is convex or empty in X;

(3) For all x in X, $f(x,\cdot)$ is quasiconvex on Y, i.e., for all $\lambda \in \mathbb{R}$,

$$\{y : y \in Y, f(x,y) \leq \lambda\}$$

is convex or empty in Y;

(4) For all y in Y, $f(\cdot,y)$ is upper semicontinuous on X and for all x in X, $f(x,\cdot)$ is lower semicontinuous on Y.

Then $(*)$ holds.

Theorem 1.2 [Fan] *Assume that*

(1) X and Y are compact convex sets in topological vector spaces, not necessarily the same space;

(2) For all y in Y, $f(\cdot,y)$ is quasiconcave on X;

(3) For all x in X, $g(x,\cdot)$ is quasiconvex on Y;

(4) For all y in Y, $g(\cdot,y)$ is upper semicontinuous on X and for all x in X, $f(x,\cdot)$ is lower semicontinuous on Y.

Then $(**)$ holds.

Remark. It was observed in [32] and [34] that the compactness of Y is not necessary and in (4), only need to assume that for all x in X, $f(x, \cdot)$ is lower semicontinuous on line segments in Y.

In Fan's two functions minimax theorem, it is assumed that the "lower" function f is "concave" in a certain sense and that the "upper" function g is "convex" in a certain sense. We shall call these kind of two functions minimax theorems type (A).

Another type of two functions minimax theorem was obtained by S. Simons [34] in extending minimax theorems of K. Fan [5] and H. König [21] to two functions.

Theorem 1.3 [Fan] *Assume that*

(1) X is a compact space and Y is a nonempty set;

(2) f is concavelike on X, i.e., for all x_1, x_2 in X and for all s in $[0, 1]$, there exists x_0 in X such that for all y in Y,

$$f(x_0, y) \geq sf(x_1, y) + (1 - s)f(x_2, y) ;$$

(3) f is convexlike on Y, i.e., for all y_1, y_2 in Y and for all t in $[0, 1]$, there exists y_0 in Y such that for all x in X,

$$f(x, y_0) \leq tf(x, y_1) + (1 - t)f(x, y_2) ;$$

(4) For all y in Y, $f(\cdot, y)$ is upper semicontinuous on X.

Then $(*)$ holds.

Theorem 1.4 [König] *Assume that*

(1) X is a compact space and Y is a nonempty set;

(2) There exists $s \in (0, 1)$ such that f is s-concavelike on X, i.e., for all x_1, x_2 in X, there exists x_0 in X such that for all y in Y,

$$f(x_0, y) \geq sf(x_1, y) + (1 - s)f(x_2, y) ;$$

(3) There exists $t \in (0, 1)$ such that f is t-convexlike on Y, i.e., for all y_1, y_2, there exists y_0 in Y such that for all x in X,

$$f(x, y_0) \leq tf(x, y_1) + (1 - t)f(x, y_2) ;$$

(4) For all y in Y, $f(\cdot, y)$ is upper semicontinuous on X.

Then $(*)$ holds.

Theorem 1.5 [Simons] *Assume that*

(1) X is a compact space and Y is a nonempty set;
(2) For all y in Y, $g(\cdot,y)$ is $\frac{1}{2}$−concavelike on X;
(3) For all x in X, $f(x,\cdot)$ is $\frac{1}{2}$−convexlike on Y;
(4) For all y in Y, $g(\cdot,y)$ is upper semicontinuous on X.

Then (**) holds.

In Simons' two functions minimax theorem, it is assumed that the "lower" function f is "convex" in a certain sense and the "upper" function g is "concave" in a certain sense. We shall call these kind of two functions minimax theorems type B. Observed that type A and type B have "oppositive" geometric pictures.

Another type of two functions minimax theorem was obtained by Bor-Luh Lin and Xiu-Chi Quan in generalizing an one function symmetric minimax theorem of Lin-Quan [24].

Theorem 1.6 [24] *Assume that*

(1) X is a compact space and Y is a nonempty set;
(2) There exists s in $(0,1)$ such that f is s-concave on X, i.e., for all x_1, x_2 in X, there exists x_0 in X such that for all y in Y,

$$f(x_0,y) \geq s\max\{f(x_1,y), f(x_2,y)\} + (1-s)\min\{f(x_1,y), f(x_2,y)\} \ ;$$

(3) There exists t in $(0,1)$ such that f is t-convex on Y, i.e., for all y_1, y_2 in Y, there exists y_0 in Y such that for all x in X,

$$f(x,y_0) \leq t\max\{f(x,y_1), f(x,y_2)\} + (1-t)\min\{f(x,y_1), f(x,y_2)\} \ ;$$

(4) For all y in Y, $f(\cdot,y)$ is upper semicontinuous on X.

Then (*) holds.

Theorem 1.7 [25] *Assume that,*

(1) X is a compact space and Y is a nonempty set;
(2) There exists s in $(0,1)$ such that for all x_1, x_2 in X, there exists x_0 in X such that for all y in Y,

$$f(x_0,y) \geq s\max\{f(x_1,y), g(x_2,y)\} + (1-s)\min\{f(x_1,y), g(x_2,y)\} \ ;$$

(3) There exists t in $(0,1)$ such that for all y_1, y_2 in X, there exists y_0 in Y such that for all x in X,

$$g(x,y_0) \leq t\max\{f(x,y_1), g(x,y_2)\} + (1-t)\min\{f(x,y_1), g(x,y_2)\} \ ;$$

(4) For all y in Y, $f(\cdot,y)$ and $g(\cdot,y)$ are upper semicontinuous on X.
Then $(**)$ holds.

Observed in conditions (2) and (3) of Theorem 1.7, instead of assuming either f or g is s-concave or t-convex, the conditions involved both f and g. We shall call such two functions minimax theorems of mixed-type.

2. Nonlinear Minimax Theorems

We shall concentrate our study to minimax theorems without linear structure. In this section, we list the minimax theorems that we shall consider for possible extending to two functions minimax theorems. We list the theorems in chronological order.

Theorem 2.1 [5] *Assume that*
(1) X is a compact space and Y is a nonempty set;
(2) f is concavelike on X;
(3) f is convexlike on Y;
(4) For all y in Y, $f(\cdot,y)$ is upper semicontinuous on X.
Then $(*)$ holds.

Theorem 2.2 [5] *Assume that*
(1) X and Y are nonempty sets;
(2) f is concavelike on X;
(3) f is convexlike on Y;
(4) f is almost periodic on $X \times Y$.
Then $(*)$ holds.

Recall that a bounded function f is right (resp. left) almost periodic on $X \times Y$ if for all $\epsilon > 0$ there exists a finite covering of Y (resp. X), $Y = \cup_{i=1}^{m} Y_i$ (resp. $X = \cup_{j=1}^{n} X_j$) such that $|f(x,y') - f(x,y'')| < \epsilon$ for all x in X and for all y', y'' which are elements in Y_i for some i (resp. $|f(x',y) - f(x'',y)| < \epsilon$ for all y in Y and for all x', x'' which are elements in X_j for some j). It is easy to show that f is right almost periodic on $X \times Y$ if and only if f is left almost periodic on $X \times Y$ and we shall simply say that f is almost periodic on $X \times Y$.

Theorem 2.3 [21] *Assume that*
(1) X is a compact space and Y is a nonempty set;
(2) There exists s in $(0,1)$ such that f is s-concavelike on X;

(3) There exists t in $(0,1)$ such that f is t-convexlike on Y;
(4) For all y in Y, $f(\cdot, y)$ is upper semicontinuous on X.
Then $(*)$ holds.

Theorem 2.4 [7] *Assume that*
(1) X is a compact space and Y is a nonempty set;
(2) For any $y_1, y_2, \ldots, y_n$ in Y and for all $\lambda \in \mathbb{R}$, $\cap_{i=1}^n \{x : x \in X, f(x, y_i) \geq \lambda\}$ is connected or empty;
(3) There exists t in $(0,1)$ such that f is t-convex on Y;
(4) For all y in Y, $f(\cdot, y)$ is upper semicontinuous on X.
Then $(*)$ holds.

Theorem 2.5 [8] *Assume that*
(1) X is a compact space and Y is a topological space;
(2) For any $y_1, y_2, \ldots, y_n$ in Y and for all λ in $\mathbb{R}$, $\cap_{i=1}^n \{x : x \in X, f(x, y_i) \geq \lambda\}$ is connected or empty;
(3) f is submaximum on Y, i.e., for any y_1, y_2 in Y, there exists a continuous map $S : [0,1] \to Y$ such that $S(0) = y_1$, $S(1) = y_2$ and for any $b \in [a, c] \subset [0,1]$, $f(x, S(b)) \leq \max\{f(x, S(a)), f(x, S(c))\}$ for all x in X;
(4) For all y in Y, $f(\cdot, y)$ is upper semicontinuous on X.
Then $(*)$ holds.

Theorem 2.6 [10] *Assume that*
(1) X is a compact space and Y is a topological space;
(2) There exists s in $(0,1)$ such that f is s-concave on X;
(3) f is submaximum on Y;
(4) For all y in Y, $f(\cdot, y)$ is upper semicontinuous on X.
Then $(*)$ holds.

Theorem 2.7 [24] *Assume that*
(1) X is a compact space and Y is a nonempty set;
(2) There exists s in $(0,1)$ such that f is s-concave on X;
(3) There exists t in $(0,1)$ such that f is t-convex on Y;
(4) For all y in Y, $f(\cdot, y)$ is upper semicontinuous on X.
Then $(*)$ holds.

Theorem 2.8 [19] *Assume that*
(1) X is a compact space and Y is a topological space;
(2) For any $y_1, y_2, \ldots, y_n$ in Y and for any $\lambda_1, \lambda_2, \ldots, \lambda_n$ in $\mathbb{R}$, $\cap_{i=1}^n \{x : x \in X, f(x, y_i) \geq \lambda_i\}$ is connected or empty;

(3) For any y_1, y_2 in Y, there exists a connected subset C in Y containing y_1, y_2 and for any y in C and any x in X.

$$f(x,y) \leq \max\{f(x,y_1, f(x,y_2)\} \ ;$$

(4) For any closed set A in X and for any $\lambda < \inf_Y \sup_A f(x,y)$, the level map

$$L_\lambda(y) = \{x : x \in A, f(x,y) \geq \lambda\}$$

is quarter-continuous on A, i.e., for all y in Y and for any open set G in A, if $L_\lambda(y) \subset G$ then there is a neighborhood U of y in Y such that $L_\lambda(u) \cap G \neq \phi$ for all u in U.

Then $(*)$ holds.

Theorem 2.9 [36] *Assume that*

(1) X is a compact space and Y is a nonempty set;

(2) For any $y_1, y_2, \cdots, y_n$ in Y and for all λ in $\mathbb{R}$, $\cap_{i=1}^n \{x : x \in X, f(x,y_i) \geq \lambda_i\}$ is connected or empty;

(3) f is upward on Y, i.e., for any $\epsilon > 0$, there exists $\delta > 0$ such that for all y_1, y_2 in Y, there is y_0 in Y with the properties that for all x in X.

$$f(x,y_0) \leq \max\{f(x,y_1), f(x,y_2)\} \ ,$$

and if

$$|f(x,y_1) - f(x,y_2)| \geqq \epsilon \text{ then } f(x,y_0) \leq \max\{f(x,y_1), f(x,y_2)\} - \delta \ ;$$

(4) For all y in Y, $f(\cdot, y)$ is upper semicontinuous on X.

Then $(*)$ holds.

Theorem 2.10 [35] *Assume that*

(1) X is a compact space and Y is a set;

(2) f is downward on X, i.e., for any $\epsilon > 0$, there exists $\delta > 0$ such that for all x_1, x_2 in X, there is x_0 in X with the properties that for all y in Y,

$$f(x_0,y) \geq \min\{f(x_1,y), f(x,y_2)\} \ ,$$

and if

$$|f(x_1,y) - f(x_2,y)| \geq \epsilon \text{ then } f(x_0,y) \geq \min\{f(x_1,y), f(x_2,y)\} + \delta \ ;$$

(3) f is upward on Y ;

(4) For all y in Y, $f(\cdot, y)$ is upper semicontinuous on X.

Then $(*)$ holds.

Theorem 2.11 [37] *Assume that*

(1) X *is a compact space and* Y *is a nonempty set;*

(2) *For all* $a < b < p$, *there exists an* (a, b, p)*-descending staircase for* X, *i.e., there exist* $\{b(0), \cdots, b(N)\}$ *in* $\mathbb{R}$ *such that*
(i) $p = b(0) > \cdots > b(N) = b$;
(ii) *for all* y_1, y_2 *in* Y, *there is* y_0 *in* Y *with the properties that for all* x *in* X.

$$f(x, y_0) \leq \max\{f(x, y_1), f(x, y_2)\}$$

and if for some x *in* X *and some* $n \leq N$,

$$\min\{f(x, y_1), f(x, y_2)\} \leq a < b < \max\{f(x, y_1), f(x, y_2)\} \leq b(n-1)$$

then $f(x, y_0) \leq b(n)$,

(3) *For all* $q < a < b$, *there exists a* (q, a, b)*-ascending staircase for* f, *i.e., there exists* $\{a(0), \cdots, a(N)\}$ *in* $\mathbb{R}$ *such that*
(i) $q = a(0) < \cdots < a(N) = a$;
(ii) *for all* x_1, x_2 *in* X, *there is* x_0 *in* X *with the properties that for all* y *in* Y,

$$f(x_0, y) \geq \min\{f(x_1, y), f(x_2, y)\}$$

and if for some y *in* Y *and some* $n \leq N$,

$$a(n-1) \leq \min\{f(x_1, y), f(x_2, y)\} < a < b \leq \max\{f(x_1, y), f(x_2, y)\}$$

then $a(n) \leq f(x_0, y)$

(4) *For all* y *in* Y, $f(\cdot, y)$ *is upper semicontinuous on* X.

Then $(*)$ *holds.*

Theorem 2.12 [22] *Assume that*

(1) X *is a compact space and* Y *is connected;*

(2) *For all* $y_1, y_2, \ldots, y_n$ *in* Y *and for all* $\lambda \in \mathbb{R}$, $\cap_{i=1}^{n}\{x : x \in X, f(x, y_i) \geq \lambda\}$ *is connected or empty;*

(3) *For any subset* H *in* X *and for all* $\lambda \in \mathbb{R}$, $\cap_{x \in H}\{y : y \in Y, f(x, y) \leq \lambda\}$ *is connected or empty;*

(4) *For all* y *in* Y, $f(\cdot, y)$ *is upper semicontinuous on* X *and for all* x *in* X, $f(x, \cdot)$ *is lower semicontinuous on* Y.

Then $(*)$ *holds.*

Theorem 2.13 [33] *Assume that*

(1) *There exists a continuous bijection from* X *onto* $[0, 1]$ *and* Y *is a topological space;*

(2) *For all* y *in* Y *and for all* $\lambda > \sup_X \inf_Y f(x, y)$, $\{x : x \in X, f(x, y) > \lambda\}$ *is connected or empty;*

(3) For all x in X and for all $\lambda > \sup_X \inf_Y f(x,y)$, $\{y : y \in Y, f(x,y) \leq \lambda\}$ is connected or empty;

(4) For all y in Y, $f(\cdot,y)$ is upper semicontinous on X and for all x in X, $f(x,\cdot)$ is lower semicontinuous on Y (or f is upper semicontinuous on $X \times Y$, or f is lower semicontinuous on $X \times Y$ and Y is compact),

Then $(*)$ holds.

Theorem 2.14 [17] *Assume that there exists $\lambda \in \mathbb{R} \cup \{\pm\infty\}$ and*

(1) X is a nonempty set and Y is a compact space;

(2) For all x_1, x_2 in X, there exists x_0 in X such that for all y in Y,

$$f(x_0, y) \geq median(f(x_1, y), f(x_2, y), \lambda)$$

where median $(f(x_1, y), f(x_2, y), \lambda)$ is the middle number of $f(x_1, y), f(x_2$ and λ;

(3) For all y_1, y_2 in Y, there exists y_0 in Y such that for all x in X.

$$f(x, y_0) \leq median(f(x, y_1), f(x, y_2), \lambda) \ ;$$

(4) For all x in X, $f(x,\cdot)$ is lower semicontinuous on X.

Then $(*)$ holds.

In the rest of the section, we discuss some minimax theorems involving the interval spaces introduced by L. L. Stacho in 1980 [39]. A topological space X is called an interval space if for all x_1, x_2 in X, this is a connected set $[x_1, x_2]$ in X, called interval in X, such that $x_i \in [x_1, x_2] = [x_2, x_1]$, $i = 1, 2$. If, in addition that $[x_1, x_2]$ is pathwise connected then X is called a strong interval space. A set K in X is convex if $[x_1, x_2] \subset K$ for all x_1, x_2 in K. An interval space X is said to be Dedekind complete if for all x_1, x_2 in X and for any convex sets K_1 and K_2 in X with $x_i \in K_i$, $i = 1, 2$ and $[x_1, x_2] \subset K_1 \cup K_2$, then there exists $x \in K_i$ for some $i = 1, 2$ such that $[x_j, x] \backslash \{x\} \subset K_j$ for $j \neq i$ and $j = 1, 2$. A map $f : X \to \mathbb{R}$ is called quasiconcave (resp. quasiconvex) if for all λ in $\mathbb{R}$, $\{x : x \in X, f(x) \geq \lambda\}$ (resp. $\{x : x \in X, f(x) \leq \lambda\}$) is convex. For minimax theorems on interval spaces, we refer to [18,20,39]. We list only the following result on interval spaces.

Theorem 2.15 [3] *Let $f : X \times Y \to Z$. Assume that*

(1) X is an interval space, Y is a strong interval space and Z is an order complete order dense linear space such that there exists $z_0 < \inf_Y \sup_X f(x,y)$ and y_0 in Y with the property that $\{x : x \in X, f(x, y_0) \geq z_0\}$ is compact in X.

(2) *For all* $y_1, y_2, \ldots, y_n$ *in* Y *and for all* z *in* Z, $\cap_{i=1}^n \{x : x \in X, f(x, y_i) > z\}$ *is connected or empty;*
(3) *For all* x *in* X, $f(x, \cdot)$ *is quasiconvex on* X;
(4) *For all* y *in* Y, $f(\cdot, y)$ *is upper semicontinuous on* X *and for all* x *in* X, $f(x, \cdot)$ *is lower semicontinuous on all intervals of* Y,

Then $(*)$ *holds.*

3. Two Functions Minimax Theorems of Type A/Type B

There are not too many two functions minimax theorems of type A/type B. We shall first discuss minimax theorems that have two functions extension of type A/type B and then give examples to show that most of the minimax theorems in section II cannot be extended to two functions minimax theorems of type A/type B.

In the introduction, we have stated Fan's generalization of Sion's theorem to two functions minimax theorem. The best result in this direction is the following theorem due to Fon-Che Liu [32].

Theorem 3.1 *Assume that*
(1) X *is a compact convex set of a topological vector space and* Y *is a convex set of a vector space;*
(2) *For all* y *in* Y, $f(\cdot, y)$ *is quasiconcave on* X;
(3) *For all* x *in* X, $g(x, \cdot)$ *is quasiconvex on* Y;
(4) *For all* y *in* Y, $g(\cdot, y)$ *is upper semicontinuous on* X *and for all* x *in* X, $f(x, \cdot)$ *is lower semicontinuous on convex hull of all finite subset in* Y.

Then $(**)$ *holds.*

In the introduction, we have stated Simons' generalization of Fan and König's minimax theorems on concavelike-convexlike.

The following two theorems are recently extensions to two functions minimax theorems on minimax theorems involving concavelike and convexlike.

Theorem 3.2 [1] *Assume that*
(1) X *is a topological space and* Y *is a set such that there exists* $\lambda_0 \leq \inf_Y \sup_X f(x, y)$ *such that for all* y *in* Y, $\{x : x \in X, g(x, y) \geq \lambda_0\}$ *is compact,*
(2) *There exists* $s \in (0, 1)$ *such that* g *is* s*-concavelike on* X;
(3) *There exists* $t \in (0, 1)$ *such that* f *is* t*-convexlike on* Y;

(4) For all y in Y, $g(\cdot, y)$ is upper semicontinuous on X.
Then $(**)$ holds.

Theorem 3.3 [1] *Assume that*
(1) X and Y are sets;
(2) There exists $s \in (0,1)$ such that g is s-concavelike on X;
(3) There exists $t \in (0,1)$ such that f is t-convexlike on Y;
(4) f and g are almost periodic on $X \times Y$,
Then $(**)$ holds.

In the Erice Workshop, J. Kindler has suggested that functions with certain double limits properties may possess two functions minimax theorems on concavelike-convexlike. We shall now present some results of this kind. The authors wish to thank Professor J. Kindler for calling attention to this kind of problems.

Let X and Y be nonempty sets and let $f, g : X \times Y \to \mathbb{R}$ with the property that $f \leqq g$ on $X \times Y$. We write

$$f^* = \inf_Y \sup_X f(x,y) \ , \qquad f_* = \sup_X \inf_Y f(x,y) \ ,$$
$$\bar{f}^* = \sup_{A \subset X} \inf_Y \sup_{x \in A} f(x,y) \ , \bar{f}_* = \inf_{b \subset Y} \sup_X \inf_{y \in B} f(x,y) \ ,$$

where A, B is finite subset in X and Y respectively. It is clear that $f_* \leq \min(\bar{f}_*, \bar{f}^*) \leq \max(\bar{f}_*, \bar{f}^*) \leq f^*$. As in the proof of Theorem 2 in [1], using the classical von Neumann minimax theorem, the following Lemmas can be proved.

Lemma 1. If g is s-concavelike on X and f is t-convexlike on Y for some $s,\ t \in (0,1)$, then $\bar{f}^* \leq \bar{g}_*$. In particular, if f is s-concavelike on X and is t-convexlike on Y for some $s,\ t \in (0,1)$, then $\bar{f}^* \leq f_* \leq f^* \leq \bar{f}_*$.

Lemma 2. If f is almost periodic, then $\bar{f}_* \leq f_*$ and $f^* \leq \bar{f}_*$.

Lemma 3. If for all sequences $\{x_i\}$ in X and $\{y_j\}$ in Y,

$$\liminf_j \liminf_i f(x_i, y_j) \leq \limsup_i \limsup_j g(x_i, y_j)$$

then $\bar{f}_* \leq \bar{g}^*$.

Proof. Suppose that $\bar{f}_* > \beta > \alpha > \bar{g}^*$. Then for finite subsets $A \subset X$ and $B \subset Y$, we have

(1) $\sup_{x\in X}\inf_{y\in B} f(x,y) > \beta$,

(2) $\alpha > \inf_{y\in Y}\sup_{x\in A} g(x,y)$.

Take $A = \{x_1\}$. By (2) there exists $y_1 \in Y$ such that $g(x_1, y_1) < \alpha$. Take $B = \{y_1\}$. By (1) there exists $x_2 \in X$ such that $f(x_2, y_1) > \beta$. Take $A = \{x_1, x_2\}$ and continue by induction, we have sequenced $\{x_i\} \subset X$ and $\{y_j\} \subset Y$ such that $g(x_i, y_j) < \alpha(i \le j)$ and $f(x_i, y_j) > \beta(i > j)$. Thus

$$\lim_j \inf \lim_i \inf f(x_i, y_j) \ge \beta > \alpha \ge \lim_i \sup \lim_j \sup g(x_i, y_j) .$$

This contradicts to the hypothesis and completes the proof. □

Lemma 4.

(i) If for all sequences $\{x_i\} \subset X$ and $\{y_j\} \subset Y$, $\lim_j \inf f(x_j, y_j) \le \lim_i \sup \lim_j \sup g(x_i, y_j)$, then $f^* \le \bar{g}^*$;

(ii) If for all sequences $\{x_i\} \subset X$ and $\{y_j\} \subset Y$, $\lim_j \inf \lim_i \inf f(x_i, y_j) \le \lim_i \sup g(x_i, y_i)$, then $\bar{f}_* \le g_*$.

Proof.

(i) Suppose that $f^* > \beta > \alpha > \bar{g}^*$. Then for all finite sets $A \subset X$, we have

$$\inf_{y\in Y}\sup_{x\in A} g(x,y) < \alpha \ , \tag{3}$$

$$\beta < \inf_Y \sup_X f(x,y) \ . \tag{4}$$

Take $A = \{x_0\}$. By (3), there is $y_1 \in Y$ with $g(x_0, y_1) < \alpha$. For y_1, by (4), there is $x_1 \in X$ such that $f(x_1, y_1) > \beta$. Continue by induction, there exist sequences $\{x_i\} \subset X$ and $\{y_j\} \subset Y$ such that $g(x_i, y_j) < \alpha(i < j)$ and $f(x_j, y_j) > \beta(j \in \mathbb{N})$. Thus

$$\lim_i \sup \lim_j \sup g(x_i, y_j) \le \alpha < \beta < \lim_j \inf f(x_j, y_j) .$$

This contradicts to the assumption. Thus $f^* \le \bar{g}^*$.

The proof of (ii) is similar. We omit it. □

Theorem 3.4. *Assume that,*

(1) g is s-concavelike on X and f is t-convexlike on Y for some s, t in $(0,1)$;

(2) f satisfies one of the following conditions:

(a_1) f is almost periodic on $X \times Y$;

(a_2) for all sequences $\{x_i\} \subset X$ and $\{y_j\} \subset Y$,

$$\limsup_i \limsup_j f(x_i, y_j) \geq \liminf_j f(x_j, y_j) ;$$

(a_3) Y is compact and for all x, $f(x, \cdot)$ is lower semicontinuous on Y;

g satisfies one of the following conditions:

(b_1) g is almost periodic;

(b_2) for all sequences $\{x_i\} \subset X$ and $\{y_j\} \subset Y$,

$$\liminf_j \liminf_i g(x_i, y_j) \leq \limsup_i g(x_i, y_i) ;$$

(b_3) X is compact and for all $y \in Y$, $g(\cdot, y)$ is upper semicontinuous on X.

Then ($**$) holds.

Proof. $\bar{f}^* \leq \bar{g}_*$ follows from (1) and Lemma 1. $f^* \leq \bar{f}^*$ follows from (a_1) and Lemma 2, or from (a_2) and Lemma 4, or from (a_3). $\bar{g}_* \leq g_*$ follows from (b_1) and Lemma 2, or from (b_2) and Lemma 4, or from (b_3). Thus $f^* \leq \bar{f}^* \leq \bar{g}_* \leq g_*$. □

Using the Lemmas, the following one function minimax theorems and two functions minimax theorems can be proved:

Theorem 3.5 *Assume that*

(1) X and Y are nonempty sets;

(2) There exist s and t in $(0, 1)$ such that f is s-concavelike on X and t-convexlike on Y;

(3) f satisfies any one of the following six conditions:

(a_1), (a_2), (a_3) or

(a_4) for all sequences $\{x_i\}$ in X and $\{y_j\}$ in Y,

$$\liminf_j \liminf_i f(x_i, y_j) \leq \limsup_i f(x_i, y_i) ;$$

(a_5) X is compact and for all y in Y, $f(\cdot, y)$ is upper semicontinuous on X;

(a_6) for all sequences $\{x_i\}$ in X and $\{y_j\}$ in Y,

$$\liminf_j \liminf_i f(x_i, y_j) \leq \limsup_i \limsup_j f(x_i, y_j) .$$

Then ($*$) holds.

Theorem 3.6 *Assume that*

(1) *X and Y are nonempty sets;*

(2) *There exist s and t in $(0,1)$ such that g is s-concavelike on X and t-convexlike on Y,*

(3) *f satisfies one of (a_1), (a_2) or (a_3).*

Then $(**)$ holds.

Theorem 3.7 *Assume that*

(1) *There exist s and t in $(0,1)$ such that f is s-concavelike on X and t-convexlike on Y;*

(2) *g satisfies one of (b_1), (b_2) or (b_3).*

Then $(**)$ holds.

Theorem 3.8 *Assume that*

(1) *There exist s_1, s_2, and t_1, t_2 in $(0,1)$ such that f is s_1-concavelike on X and t_1-convexlike on Y and g is s_2-concavelike on X and t_2-convexlike on Y;*

(2) *For all sequences $\{x_i\}$ in X and $\{y_j\}$ in Y,*

$$\liminf_j \liminf_i f(x_i, y_j) \leq \limsup_i \limsup_j g(x_i, y_j) .$$

Then $(**)$ holds.

* * * * *

In [33], B. Ricceri gave an alternative principle for multifunctions and used it to establish some minimax theorems. Using the analogous argument, some two functions version of these minimax theorems remain true. We give only the following theorem that generalizes Theorem 2.13 in (II).

Theorem 3.9 *Assume that*

(1) *X and Y are topological spaces with X admitting a continuous bijection onto $[0,1]$;*

(2) *For each $\lambda > g_*$, the sets $\{x : x \in X, f(x,y) > \lambda\}$ are connected;*

(3) *For each $\lambda > g_*$, the sets $\{y : y \in Y, g(x,y) < \lambda\}$ are connected;*

(4) *One of the following conditions is satisfied:*

(i) for all $x \in X$, $f(x,\cdot)$ is lower semicontinuous on Y and for all y in Y, $g(\cdot,y)$ is upper semicontinuous on X;

(ii) f is upper semicontinuous on $X \times Y$ and for all y in Y, $g(\cdot,y)$ is upper semicontinuous on X;

(iii) Y is compact and for all x in X, $f(x,\cdot)$ is lower semicontinuous on Y, g is lower semicontinuous on $X \times Y$.
Then (**) holds.

Proof. Suppose that $f^* > g_*$. Choose α, β in $\mathbb{R}$ such that $f^* > \beta > \alpha > g_*$. Let

$$S = \begin{cases} \{(x,y) : g(x,y) < \alpha\} & \text{if (i) or (ii) is satisfied,} \\ \{(x,y) : g(x,y) \leq \alpha\} & \text{if (iii) is satisfied,} \end{cases}$$

$$T = \begin{cases} \{(x,y) : f(x,y) > \beta\} & \text{if (i) or (ii) is satisfied,} \\ \{(x,y) : f(x,y) \geq \beta\} & \text{if (iii) is satisfied .} \end{cases}$$

Apply Ricceri's alternative principle for multifunctions to complete the proof. □

* * * * *

J. Kindler's Dini-Dax minimax theorem, i.e., Theorem 2.14 in (II) allows four different two functions minimax theorems of type (A)/(B).

Theorem 3.10 *Assume that there exists $\lambda \in \mathbb{R} \cup \{\pm\infty\}$ and*
(1) X is a nonempty set and Y is a compact space;
(2) For all x_1, x_2 in X, there exists x_0 in X such that for all y in Y,

$$f(x_0, y) \geq \text{ median } \{f(x_1,y), f(x_2,y), \lambda\} \ ;$$

(3) For all y_1, y_2 in Y, there exists y_0 in Y such that for all x in X.

$$g(x, y_0) \leq \text{ median } \{g(x,y_1), g(x,y_2), \lambda\} \ ;$$

(4) For all x in X, $f(x,\cdot)$ and $g(x,\cdot)$ are lower semicontinuous on Y.
Then (**) holds.

Proof. Suppose $f^* > g_*$. Choose $\alpha \in \mathbb{R}$ such that $f^* > \alpha > g_*$. If $\lambda > \alpha$, we shall show that for any finite subset A in X, $Y(A, f, \leq, \alpha) = \{y : y \in Y,\ f(x,y) \leq \alpha \text{ for all } x \in A\} \neq \phi$.

In fact, if card $A = 1$, since $\alpha > g_* \geq f_*$, $Y(A, f, \leq, \alpha) \neq \phi$. Suppose $Y(A, f, \leq, \alpha) \neq \phi$ for all card $A \leq n$ and in case card $A = n+1$, let $A = B \cup \{x_1, x_2\}$ and $x_1 \neq x_2$, $x_i \notin B, i = 1,2$. Take $x_0 \in X$ according to (2) and let $y_0 \in Y(B \cup \{x_0\}, f, \leq, \alpha)$. Then, since $\lambda > \alpha$, we have

$$\begin{aligned} \alpha \geq f(x_0, y_0) &\geq \text{median } \{f(x_1,y_0), f(x_2,y_0), \lambda\} \\ &\geq \max\{f(x_1,y_0), f(x_2,y_0)\} \ . \end{aligned}$$

Thus $y_0 \in Y(A, f, \leq, \alpha)$. Since Y is compact and $f(x, \cdot)$ is lower semicontinuous, we have $Y(X, f, \leq, \alpha) \neq \phi$. Then $f^* \leq \alpha$, contradicts to $f^* > \alpha$.

If $\lambda \leq \alpha$, we shall show that $Y(A, g, \leq, \alpha) \neq \phi$ for all finite subset A in X.

In fact, if card $A = 1$, since $\alpha > g_*$, $Y(A, g, \leq, \alpha) \neq \phi$. Suppose $Y(A, g, \leq, \alpha) \neq \phi$ for all A, card $A \leq n$. Let $A \subset X$ and card $A = n+1$. Let $A = B \cup \{x_1, x_2\}$, $x_1 \neq x_2$ and $x_i \notin B$, $i = 1, 2$. Take $y_i \in Y(B \cup \{x_i\}, g, \leq, \alpha)$, $i = 1, 2$ and take $y_0 \in Y$ according to (3). Then $y_0 \in Y(A, g, \leq, \alpha)$. Since Y is compact and $g(x, \cdot)$ is lower semicontinuous for all x in X, we have $Y(X, g, \leq, \alpha) \neq \phi$. Then $\alpha \geq g^* \geq f^*$, contradicts to $f^* > \alpha$. □

Using similar argument as in Theorem 3.10, the following result can be proved:

Theorem 3.11 *Assume that in Theorem 3.10, (2) and (3) are replaced by*

(2') For all x_1, x_2 in X, there exists x_0 in X such that for all y in Y,

$$g(x_0, y) \geq \text{ median } \{g(x_1, y), g(x_2, y), \lambda\} \ ;$$

(3') For all y_1, y_2 in Y, there exists y_0 in Y such that for all x in X.

$$f(x, y_0) \leq \text{ median } \{f(x, y_1), f(x, y_2), \lambda\} \ .$$

Then (**) holds.

In case that X is also compact, by replacing $g(x, \cdot)$ lower semicontinuous by $g(\cdot, y)$ upper semicontinuous for all y in Y. Using the similar arguments, the following results holds:

Theorem 3.12 *Assume that there exists $\lambda \in \mathbb{R} \cup \{\pm\infty\}$ and*

(1) X and Y are compact spaces;

(2) and (3) in Theorem 3.10 or (2') and (3'), in Theorem 3.11;

(3) For all x in X, $f(x, \cdot)$ is lower semicontinuous on Y and for all y in $Y, g(\cdot, y)$ is upper semicontinuous on X.

Then (**) holds.

* * * * *

In the rest of the section, we shall examine the example given by I. Joo and G. Kassay in [13]. I. Joo and G. Kassay gave an example to show that Simons' upward-downward minimax theorem cannot be extended to two functions minimax theorem in type (B). In fact, we shall show that their example can be used to show that several minimax theorems in (II) cannot be extended to two functions minimax theorems of type (B).

Example 1. [13] Let $X = Y = [-1, 1]$ and let $f, g : X \times Y \to \mathbb{R}$ given by

$$f(x,y) = \begin{cases} (y-2)x - 1, & \text{if } (x,y) \in [-1,0] \times [-1,0] \\ (3y-2)x - 1, & \text{if } (x,y) \in [-1,0] \times [0,1] \\ (3y+2)x - 1, & \text{if } (x,y) \in (0,1] \times [-1,0] \\ (y+2)x - 1, & \text{if } (x,y) \in (0,1] \times [0,1] \ , \end{cases}$$

$$g(x,y) = \begin{cases} (3x+2)y + 1, & \text{if } (x,y) \in [-1,0] \times [-1,0] \\ (x-2)y + 1, & \text{if } (x,y) \in [-1,0] \times (0,1] \\ (x+2)y + 1, & \text{if } (x,y) \in (0,1] \times [-1,0] \\ (3x-2)y + 1, & \text{if } (x,y) \in (0,1] \times (0,1] \ . \end{cases}$$

Clearly $f \le g$ on $X \times Y$ and
(i) X and Y are compact spaces,
(ii) for all $x \in X$ and for all $y \in Y$, $f(x,\cdot)$ and $g(\cdot,y)$ are strictly monotone or constant, hence for all $\lambda \in \mathbb{R}$ and for all $A \subset X$, $B \subset Y$, $Y(A, f, \le, \lambda)$ and $X(B, g, \ge, \lambda)$ are connected or empty;
(iii) f and g are continuous on $X \times Y$.
We claim that
(iv) f is $\frac{3}{4}$-convex on Y. Hence f is upward on Y.
It is straight forward to check that for any given y_1, y_2 in Y, let $y_0 = \frac{y_1+y_2}{2}$, then for all $x \in X$,

$$f(x, y_0) \le \frac{3}{4} \max\{f(x,y_1), f(x,y_2) + \frac{1}{4} \min\{f(x,y_1), f(x,y_2)\} \ .$$

Similarly, it can be proved that g is $\frac{1}{4}$-concave on X and so g is downward on X.

However $f^* = 1$ and $g_* = -1$. Thus Example 1 shows that Theorem 2.4 [7], Theorem 2.7 [24], Theorem 2.9 [36], Theorem 2.10 [35], Theorem 2.12 [22] and Theorem 2.13 [33] cannot be extended to two functions minimax theorems of type B.

Example 2. Let $X = Y = [0, 1]$ and $f, g : X \times Y \to \mathbb{R}$ be given by

$$f(x,y) = \begin{cases} \frac{1}{2}(1-y) + x & \text{if } x \in [0,1] \text{ and } x \le y \le 1 \\ \frac{1}{2}(1-x) + y & \text{if } x \in [0,1] \text{ and } x \ge y \ge 0 \end{cases}$$

$$g(x,y) = \begin{cases} \frac{1}{2}(1-y) + x & \text{if } x \in [0,\frac{1}{4}] \text{ and } y \in [0,1] \\ \frac{1}{2}(1-x) + y & \text{if } x \in (\frac{1}{4},1] \text{ and } y \in [0,1] \ . \end{cases}$$

clearly $f \le g$ on $X \times Y$ and

(i) X and Y are compact spaces;

(ii) for all $x \in X$, $g(x, \cdot)$ is linear and so is convex and upward on Y and for all $\lambda \in \mathbb{R}$ and for any $A \subset X$, $Y(A, g, \le, \lambda)$ is connected or empty;

(iii) for all $y \in Y$, $f(\cdot, y)$ is concave and so is $\frac{1}{2}$- concave and downward on X;

(iv) f is continuous on $X \times Y$ and for all $x \in X$, $g(x, \cdot)$ is lower semicontinuous on Y.

However $f^* = \frac{1}{2}$ and $g_* = \frac{3}{8}$. Example 2 shows that Theorem 2.4[7], Theorem 2.7[24] and Theorem 2.9[36] cannot be extended to two functions minimax theorems of type A.

We believe only very few minimax theorems can be extended to two functions minimax theorems of either type A or type B.

4. Two Functions Minimax Theorems of Mixed Type

The first two functions minimax theorem of mixed type was given by [25], as stated in Theorem 1.7 in extending Theorem 2.7[24]. Subsequently, Theorem 2.8[19] was extended in [27], Theorem 2.11[37] in [26] and recently, Theorem 2.5 and Theorem 2.15 in [29]. In [29] we have showed that Theorem 2.4[7] can be extended to two functions minimax theorems of mixed type.

To conclude this paper, we remark that the Dini-Dax minimax theorem, Theorem 2.14[17] can be extended to a two functions minimax theorem of mixed type.

Theorem 4.1 *Assume that there exists $\lambda \in \mathbb{R} \cup \{\pm\infty\}$ and*

(1) X is a nonempty set and Y is a compact space;

(2) For all x_1, x_2 in X, there exists $x_0 \in X$ such that for all $y \in Y$,

$$f(x_0, y) \ge \textit{ median } (f(x_1, y), g(x_2, y), \lambda) \;;$$

(3) For all y_1, y_2 in Y, there exists $y_0 \in Y$ such that for all $x \in X$,

$$g(x, y_0) \le \textit{ median } (f(x, y_1), g(x, y_2), \lambda) \;;$$

(4) For all $x \in X$, $f(x, \cdot)$ and $g(x, \cdot)$ are lower semicontinuous on Y. Then $(**)$ holds.

Observe that since $f \leq g$, the inequality in (2) and in (3) imply, respectively, the following inequalities:

$$f(x_0, y) \geq \text{ median } (f(x_1, y), f(x_2, y), \lambda)$$

and

$$g(x, y_0) \leq \text{ median } (g(x, y_1), g(x, y_2), \lambda) .$$

Hence Theorem 4.1 is a consequence of Theorem 3.10.

References

1. Cao-Zong Cheng and Bor-Luh Lin, *Nonlinear, noncompact minimax theorems,* Bull. Inst. Math. Academia Sinica.
2. Cao-Zong Cheng and Bor-Luh Lin, *A two functions, noncompact topological minimax theorems,* Acta Math. Hungar. **73** (1996), 65-69.
3. Cao-Zong Cheng and You-Hao Lin, *A generalization of minimax theorems of fixed-point type,* Acta Math. Sinica **34** (1991), 502-507 (Chinese).
4. A. Chinnì , *A two-function minimax theorem,* Arch. Math. **66** (1996), 214-218.
5. Ky Fan, *Minimax theorems,* Proc. Nat. Acad. Sci., USA **39** (1953), 42-47.
6. Ky Fan, *Sur un théoréme minimax,* C.R. Acad. Sci. Paris **259** (1964), 3925-3928.
7. M. A. Geraghty and Bor-Luh Lin, *On a minimax theorem of Terkelsen,* Bull. Inst. Math. Acad. Sinica **11** (1983), 343-347.
8. M. A. Geraghty and Bor-Luh Lin, *Topological minimax theorems,* Proc. Amer. Math. Soc. **91** (1984), 377-380.
9. M. A. Geraghty and Bor-Luh Lin, *Minimax theorems without linear structure,* Linear and Multilinear Algebra **17** (1985), 171-180.
10. M. A. Geraghty and Bor-Luh Lin, *Minimax theorems without convexity,* Contemporary Math. **52** (1986), 102-108.
11. I. Irle, *Minimax theorems in convex situations,* Game Theory and Mathematical Economics, O. Moeschlin and D. Pallaschke, eds., North-Holland (1981), 321-331.
12. I. Irle, *A general minimax theorem,* Zeitschrift für Operations Research **29** (1985), 229-247.
13. I. Joo and G. Kassay, *Convexity, minimax theorems and their applications,* (preprint).
14. J. Kindler, *Some consequences of a double limit condition,* Game Theory and Mathematical Economics, O. Moeschlin and D. Pallaschke, eds., North-Holland (1981), 73-82.
15. J. Kindler, *Equilibrium point theorems for two-person games,* Siam. J. Control. Opt. **22** (1984), 671-683.
16. J. Kindler, *On a minimax theorem of Terkelsen's,* Arch. Math. **55** (1990), 573-583.
17. J. Kindler, *A Dini-Dax theorem,* Acta Math. Hungar. **63** (1994), 53-55.
18. J. Kindler and R. Trost, *Minimax theorems for interval spaces,* Acta. Math. Hungar. **54** (1989), 39-49.
19. H. Komiya, *A minimax theorem without linear structure,* Hiyoshi Review of Natural Science **8** (1990), 74-78.
20. V. Komornik, *Minimax theorems for upper semicontinuous functions,* Acta. Math. Acad. Sci. Hungar. **40** (1982), 159-163.
21. H. König, *Über das von Neumannsche Minimax-Theorem,* Arch. Math. **19** (1968), 482-487.
22. H. König, *A general minimax theorem based on connectedness,* Arch. Math. **59** (1992), 55-64.
23. H. König, *Addendum to "A general minimax theorem based on connectedness",* Arch. Math. **64** (1995), 139-143.

24. Bor-Luh Lin and Xiu-Chi Quan, *A symmetric minimax theorem without linear structure,* Arch. Math. **52** (1989), 367-370.
25. Bor-Luh Lin and Xiu-Chi Quan, *A two functions symmetric nonlinear minimax theorem,* Arch. Math. **57** (1991), 75-79.
26. Bor-Luh Lin and Xiu-Chi Quan, *Two functions minimax theorem with staircase,* Bull. Inst. Math. Acad. Sinica **19** (1991), 279-287.
27. Bor-Luh Lin and Xiu-Chi Quan, *A two functions nonlinear minimax theorem,* Fixed Point Theory and Applications, M. A. Thera and J. B. Baillon, eds., Pitman Research Notes 252 (1991), 321-325.
28. Bor-Luh Lin and Xiu-Chi Quan, *A noncompact topological minimax theorem,* J. Math. Anal. Appl. **161** (1991), 587-590.
29. Bor-Luh Lin and Feng-Shuo Yu, *A two functions version of generalized Terkelsen Minimax theorem,* to appear.
30. Bor-Luh Lin and Feng-Shuo Yu, *A two functions generalization of Horvath's minimax theorem,* to appear.
31. Bor-Luh Lin and Feng-Shuo Yu, *How far can we go in generalizing nonlinear minimax theorems to two functions minimax theorems,* to appear.
32. Fon-Che Liu, *A note on the von Neumann-Sion minimax principle,* Bull. Inst. Math. Acad. Sinica **6** (1978), 517-524.
33. B. Ricceri, *Some topological minimax theorems via an alternative principle for multifunctions,* Arch. Math. **60** (1993), 367-377.
34. S. Simons, *Two-function minimax theorems and variational inequalities for functions on compact and noncompact sets, with some comments on fixed-points theorems,* Proc. Symp. in Pure Math. **45** (1986), 377-392.
35. S. Simons, *An upward-downward minimax theorem,* Arch. Math. **55** (1990), 275-279.
36. S. Simons, *On Terkelsen's minimax theorem,* Bull. Inst. Math. Acad. Sinica **18** (1990), 35-39.
37. S. Simons, *Minimax theorems with staircases,* Arch. Math. **57** (1991), 169-179.
38. M. Sion, *On general minimax theorems,* Pacif. J. Math. **8** (1958), 171-176.
39. L. L. Stacho, *Minimax theorems beyond topological vector spaces,* Acta Sci. Math. **42** (1980), 157-164.
40. J. von Neumann, *Über ein ökonomisches Gleichungssystem und eine Verallgemeinerung des Brouwerschen, Fix punkt stages,* Ergeb. Math. Kolloq. Wein **8** (1937), 73-83.

WEAKLY UPWARD-DOWNWARD MINIMAX THEOREM

CAO-ZONG CHENG
Department of Mathematics
Computer Institute
Beijing Polytechnic University
Beijing 100044, China

BOR-LUH LIN AND FENG-SHUO YU
Department of Mathemtics
University of Iowa
Iowa City, IA 52242, U.S.A.
e-mail: bllin@math.uiowa.edu

Let X and Y be nonempty sets and let f be a real-valued function defined on $X \times Y$. For $S \subset X$ and $T \subset Y$, let

$$f^*(S,T) = \inf_T \sup_S f(x,y), \quad f_*(S,T) = \sup_S \inf_T f(x,y).$$

In particular, let $f^* = f^*(X,Y)$ and $f_* = f_*(X,Y)$. For $E \subset X$ and $\lambda \in R$, let $Y(E,f,\leq,r) = \cap_{x \in E}\{y \in Y : f(x,y) \leq r\}$.

f is said to be downward on X if for every $\epsilon > 0$, there exists $\delta > 0$ such that for all x_1, x_2 in X there exists an element x_0 in X with the properties that for all y in Y,

$$f(x_0,y) \geq \min\{f(x_1,y), f(x_2,y)\},$$

and

$$|f(x_1,y) - f(x_2,y)| \geq \epsilon \Rightarrow f(x_0,y) \geq \min\{f(x_1,y), f(x_2,y)\} + \delta.$$

Similarly, f is said to be upward on Y if for every $\epsilon > 0$, there eixsts $\delta > 0$ such that for all y_1, y_2 in Y, there exists an element y_0 in Y with the properties that for all x in X,

B. Ricceri and S. Simons (eds.), Minimax Theory and Applications, 21–28.

$$f(x, y_0) \leq \max\{f(x, y_1), f(x, y_2)\},$$

and

$$|f(x, y_1) - f(x, y_2)| \geq \epsilon \Rightarrow f(x, y_0) \leq \max\{f(x, y_1), f(x, y_2)\} - \delta.$$

S. Simons introduced the upward-downward functions to include the t-convex and s-concave functions considered in [3]. He proved the following two theorems.

Theorem A [7]. *Let X be a nonempty set and let Y be a compact topological space. Suppose $f : X \times Y \to R$ such that $f(x, \cdot)$ is lower semicontinuous on Y for any $x \in X$. Suppose that*

(1) f is downward on X,

(2) $Y(E, f, \leq r)$ is connected or empty for any finite subset E of X and $r \in R$.

Then,

$$f^* = f_*.$$

Theorem B [8]. *Let X be a nonempty set and let Y be a compact topological space. Suppose $f : X \times Y \to R$ such that $f(x, \cdot)$ is lower semicontinuous on Y for any $x \in X$. Suppose that*

(3) f is downward on X,

(4) f is upward on Y.

Then,

$$f^* = f_*.$$

Theorem A generalizes the minimax theorem in [3] and a Terkelsen minimax theorem [11]. See [4] for other extensions of Terkelsen minimax theorems. Theorem B generalizes the minimax theorem in [6] and the minimax theorem of H. König [5] and K. Fan [2].

In the minimax Workshop held at Erice, Italy in September, 1996, Simons raised the question whether there are minimax theorems with the conditions on upward-downward on f be weaken, the condition on continuity of f be strengthen. In this note, we answer the question in certain ways.

Definition 1. *Let X and Y be nonempty sets. A real-valued function f defined on $X \times Y$ is said to be weakly downward on X if for all x_1, x_2 in X there exists an x_0 in X such that for all y in Y,*

$$f(x_0, y) \geq \min\{f(x_1, y), f(x_2, y)\},$$

and

$$f(x_1,y) \neq f(x_2,y) \Rightarrow f(x_0,y) > \min\{f(x_1,y), f(x_2,y)\}.$$

Similarly, f is said to be weakly upward on Y if for all y_1, y_2 in Y there exists y_0 in Y such that for all x in X,

$$f(x,y_0) \leq \max\{f(x,y_1), f(x,y_2)\},$$

and

$$f(x,y_1) \neq f(x,y_2) \Rightarrow f(x,y_0) < \max\{f(x,y_1), f(x,y_2)\}.$$

Definition 2 [9]. *We say that sets H_0 and H_1 are joined by a set H if $H \cap H_0 \neq \emptyset$, $H \cap H_1 \neq \emptyset$ and $H \subset H_0 \cup H_1$. We say that a family of sets $\mathcal{H}$ is pseudoconnected if for $H_0, H_1, H \in \mathcal{H}$ and H_0 and H_1 are joined by H implies $H_0 \cap H_1 \neq \emptyset$.*

The following example shows that in case that X is a nonempty set and Y is compact, even when $f(x,\cdot)$ is affine on X for all x in X and $Y = [-1,1]$, Theorem A and B fail to be true when the conditions that f is upward-downward are replaced by f is weakly upward-downward.

Example. Let $X = \{\pm\frac{1}{n} : n \in N\}$ and $Y = [-1,1]$. Define $f : X \times Y \to R$ by

$$f(\frac{1}{n}, y) = -y - \frac{1}{n}$$

and

$$f(-\frac{1}{n}, y) = y - \frac{1}{n}, \quad n \in N \text{ and } y \in Y.$$

Obviously, Y is compact and $f(x,\cdot)$ is affine for all x in X, hence f is upward on Y and $L_f^\lambda(F)$ is connected for all finite set F in X and for all $\lambda \in R$. To see that f is weakly downward on X, for $x_1 = \frac{1}{n}$, $x_2 = \frac{1}{m}$, takes $x_0 = \frac{1}{n+m}$; for $x_1 = -\frac{1}{n}$, $x_2 = -\frac{1}{m}$, takes $x_0 = -\frac{1}{n+m}$ and for $x_1 = \frac{1}{n}$, $x_2 = -\frac{1}{m}$, takes $x_0 = -\frac{1}{n+1}$. It is easy to see that $f^* = 0$ and $f_* = -1$.

However, Simons' conjecture is corrected when both X and Y are compact. We may prove the following results.

Theorem 1. *Let X and Y be compact topological spaces. Let $f : X \times Y \to R$ such that $f(\cdot,y)$ is upper semicontinuous on X for any $y \in Y$, and $f(x,\cdot)$ is lower semicontinuous on Y for any $x \in X$. Suppose that*

(5) f is weakly downward on X,

(6) the family $\{Y(x,f <,r) \cap Y(E,f,\leq,r)\}_{x \in X}$ is pseudoconnected for any finite subset E of X and any r in R.

Then,

$$f^* = f_*.$$

Proof. We first prove $f^*(A, Y) \leq f_*$ for any finite subset A of X by the induction.

When $|A| = 1$, obviously $f^*(A, Y) \leq f_*$. Suppose that when $|A| \leq n$, $f^*(A, Y) \leq f_*$. For $|A| = n + 1$, let $A = E \cup \{s, t\}, s \neq t$ and $s, t \notin E$. Hence $|E| = n - 1$. Let $r \in R$ with $f_* < r$. For convenience, we write

$$W = Y(E, f, \leq, r),$$

$$G(x) = \{y \in W : f(x, y) < r\},$$

$$h(x) = \min\{f(x, y) : y \in W\}, \quad x \in X.$$

Note that $W \neq \emptyset$ and $G(x) \neq \emptyset$ for any $x \in X$ by the induction hypothesis. Put

$$\mathcal{G}(s) = \{G(x) : x \in X, G(x) \subset G(s)\},$$

$$\mathcal{G}(t) = \{G(x) : x \in X, G(x) \subset G(t)\}.$$

Obviously $\mathcal{G}(s)$ is a nonempty partially ordered set with the inclusion relation of subset of Y. It is easy to see that any totally ordered subset of $\mathcal{G}(s)$ has a lower bound since X is compact and f is upper semicontinuous on X. By Zorn's Lemma (Selection Axiom), there exists a minimal element $G(x^1) \in \mathcal{G}(s)$, this is, there exists an $x^1 \in X$ such that $G(x^1) \subset G(s)$, and $G(x) = G(x^1)$ if $G(x) \subset G(x^1)$ for any $x \in X$.

Put $X_1 = \{x \in X : G(x) = G(x^1)\}$.

It is easy to verify that X_1 is a closed subset, and so compact subset of X.

In fact, let $x_i \in X_1$, $x_0 \in X$ and $x_i \to x_0$. For any $y \in G(x_0)$, it follows that $y \in W$ and $f(x_0, y) < r$. Since $f(\cdot, y)$ is upper semicontinuous on X, there exists $i \in N$ such that $f(x_i, y) < r$. It follows that $y \in G(x_i) = G(x^1)$. Therefore $G(x_0) \subset G(x^1)$, and so $G(x_0) = G(x^1)$, i.e., $x_0 \in X_1$.

We remark that $h(x)$ is upper semicontinuous on X since $f(\cdot, y)$ is upper semicontinuous. Hence there exists some $u \in X_1$ such that

$$h(u) = \max_{x \in X_1} h(x).$$

For $\mathcal{G}(t)$, similarly there exists a minimal element $G(x^2) \in \mathcal{G}(t)$.

Put $X_2 = \{x \in X : G(x) = G(x^2)\}$.

Take $v \in X_2$ such that $h(v) = \max_{x \in X_2} h(x)$.

Since f is weakly downward, for $u, v \in X$, there exists $\overline{x} \in X$ such that

$$f(\overline{x}, y) \geq \min\{f(t_0, y), f(t_1, y)\} \text{ for all } y \in Y,$$

$$f(u, y) \neq f(v, y) \Rightarrow f(\overline{x}, y) > \min\{f(u, y), f(v, y)\}.$$

Hence $G(\overline{x}) \subset G(u) \cup G(v)$. Next we prove

$$G(\overline{x}) \cap G(u) \neq \emptyset,$$

and

$$G(\overline{x}) \cap G(v) \neq \emptyset.$$

Suppose $G(\overline{x}) \cap G(u) = \emptyset$. For any $y \in G(\overline{x})$, $y \in W$ and

$$\min\{f(u, y), f(v, y)\} \leq f(\overline{x}, y) < r.$$

Since $y \notin G(u)$, we have $f(u, y) \geq r > f(v, y)$. It follows that

$$f(\overline{x}, y) > \min\{f(u, y), f(v, y)\} = f(v, y) \geq h(v).$$

Since $G(\overline{x}) \neq \emptyset$, there exists $\epsilon > 0$ such that $\{y \in W : f(\overline{x}, y) \leq r - \epsilon\} \neq \emptyset$.

Put

$$W_1 = \{y \in W : f(\overline{x}, y) \leq r - \epsilon\},$$

$$W_2 = \{y \in W : f(\overline{x}, y) > r - \epsilon\}.$$

It is clear that $W = W_1 \cup W_2$, $W_1 \subset G(\overline{x})$,

$$h(\overline{x}) = \min_{y \in W} f(\overline{x}, y) = \min_{y \in W_1} f(\overline{x}, y),$$

and $f(\overline{x}, y) > h(v)$ for any $y \in W_1$.

Since $f(\overline{x}, \cdot)$ is lower semicontinuous and W_1 is compact, we have $h(\overline{x}) > h(v)$. It follows that $G(\overline{x}) \setminus G(v) \neq \emptyset$, or $G(\overline{x}) \cap G(u) \neq \emptyset$.

In fact, suppose that $G(\overline{x}) \subset G(v) = G(x^2)$. Hence $G(\overline{x}) = G(x^2)$ and $\overline{x} \in X_2$. Therefore $h(\overline{x}) \leq h(v)$.

It is similarly proved that $G(\overline{x} \cap G(v) \neq \emptyset$.

By the condition (6), $G(u) \cap G(v) \neq \emptyset$. Since $G(u) = G(x^1) \subset G(s)$ and $G(v) = G(x^2) \subset G(t)$, it follows that $G(s) \cap G(t) \neq \emptyset$. Let $y_0 \in G(s) \cap G(t)$. It is clear that $\max_{x \in E \cup \{s,t\}} f(x, y_0) \leq r$. Hence $f^*(E \cup \{s, t\}, Y) \leq r$, and so $f^*(E \cup \{s, t\}, Y) \leq f_*$.

This completes the proof of the conclusion that $f^*(A, y) \leq f_*$ for any finite subset A of X.

Since Y is compact and $f(x, \cdot)$ is lower semicontinuous, it follows that $f^* \leq f_*$.

This completes the proof of Theorem 1.

□

Corollary 1. *Let X and Y be compact topological spaces. Let $f : X \times Y \to R$ such that $f(\cdot, y)$ is upper semicontinuous on X for any $y \in Y$, and $f(x, \cdot)$ is lower semicontinuous on Y for any $x \in X$. Suppose that*

(5) f is weakly downward on X,

(7) the set $\{y \in Y : f(x_i, y) < r\}$ is either connected or empty for any $n \in N$, $x_1, \ldots, x_n \in X$ and any $r \in R$.

Then,

$$f^* = f_*.$$

Lemma . *Let X be a nonempty set and Y a compact topological spaces. Let $f : X \times Y \to R$. Suppose that $f(x, \cdot)$ is lower semicontinuous on Y for any x in X and f is weakly upward on Y. Then the family $\{Y(x, f, <, r) \cap Y(E, f, \leq, r)\}_{x \in X}$ is pseudoconnected for any finite subset E of X and any r in R..* **Proof.** *Suppose that there exist some finite subset E of X and $u, v, x \in X$, and $r \in R$ such that*

(a) $Y(E, f, \leq, r) \cap Y(x, f, <, r) \subset Y(u, f, <, r) \cup Y(v, f, <, r)$,

(b) $Y(E, f, \leq, r) \cap Y(x, f, <, r) \cap Y(u, f, <, r) \neq \emptyset$,

(c) $Y(E, f, \leq, r) \cap Y(x, f, <, r) \cap Y(v, f, <, r) \neq \emptyset$,

but

(d) $Y(E, f, \leq, r) \cap Y(u, f, <, r) \cap Y(v, f, <, r) = \emptyset$.

Form (b) and (c), there exists $\epsilon > 0$ such that
$Y(E, f, \leq, r) \cap Y(x, f, \leq, r - \epsilon) \cap Y(u, f, <, r) \neq \emptyset$.
$Y(E, f, \leq, r) \cap Y(x, f, \leq, r - \epsilon) \cap Y(v, f, <, r) \neq \emptyset$.
For Convenience, we write

$$Z = Y(E, f, \leq, r) \cap Y(x, f, \leq, r - \epsilon),$$

$$Y(u) = Z \cap Y(u, f, <, r),$$

$$Y(v) = Z \cap Y(v, f, <, r).$$

Then Z is a nonempty compact subset of Y, $\emptyset \neq Y(u) \subset Z$, $\emptyset \neq Y(v) \subset Z$, and $Y(u) \cap Y(v) = \emptyset$. Put

$$\alpha = \inf_{y \in Y(u)} f(v, y),$$

and

$$\beta = \inf_{y \in Y(v)} f(u, y).$$

Since $Y(u) \cap Y(v) = \emptyset$, for $y \in Y(u)$, we have $y \notin Y(v)$ and $y \in Z$. Hence $f(v,y) \geq r$. It follows that $\alpha \geq r > -\infty$. On the other hand, $\alpha \leq f(v,y) < +\infty$ for some $y \in Y(u) \neq \emptyset$. Hence $|\alpha| < \infty$. Similarly, $|\beta| < \infty$.

For any $n \in N$, let $y_n^1 \in Y(u)$ and $y_n^2 \in Y(v)$ such that

$$\alpha \leq f(v, y_n^1) < \alpha + \frac{1}{n},$$

and

$$\beta \leq f(u, y_n^2) < \beta + \frac{1}{n}.$$

Since $y_n^1, y_n^2 \in Z(n \in N)$ and Z is compact, there exist subsequences $\{y_{n_k}^1\}$ and $\{y_{n_k}^2\}$, and $y^1, y^2 \in Z$ such that $y_{n_k}^1 \to y^1$ and $y_{n_k}^2 \to y^2$. Hence $\alpha = f(v, y^1)$ and $\beta = f(u, y^2)$.

By (a), $y^1 \in Z \subset Y(u,f,<,r) \cup Y(v,f,<,r)$. Thus $y^1 \in Y(u)$ or $y^1 \in Y(v)$.

Suppose that $y^1 \in Y(v)$, this is, $y^1 \in Z$ and $f(v,y^1) < r$. Hence there exists $y_{h_k}^1$ such that $f(v, y_{n_k}^1) < r$. It follows that $y_{n_k}^1 \in Y(v)$. This is a contradiction with $Y(u) \cap Y(v) = \emptyset$. Thus $y^1 \in Y(u)$. Similarly, $y^2 \in Y(v)$.

For y^1 and y^2, since f is weakly upward on Y, there exists y^0 in Y such that for all x in X,

$$f(x,y^0) \leq \max\{f(x,y^1), f(x,y^2)\},$$

and

$$f(x,y^1) \neq f(x,y^2) \Rightarrow f(x,y^0) < \max\{f(x,y^1), f(x,y^2)\}.$$

From y^1, $y^2 \in Z$ and (a), we have

$$y^0 \in Z \subset Y(u,f,<,r) \cup Y(v,f,<,r).$$

In case $y^0 \in Y(u)$, since $f(v,y^1) \geq r > f(v,y^2)$, we have

$$f(v,y^0) < \max\{f(v,y^1), f(v,y^2)\} = f(v,y^1) = \alpha.$$

This is a contradiction with the definition of α.

In case $y^0 \in Y(v)$, since $f(u,y^1) < r \leq f(u,y^2)$, we have

$$f(u,y^0) < \max\{f(u,y^1), f(u,y^2)\} = f(u,y^2) = \beta.$$

This is a contradiction with the definition of β.

This completes the proof.

□

Theorem 2. *Let X and Y be compact topological spaces. Let $f : X \times Y \to R$ such that $f(\cdot, y)$ is upper semicontinuous on X for any $y \in Y$ and $f(x, \cdot)$ is lower semicontinous on Y for any $x \in X$. Suppose that*
(5) f is weakly downward on X,
(8) f is weakly upward on Y.

Then,

$$f^* = f_*.$$

Proof. The conclusion follows from Lemma and Theorem 1.

□

Corollary 2. *Let X and Y be compact topological spaces. Let $f : X \times Y \to R$ be jointly continuous. Suppose that*
(5) f is weakly downward on X,
(8) f is weakly upward on Y.

Then,

$$f^* = f_*.$$

From the definition, it is not clear that downward implies weakly downward and upward implies weakly upward. For downward, y_0 is depended on ϵ and y_1, y_2. But for weakly downward, we only have $f(x, y_1) \neq f(x, y_2)$. It may happen that $|f(x, y_1) - f(x, y_2)| \to 0$ for $x \in X$.

After the note is completed, S. Simons kindly provided the authors a reprint of A. Domokos' paper [1] in which Corollary 2 is proved. However, the example b) in [1] is incorrect since f is not weakly downward on X (for instance, for $x_1 = -0.5$ and $x_2 = 1$, there is not x_3 as in the definition of the weakly downward).

References

1. A. Domokos, *A minimax theorem,* Ann. Univ. Sci. Budapest. **37** (1994), 157-163.
2. Ky Fan, *Minimax theorems,* Proc. Nat. Acad. Sci. U.S.A. **39** (1953), 42-47.
3. M. A. Geraghty and Bor-Luh Lin, *On a minimax theorem of Terkelsen,* Bull. Inst. Math. Acad. Sinica **11** (1983), 343-347.
4. J. Kindler, *On a minimax theorem of Terkelsen's,* Arch. Math. **55** (1990), 573-583.
5. H. König, *Über das von Neumansche minimax theorem,* Arch. Math. **19** (1968), 482-487.
6. Bor-Luh Lin and Xiu-Chi Quan, *A symmetric minimax theorem without linear structure,* Arch. Math. **52** (1989), 367-370.
7. S. Simons, *On Terkelsen's minimax theorem,* Bull. Inst. Math. Acad. Sinica **18** (1990), 35-39.
8. S. Simons, *An upward-downward minimax theorem,* Arch. Math. **55** (1990), 275-279.
9. S. Simons, *A flexible minimax theorem,* Acta Math. Hungar. **63** (1994), 119-132.
10. S. Simons, *Addendum to "A flexible minimax theorem",* Acta Math. Hungar. **69** (1995), 359-360.
11. F. Terkelsen, *Some minimax theorems,* Math. Scand. **31** (1972), 405-413.

A TWO-FUNCTION MINIMAX THEOREM

ANTONIA CHINNÌ
Department of Mathematics
University of Messina
98166 Sant'Agata-Messina, Italy
e-mail: chinni@riscme.unime.it

1. Introduction

Let X, Y be two non-empty sets and f, g two real functions on $X \times Y$. In this paper, we consider the problem of guaranteeing the validity of the inequality

$$\inf_{x \in X} \sup_{y \in Y} f(x,y) \leq \sup_{y \in Y} \inf_{x \in X} g(x,y).$$

In the sequel, it is implicitly assumed that $\sup_{y \in Y} \inf_{x \in X} g(x,y) < +\infty$. Several papers have already been devoted to this problem. Most of them are quoted in a very recent survey by S.Simons [2] to which we also refer for a discussion of the above inequality. A feature which is common to each of that papers is to assume the condition

$$f(x,y) \leq g(x,y) \tag{1}$$

for all $(x,y) \in X \times Y$.

We have recalled this fact just because the motivation for this paper is to present a result where, for the first time (as far as we know), condition (1) is noticeably weakened. Actually, instead of (1), we assume that

$$f(x,y) \leq \max\{g(x,y), h(x,y)\}$$

for all (x,y) in a certain subset of $X \times Y$, where $h : X \times Y \to \mathbb{R}$ is another function having suitable properties. But, that is not all. Indeed, we are also able to prove the better inequality

$$\inf_{x \in X \setminus H} \sup_{y \in Y} f(x,y) \leq \sup_{y \in Y} \inf_{x \in X} g(x,y)$$

B. Ricceri and S. Simons (eds.), Minimax Theory and Applications, 29–33

for each H in a suitable family of sets.

2. The Main Result

For the reader's convenience, we first recall a particular case of a recent result of B.Ricceri which will be the main tool used later.

If A is a subset of a product space $X \times Y$, for each $x \in X$, $y \in Y$, we put

$$A_x = \{v \in Y : (x,v) \in A\}$$

and

$$A^y = \{u \in X : (u,y) \in A\}.$$

Theorem 2.1 ([1], Theorem 5) *Let E be an infinite-dimensional Hausdorff topological vector space, $X \subseteq E$ a convex set with non-empty interior, $K \subseteq E$ a relatively compact set, $Y \subseteq \mathbb{R}$ a compact interval, and S, T two subsets of $X \times Y$. Assume that:*

(*i*) *S_x is open in Y for each $x \in X \setminus K$, and S^y is convex and with non-empty interior for each $y \in Y$;*

(*ii*) *T_x is non-empty and connected for each $x \in X \setminus K$, and either $T^y \setminus K$ is open in $X \setminus K$ for each $y \in Y$, or $T \setminus (K \times Y)$ is closed in $(X \setminus K) \times Y$.*

Then, for every set $V \subseteq X \times Y$ such that V^y is relatively compact in E for each $y \in Y$ and V_x is closed in Y for each $x \in X \setminus K$, the set $(S \setminus (V \cup (K \times Y))) \cap T$ is non-empty.

We also recall that, if A is a convex in a vector space, a function $\phi : A \to \mathbb{R}$ is said to be *quasi-convex* (resp. *quasi-concave*) if, for each $\lambda \in \mathbb{R}$, the set $\{x \in A : \phi(x) \leq \lambda\}$ (resp. $\{x \in A : \phi(x) \geq \lambda\}$) is convex.
If A is a subset of a topological space E and $\alpha \in [-\infty, +\infty[$, a function $\phi : A \to \mathbb{R}$ is said to be $\alpha - sup - compact$ with respect to E if, for each $\lambda > \alpha$, the set $\{x \in A : \phi(x) \geq \lambda\}$ is relatively compact in E.

Our main result is the following

Theorem 2.2 *Let E be an infinite-dimensional Hausdorff topological vector space, $X \subseteq E$ a convex set with non-empty interior, $K \subseteq E$ a relatively compact set, $Y \subseteq \mathbb{R}$ a compact interval, and f, g, h three real functions on $X \times Y$. Assume that:*

(*a*) *$f(x, .)$ is quasi-concave in Y for each $x \in X \setminus K$, and either f is upper semicontinuous in $(X \setminus K) \times Y$ or $f(., y)$ is lower semicontinuous in $X \setminus K$ for each $y \in Y$;*

(*b*) *$g(x, .)$ is upper semicontinuous in Y for each $x \in X \setminus K$, and $g(., y)$ is upper semicontinuous and quasi-convex in X for each $y \in Y$;*

(*c*) *$h(x,.)$ is upper semicontinuous in Y for each $x \in X \setminus K$, and $h(.,y)$ is α-sup-compact in X with respect to E for each $y \in Y$, where*

$$\alpha = \sup_{v \in Y} \inf_{u \in X} g(u,v);$$

(*d*) *$f(x,y) \leq \max\{g(x,y), h(x,y)\}$ for each $(x,y) \in (X \setminus K) \times Y$.*
Then, for every relatively compact set $H \subseteq E$, one has

$$\inf_{x \in X \setminus H} \sup_{y \in Y} f(x,y) \leq \sup_{y \in Y} \inf_{x \in X} g(x,y).$$

Proof. Arguing by contradiction, assume that there is some relatively compact set $H \subseteq E$ such that

$$\sup_{y \in Y} \inf_{x \in X} g(x,y) < \inf_{x \in X \setminus H} \sup_{y \in Y} f(x,y).$$

Fix a real number λ such that

$$\sup_{y \in Y} \inf_{x \in X} g(x,y) < \lambda < \inf_{x \in X \setminus H} \sup_{y \in Y} f(x,y). \tag{2}$$

Now, put

$$S = \{(x,y) \in X \times Y : g(x,y) < \lambda\},$$
$$T = \{(x,y) \in X \times Y : f(x,y) \geq \lambda\},$$
$$\tilde{T} = \{(x,y) \in X \times Y : f(x,y) > \lambda\},$$
$$V = \{(x,y) \in X \times Y : h(x,y) \geq \lambda\}.$$

Fix $x \in X \setminus (H \cup K)$. In view of (*b*), S_x is open in Y. In view of (2) and (*a*), T_x and $\tilde{T}_x$ are non-empty and connected. In view of (*c*), V_x is closed. Now, fix $y \in Y$. By (2) and (*b*), S^y is non-empty, convex and open in X, and hence its interior in E is non-empty. According to the alternative allowed in (*a*), either $T \setminus ((H \cup K) \times Y)$ is closed in $(X \setminus (H \cup K)) \times Y$ or $\tilde{T}^y \setminus (H \cup K)$ is open in $X \setminus (H \cup K)$. Finally, by (*c*), the set V^y is relatively compact in E. Thus, the assumptions of Theorem 2.1 are satisfied, and so the set $S \setminus (V \cup ((H \cup K) \times Y))$ intersects T or even $\tilde{T}$. Clearly, this contradicts (*d*), and, therefore, the proof is complete. □

3. Remarks and Examples Related to Theorem 2.2

Remark 3.1 In the case where $f(x,y) \leq h(x,y)$ for each $(x,y) \in (X \setminus K) \times Y$, and $h(.,y)$ is $-\infty$-sup-compact with respect to E for each $y \in Y$, the conclusion of Theorem 2.2 holds, in particular, for any constant function

g. This clearly implies that, in that case, for each relatively compact set $H \subseteq E$, one has

$$\inf_{x \in X \setminus H} \sup_{y \in Y} f(x,y) = -\infty.$$

Remark 3.2 In the case $g(x,y) \leq f(x,y)$ for each $(x,y) \in X \times Y$, the conclusion of Theorem 2.2 becomes the following: for each relatively compact set $H \subseteq E$, one has

$$\inf_{x \in X \setminus H} \sup_{y \in Y} f(x,y) = \inf_{x \in X} \sup_{y \in Y} f(x,y) = \sup_{y \in Y} \inf_{x \in X} f(x,y) =$$

$$= \inf_{x \in X \setminus H} \sup_{y \in Y} g(x,y) = \inf_{x \in X} \sup_{y \in Y} g(x,y) = \sup_{y \in Y} \inf_{x \in X} g(x,y).$$

In such an instance, one clearly sees how important it is to allow that $f(x,y) > g(x,y)$ somewhere. Actually, we are able to derive two distinct one-function minimax theorems in connection with given functions satisfying either (a) or (b).

Remark 3.3 Theorem 2.2 is no longer true if $\dim(E) < +\infty$, even if we take $H = \emptyset$. Indeed, if $\dim(E) < +\infty$, we can choose a compact convex set $X \subseteq E$ with non-empty interior, and then define f, g, h by putting $f(x,y) = h(x,y) = 1$, $g(x,y) = 0$ for all $(x,y) \in X \times Y$. Of course, apart from $\dim(E) = +\infty$, each other assumption of Theorem 2.2 is satisfied (take $K = \emptyset$). However, we have $0 = \sup_{y \in Y} \inf_{x \in X} g(x,y) < \inf_{x \in X} \sup_{y \in Y} f(x,y) = 1$.

Remark 3.4 Theorem 2.2 cannot be extended to the case where Y is any compact convex subset of a Hausdorff locally convex topological vector space. Indeed, let $(E, \langle .,. \rangle)$ be an infinite-dimensional real Hilbert space, considered with its weak topology, let $X = E$, and let Y be the closed unit ball of E. So, Y is weakly compact. Then put

$$f(x,y) = \langle x, y \rangle$$

for all $(x,y) \in X \times Y$. Finally, take $g = f$ and $h = -1$. Observe that h satisfies condition (c) since $\sup_{y \in Y} \inf_{x \in X} g(x,y) = 0$. Clearly f satisfies (a) and (b). However, we have

$$0 = \sup_{y \in Y} \inf_{x \in X} f(x,y) < \inf_{x \in X \setminus Y} \sup_{y \in Y} f(x,y) = \inf_{x \in X \setminus Y} \|x\| = 1.$$

We conclude by pointing out a simple example of application of Theorem 2.2.

Theorem 3.1 *Let E be an infinite-dimensional reflexive Banach space, $X \subseteq E$ a convex set with non-empty interior in the weak topology, $K \subseteq X$ a bounded set, with $\theta_E \in K$, $Y \subseteq \mathbb{R}$ a compact interval, and f, g two real functions on $X \times Y$, with*

$$\alpha = \sup_{y \in Y} \inf_{x \in X} g(x,y) > -\infty.$$

Assume that

(a) *$f(x,.)$ is quasi-concave in Y for each $x \in X \setminus K$, and either f is weakly upper semicontinuous in $(X \setminus K) \times Y$ or $f(.,y)$ is weakly lower semicontinuous in $X \setminus K$ for each $y \in Y$;*

(b) *$g(x,.)$ is upper semicontinuous in Y for each $x \in X \setminus K$, and $g(.,y)$ is weakly upper semicontinuous and quasi-convex in X for each $y \in Y$;*

(c) *for each $y \in Y$ there exists $\rho_y > 0$ such that*

$$f(x,y) \leq \begin{cases} g(x,y) & \text{if } x \in X \setminus K \text{ and } \|x\| > \rho_y \\ \frac{1}{\rho_y} + \alpha & \text{if } x \in X \setminus K \text{ and } \|x\| \leq \rho_y \end{cases}$$

Then, for every bounded set $H \subseteq X$, one has

$$\inf_{x \in X \setminus H} \sup_{y \in Y} f(x,y) \leq \sup_{y \in Y} \inf_{x \in X} g(x,y).$$

Proof. Apply Theorem 2.2 endowing E with its weak topology. So, since E is reflexive, the relatively weakly compact subsets of E are precisely the bounded ones. Now, for each $(x,y) \in X \times Y$, put

$$h(x,y) = \begin{cases} \frac{1}{\|x\|} + \alpha & \text{if } x \neq \theta_E \\ \alpha & \text{if } x = \theta_E \end{cases}$$

For each $\lambda > \alpha$, we have $\{x \in X : h(x,y) \geq \lambda\} = \{x \in X : 0 < \|x\| \leq \frac{1}{\lambda - \alpha}\}$. So, h satisfies condition (c) of Theorem 2.2. Finally, condition (c) clearly implies that

$$f(x,y) \leq \max\{g(x,y), h(x,y)\}$$

for each $(x,y) \in (X \setminus K) \times Y$. Therefore, the assumptions of Theorem 2.2 are satisfied, and then our conclusion follows. □

References

1. B. Ricceri, *Applications of a theorem concerning sets with connected sections,* Topol. Methods Nonlinear Anal. **5** (1995), 237–248.
2. S. Simons, *Minimax theorems and their proofs,* In Minimax and Applications, 1-23. D.-Z. Du and P.M. Pardalos (eds.), Kluwer Academic Publishers, 1995.

GENERALIZED FIXED-POINTS AND SYSTEMS OF MINIMAX INEQUALITIES

PAUL DEGUIRE
Département de Mathématiques et de Statistique
Université de Moncton
Moncton, N.-B., Canada, E1A 3E9
e-mail: deguirp@umoncton.ca

Abstract. To begin, we recall the notion of selecting families, introduced by Deguire and Lassonde in [4]. We also recall a selection theorem and a generalized fixed-point theorem for Ky Fan families of set valued maps. Applications of those results are then presented: a generalization of the Tychonoff fixed-point theorem and a version of the Ky-Fan intersection theorem for families of maps together with it's analytical formulation, a very general system of minimax inequalities.

1. Introduction

All topological vector spaces will be Hausdorff and noted t.v.s. or l.c.s. when they are locally convex. A set valued map $A : X \to Y$ associates to each $x \in X$, a subset $A(x) \subseteq Y$, the value of A at X. The fiber of A at $y \in Y$ is the subset $A^{-1}(y) \subseteq X$ defined by $A^{-1}(y) = \{x \in X \mid y \in A(x)\}$. Values and fibers of a set-valued map can be empty.

The single-valued map $s : X \to Y$ is a selection of the map $A : X \to Y$ if $s(x) \in A(x)$ for all $x \in X$. If $X = \prod_{i\in I} X_i$, the family of maps $\{A_i : X \to X_i\}_{i\in J}$ is said to be a Ky Fan family of maps if the following is verified:

i) $A_i(x)$ is convex, for all $x \in X$ and all i in I
ii) $A_i^{-1}(y_i)$ is open, for all $y_i \in X_i$ and all i in I
iii) for all $x \in X$, there exists $i \in I$ with $A_i(x) \neq \emptyset$

We say that the family $\{A_i\}_{i\in I}$ has a generalized fixed-point if there exists $x \in X$ and $i \in I$ with $x_i \in A_i(x)$. A generalized fixed-point is not a

B. Ricceri and S. Simons (eds.), Minimax Theory and Applications, 35–40.

fixed-point, but we find back fixed-points when I has only one element. In general there is no restriction on the set of indexes I.

Definition: The family of continuous singled-valued maps,

$$\{s_i : X \to X_i\}_{i\in I},$$

is called a selecting family for the family $\{A_i : X \to X_i\}_{i\in I}$ of set-valued maps if for all x in X, there exists $i \in I$ with $s_i(x) \in A_i(x)$.

One easily observes that the notion of selecting family reduces to the notion of continuous selection when I has only one element.

Theorem 1 (Deguire and Lassonde, [4]): *Let* $X = \prod_{i\in I} X_i$ *where for all* $i \in I$, X_i *is a non-empty compact and convex subset of a t.v.s.* E_i. *Then, any Ky Fan family of set-valued maps,* $\{A_i : X \to X_i\}_{i\in I}$ *admits a selecting family* $\{s_i : X \to X_i\}_{i\in I}$.

The selecting family can be chosen in such a way that, for each $i \in I$, $s_i(X) \subseteq C_i \subseteq X_i$ where C_i is the convex hull of a finite number of points in X_i and for all but a finite number of indexes $i \in I$, C_i reduces to a single point. This means that the subset of X, $C = \prod_{i\in I} C_i$, behaves like a finite dimensional compact and convex set.

The following generalized fixed-point theorem is the simplest application of theorem 1. More general results can be obtained easily but this version is all that we need for the following applications.

Theorem 2 : *Let* $X = \prod_{i\in I} X_i$ *and* $\{A_i : X \to X_i\}_{i\in I}$ *as in theorem 1. Then there exists* $x \in X$ *and* $i \in I$ *such that* $x_i \in A_i(x)$.

Proof: Let $\{s_i : X \to X_i\}_{i\in I}$ be a selecting family for $\{A_i : X \to X_i\}_{i\in I}$. If $s : X \to C$ is the continuous map defined by $s(x) = (s_i(x))_{i\in I}$, for each x in X, the restriction $s|_C : C \to C$ has a fixed-point by the Brouwer fixed-point theorem. If $s(x) = x$, for all i in I we have $x_i = s_i(x)$. Then, for some $i \in I$, we have $x_i \in A_i(x)$.•

Theorem 2 reduces to the Fan-Browder fixed-point theorem when I contains only one element. Related results can be found in Deguire [3], Deguire and Lassonde [4], Gale-Mas-Colell [7], Marchi, Martinez-Legaz [10] and Toussaint [13].

2. Applications

The following generalized fixed-point theorem reduces to the Tychonoff fixed-point theorem when I contains only one element.

Theorem 3 : *Let* $X = \prod_{i \in I} X_i$ *where for all* $i \in I$, X_i *is a non empty compact and convex subset of a l.c.s.* E_i *and let* $\{O_i\}_{i \in I}$ *be an open covering of* X.

If, for all $i \in I$, $f_i : O_i \to X_i$ *is a continuous map, then there exists* $x \in X$ *such that for all* i *in* I *satisfying* $x \in O_i$ *we have* $x_i = f_i(x)$.

Proof: For all $i \in I$, let $\left\{p_j^i\right\}_{j \in J_i}$ the set of all continuous seminorms in E_i. We define a family of maps $\{A_i : X \to X_i\}_{i \in I}$:

For all $i \in I$ and all $x \in X$,

$$A_i(x) = \begin{cases} \emptyset & \text{if } x \notin O_i \\ \{y_i \in X_i : & \sum_{j \in J_i} p_j^i(x_i - f_i(x)) > \sum_{j \in J_i} p_j^i(y_i - f_i(x))\} \text{ if } x \in O_i \end{cases}$$

For all $i \in I$, the map $A_i : X \to X_i$ has convex values and open fibers but, $\{A_i\}_{i \in I}$ do not have a generalized fixed point because $x_i \in A_i(x)$ leads to the impossible inequality $\sum_{j \in J_i} p_j^i(x_i - f_i(x)) > \sum_{j \in J_i} p_j^i(x_i - f_i(x))$. Then, by theorem 2, the family $\{A_i\}_{i \in I}$ cannot be a Ky Fan family of maps. Therefore, there is some $x \in X$ such that for all $i \in I$, $A_i(x) = \emptyset$. This means that for all i for which $x_i \in O_i$ we have for all $y_i \in X_i$:

$$\sum_{j \in J_i} p_j^i(x_i - f_i(x) \leq \sum_{j \in J_i} p_j^i(y_i - f_i(x).$$

By taking $y_i = f_i(x)$ in the above inequality, one concludes that $\sum_{j \in J_i} p_j^i(x_i - f_i(x) = 0$ and that is equivalent to the conclusion of the theorem.•

We now introduce some notation that will be needed in our last application.

If $\{X_i\}_{i \in I}$ is a family of sets with $X = \prod_{i \in I} X_i$, we will write $X^i = \prod_{j \in I - \{i\}} X_j$, $(x^i, x_i) = x \in X$, $x_i \in X_i$ and $x^i \in X^i$.

In the more complex situation where for each $k \in K$ we have a family of sets $\{X_{k,i}\}_{i \in I}$, we will write

$$X = \prod_{k \in K} X_k \ , \ X^k = \prod_{j \in K - \{k\}} X_j \ , \text{ and } X_k = \prod_{i \in I} X_{k,i}$$

The following result is a known version of Ky Fan intersection theorem.

Theorem 4 : *Let $X = \prod_{i \in I} X_i$, where each X_i is a non-empty compact and convex subset of a t.v.s. E_i and let $\{A_i : X^i \to X_i\}_{i \in I}$ a family of maps such that for all $i \in I$ we have:*

i) $A_i(x^i)$ is convex and non empty, for all $x^i \in X^i$

ii) $A_i^{-1}(x_i)$ is open, for all $x_i \in X_i$

Then there exists $x \in X$ such that $x \in \bigcap_{i \in I} \Gamma A_i$ where ΓA_i, the graph of A_i, is the set of $x \in X$ with $x_i \in A_i(x^i)$.

The intersection theorem of Ky Fan has been proved in the case where I is a finite set and has been extended by Ma to the infinite case. When I has only 2 elements, this is the well known coincidence theorem of Fan. (see Fan [6] and Ma [9].).

The following result is an extension of theorem 4 in which set-valued maps are replaced by families of set-valued maps.

Theorem 5 : *Let K and I be sets of indexes with $X_{k,i}$ a non-empty compact and convex subset of a l.c.s. $E_{k,i}$, for all $k \in K$ and $i \in I$. If for each $k \in K$, $\left\{A_{k,i} : X^k \to X_{k,i}\right\}_{i \in I}$ is a Ky Fan family of maps, then there exists $x \in X$ such that for all $k \in K$, there exists $i \in I$ with $x_{k,i} \in A_{k,i}(x^k)$.*

Proof: By theorem 1, for each fixed k in K, the family of maps

$$\left\{A_{k,i} : X^k \to X_{k,i}\right\}_{i \in I}$$

admits a selecting family $\left\{s_{k,i} : X^k \to X_{k,i}\right\}_{i \in I}$. For each $k \in K$, let $s_k : X^k \to X_k$ be the map defined by $s_k(x^k) = (s_{k,i}(x^k))_{i \in I}$ and let $s : X \to X$ be the map defined by $s(x) = (s_k(x^k))_{k \in K}$. By the Tychonoff fixed-point theorem, the map s has a fixed-point. Let $x = s(x)$. Then for each $k \in K$, we have $x_k = s_k(x^k)$ and for each $i \in I$, we have $x_{k,i} = s_{k,i}(x^k)$. For all $k \in K$ there exists $i \in I$ with $s_{k,i}(x^k) \in A_{k,i}(x^k)$ by definition of selecting family. We have proved that there exists $x \in X$ such that for all $k \in K$, there exists $i \in I$ with $x_{k,i} \in A_{k,i}(x^k)$ and the proof is complete.•

When I has only one element, theorem 5 reduces to theorem 4.
Moreover, the families of maps $\left\{A_{k,i} : X^k \to X_{k,i}\right\}_{i \in I}$ does not have to be indexed by the same set I for all k in K. But this is quite irrelevant as empty maps can be considered.

Our next result is the analytical formulation of theorem 5. When K has two elements, it is a system of minimax inequalities presented in Deguire-Lassonde (see [4]). This result also contains some well known results including the Liu minimax inequality (see [8]) and the von Neumann-Sion minimax equality(see [12]).

We recall that a numerical map $f : X \to \Re$ defined on a convex subset X of a t.v.s. E is said to be quasi concave (respectively lower semi-continuous) if for all $\lambda \in \Re$, the set $\{x \in X \mid f(x) > \lambda\}$ is convex (respectively open) and it is said to be quasi convex (respectively upper semi-continuous) when $-f$ is quasi concave (respectively lower semi-continuous).

Theorem 6 : *Let K, I, X, X^k and $X_{k,i}$ be as in theorem 5 and let K be divided in two subsets $K = K^+ \cup K^-$. If f for any k in K we have a family of numerical maps $\left\{f_{k,i} : X^k \times X_{k,i} \to \Re\right\}_{i\in I}$ such that the following three statements are satisfied:*

1) for all k in K^+, $f_{k,i}$ is quasi concave in $y_{k,i}$, for all x^k in X^k and lower semi-continuous in x^k, for all $y_{k,i}$ in $X_{k,i}$

2) for all k in K^-, $f_{k,i}$ is quasi convex in $y_{k,i}$, for all x^k in X^k and upper semi-continuous in x^k, for all $y_{k,i}$ in $X_{k,i}$

3) for a given $(\lambda_i)_{i\in I}$ in $\Re^I$ we have that for all x in X,

$$\left\{k \in K^+ : \sup_I f_{k,i}(x^k, x_{k,i}) \leq \lambda_i\right\} \bigcup \left\{k \in K^- : \inf_I f_{k,i}(x^k, x_{k,i}) \geq \lambda_i\right\} \neq \emptyset$$

Then we have the following system of minimax inequalities:

$$\left\{k \in K^+ : \inf_{X^k} \sup_{X_{k,i}} \sup_I f_{k,i}(x^k, y_{k,i}) \leq \lambda_i\right\} \bigcup$$
$$\left\{k \in K^- : \sup_{X^k} \inf_{X_{k,i}} \inf_I f_{k,i}(x^k, y_{k,i}) \geq \lambda_i\right\} \neq \emptyset$$

Sketch of the proof: for all k in K^+ and i in I, let the map $A_{k,i} : X^k \to X_{k,i}$ be defined by $A_{k,i}(x^k) = \left\{y_{k,i} \in X_{k,i} : f_{k,i}(x^k, y_{k,i} > \lambda_i\right\}$ and for all k in K^- and i in I, let the map $A_{k,i} : X^k \to X_{k,i}$ be defined by $A_{k,i}(x^k) = \left\{y_{k,i} \in X_{k,i} : f_{k,i}(x^k, y_{k,i}) < \lambda_i\right\}$.

Then, the hypothesis 3) states that the conclusion of theorem 5 is not satisfied. The conclusion follows from the fact that for some k in K, the family $\{A_{k,i}\}_{i\in I}$ is not a Ky Fan family of maps. •

Remark: All results from theorem 2 to theorem 6 can be easily extended in two natural ways. First, larger classes or families of maps can be used and secondly, the compactness hypothesis can be replaced by weaker assumptions. The proofs would remain practically unchanged and the ideas involved would be the same as those presented in this text. Related results using those improvements on maps or on compactness can be found in Deguire [3], Deguire, Lassonde [4] and Ben El Mechaïekh, Deguire and Granas [12].

Remark: Theorem 2 and theorem 4 both leads to the Nash theorem on equilibrium. As theorem 5 generalises both results, on can possibly search for a generalization of Nash theorem, concerning family of equilibria, using theorem 5.

References

1. H. Ben El Mechaïekh, P. Deguire and A. Granas, *Points fixes et coïncidences pour les applications multivoques* (applications de Ky Fan), C. R. Acad. Sci. Paris Série I Math. **295** (1982), 337-340.
2. H. Ben El Mechaïekh, P. Deguire et A. Granas, *Points fixes et coïncidences pour les applications multivoques* (applications de types M et M*), C. R. Acad. Sci. Série I, **305** (1987), 381-384.
3. P. Deguire, *Browder-Fan Fixed-Point theorem and related results,* Discussiones Mathematicae, Differential Inclusions **15** (1995), 149-162.
4. P. Deguire and M. Lassonde, *Familles sélectantes,* Topol. Methods Nonlinear Anal. **5** (1995), 261-269.
5. K. Fan, *A generalization of Tychonoff's fixed point theorem,* Math. Ann. **142** (1961), 305-310.
6. K. Fan, *Applications of a theorem concerning sets with convex sections,* Math. Ann. **163** (1966), 189-203.
7. D. Gale and A. Mas-Colell, *An equilibrium existence theorem for a general model without ordered preferences,* J. Math. Econom. **2** (1975), 9-15; Erratum, ibid. 6 (1979), 297-298.
8. F.C. Liu, *A note on the von Newmann-Sion minimax principle,* Bull. Inst. Math. Acad. Sinica **6** (1978), 517-524.
9. T.W. Ma, *On sets with convex sections,* J. Math. Anal. Appl. **27** (1969), 413-416.
10. E. Marchi and J.-E. Martinez-Legaz, *A generalization of Fan-Browder's fixed point theorem and its applications,* Topol. Methods Nonlinear Anal. **2** (1993), 277-291.
11. J. Nash, *Equilibrium points in n-person games,* Proc. Nat. Acad. Sci. U.S.A. **36** (1950), 48-49.
12. M. Sion, *On general minimax theorems,* Pacific J. Math. **8** (1958), 171-176.
13. S. Toussaint, *On the existence of equilibria in economies with infinitely many commodities and without ordered preferences,* J. Econom. Theory **33** (1984), 98-115.

A MINIMAX INEQUALITY FOR MARGINALLY SEMICONTINUOUS FUNCTIONS

GABRIELE H. GRECO AND MARIA PIA MOSCHEN
Dipartimento di Matematica
Università di Trento
38050 Povo, Italy
e-mail: greco@science.unitn.it

The purpose of this short note is to give a "finite-dimensional minimax inequality" which includes Brouwer Fixed Point Theorem, Fan Fixed Point Theorem, KKM Theorem, Fan Minimax Inequality, Liu Minimax Inequality, Sion Minimax Theorem and other minimax theorems due to first author and his collaborators.

In the sequel X, Y will always denote non-empty convex subsets of real separated topological vector spaces; X and Y will be equipped with the induced topology which, clearly, coincides with the Euclidean topology, if they are finite-dimensional.

A function $f : X \times Y \to \overline{\mathbb{R}}$ is said to be *u.s.c./l.s.c.* (resp. *quasi concave-convex*), if for every $x_0 \in X, y_0 \in Y$ and $r \in \mathbb{R}$ the sets $\{x \in X : f(x, Jy_0) \geq r\}$ and $\{y \in Y : f(x_0, y) \leq r\}$ are closed (resp. convex) in X (resp. Y). Moreover, f is said to have a *saddle value*, if $\inf_Y \sup_X f = \sup_X \inf_Y f$.

A function $f : X \times Y \to \overline{\mathbb{R}}$ is said to be *marginally u.s.c. on X* (resp. *marginally l.s.c. on Y*), if for every open subset V of Y (resp. U of X)

$$x \mapsto \inf_{y \in V} f(x, y) \quad \textit{is u.s.c. on } X \quad (\textit{resp.} \quad y \mapsto \sup_{x \in U} f(x, y) \quad \textit{is l.s.c. on } Y).$$

Moreover, if f is both marginally u.s.c. on X and marginally l.s.c. on Y, f is said to be *marginally closed.* Obviously, every u.s.c./l.s.c. function is marginally closed, because u.s.c. on X (resp. l.s.c. on Y) implies marginal u.s.c. on X (resp. marginal l.s.c. on Y). It was proved in Greco-Bassanezi [1] that every marginally closed, quasi concave-convex function on $X \times Y$

B. Ricceri and S. Simons (eds.), Minimax Theory and Applications, 41–51.

has saddle value, whenever X and Y are convex subsets of Hausdorff locally convex spaces and either X or Y is both compact and finite-dimensional.

Marginally closed functions are the functional counterpart of "lower semicontinuous multifunctions"; they appear in Greco [2] and are strictly related to "topologically closed functions" which were investigated by the first author from the point of view of minimax theorems [3] and of the convergences of functions preserving saddle points and saddle values [2]. A function $f : X \times Y \to \overline{\mathbb{R}}$ is called *topologically closed*, if for every $(x, y) \in X \times Y$ and for every open subset U of X and open subset V of Y:

$$\inf_{y' \in V} f(x, y') = \inf_{y' \in V} \limsup_{x' \to x} f(x', y') \; and \; \sup_{x' \in U} f(x', y) = \sup_{x' \in U} \liminf_{y' \to y} f(x', y').$$

Topologically closed functions are marginally closed. It was proved in Greco [3] that every quasi concave-convex toplogically closed function has saddle value, whenever either X or Y is compact.

A third class of "minimax functions" is given by the "quasi-topologically closed functions". A function $f : X \times Y \to \overline{\mathbb{R}}$ is called *quasi-topologically closed*, if for every $(x, y) \in X \times Y$ and for every open subset U of X and open subset V of Y:

$$\limsup_{x' \to x} \inf_{y' \in V} f(x', y') = \inf_{y' \in V} \limsup_{x' \to x} f(x', y') \qquad and$$
$$\liminf_{y' \to y} \sup_{x' \in U} f(x', y') = \sup_{x' \in U} \liminf_{y' \to y} f(x', y').$$

Using the minimax theorem for topologically closed functions, it is easy to check that every quasi convex-concave, quasi-topologically closed function has saddle value, whenever either X or Y is compact.

Let the symbols $\mathcal{UL}$, $\mathcal{T}$, $\mathcal{T}_q$ and $\mathcal{M}$ denote the families of all functions on $X \times Y$ which are u.s.c./l.s.c., topologically closed, quasi-topologically closed and marginally closed, respectively. Then one has that

$$\mathcal{UL} \subset \mathcal{T} = \mathcal{T}_q \cap \mathcal{M}.$$

In order to gain insigths into topologically closed and marginally closed functions and to perceive how using them, the reader can start to consider functions which are either indicator or characteristic functions of subsets of $X \times Y$.

For example, let $\Gamma : X \twoheadrightarrow Y$ be a *multifunction* and let Γ^* denote its *dual*, that is: $\Gamma^* : Y \twoheadrightarrow X$ is defined, for every $y \in Y$, by $\Gamma^* y := \{x \in X : y \notin \Gamma x\}$. And let $\psi_\Gamma : X \times Y \to \{0, +\infty\}$ denote the *indicator function* of

Γ which is defined as equal 0 on Γ and $+\infty$ otherwise. Then the following equivalences

$$\psi_\Gamma \textit{ is u.l.s./l.s.c.} \iff \textit{the values of } \Gamma \textit{ and } \Gamma^* \textit{ are closed} \quad (1)$$

$$\psi_\Gamma \textit{ is topologically closed} \iff \left\{ \begin{array}{ll} \cap_{x\in U}\, \Gamma x = \cap_{x\in U}\overline{\Gamma x} & \forall U \textit{ open} \subset X \\ \cap_{y\in V}\, \Gamma^* y = \cap_{y\in V}\overline{\Gamma^* y} & \forall V \textit{ open} \subset Y \end{array} \right\} \quad (2)$$

$$\psi_\Gamma \textit{ is marginally closed} \iff \left\{ \begin{array}{lll} \cap_{x\in U}\, \Gamma x & \textit{is closed} & \forall\, U \textit{ open} \subset X \\ \cap_{y\in V}\, \Gamma^* y & \textit{is closed} & \forall\, V \textit{ open} \subset Y \end{array} \right\} \quad (3)$$

$$\psi_\Gamma \textit{ is quasi-topol. closed} \iff \left\{ \begin{array}{ll} \overline{\cap_{x\in U}\Gamma x} = \cap_{x\in U}\overline{\Gamma x} & \forall\, U \textit{ open} \subset X \\ \overline{\cap_{y\in V}\Gamma^* y} = \cap_{y\in V}\overline{\Gamma^* y} & \forall\, V \textit{ open} \subset Y \end{array} \right\} \quad (4)$$

hold. Each of the equalities:

$$\cap_{x\in U}\Gamma x = \cap_{x\in U}\overline{\Gamma x}, \quad \cap_{x\in U}\Gamma x \textit{ is closed}, \quad \overline{\cap_{x\in U}\Gamma x} = \cap_{x\in U}\overline{\Gamma x}$$

is a weaker form than the usual closedness of the values of Γ. Hence, working with either topologically closed, marginally closed or quasi-topologically closed functions amounts to work with multifunctions whose values (not necessarily closed) verify some weak form of closedness. In the sequel, the multifunctions Γ such that:

$$\cap_{x\in U}\Gamma x \quad \textit{is closed} \quad \forall\, U \textit{ open} \subset X,$$

will play an important role; they are called *multifunctions with marginally closed values.*

To complete the scheme offered by (1)-(4), it is worth recalling these obvious equivalences:

$$\psi_\Gamma \textit{ is quasi concave-convex} \iff \textit{the values of } \Gamma \textit{ and } \Gamma^* \textit{ are convex} \quad (5)$$

$$\psi_\Gamma \textit{ has saddle value} \iff \cap_{x\in X}\Gamma x \neq \emptyset \textit{ or } \cap_{y\in Y}\, \Gamma^* y \neq \emptyset \quad (6)$$

1. Example. Let $h : X \to Y$ be a continuous function and $\Omega_1 = \{(x, y) \in X \times Y : y = h(x)\}$. Its indicator function ψ_{Ω_1} (equal 0 on Ω_1, and $+\infty$ otherwise) is marginally closed, but not topologically closed. $\diamond$

2. Example. Let $X = Y = [0, 1]$. Let $\Omega_2 = \{(x, y) \in X \times Y : y \geq 2x \;\; \textit{and} \;\; 0 \leq x < 1/2\} \cup \{(x, y) \in X \times Y : y < 2(x - 1/2) \;\; \textit{and} \;\; 1/2 <$

$x \leq 1\}$. Its indicator function ψ_{Ω_2} is quasi concave-convex; moreover, ψ_{Ω_2} is topologically closed (hence, marginally closed), but neither u.s.c. on X nor l.s.c. on Y (hence ψ_{Ω_2} is not u.s.c./l.s.c.). ◇

3. Example. Let $f : X \times Y \to \overline{\mathbb{R}}$ be a function. Define the functions f^- (the *lower regularization* of f) and f^+ (the *upper regularization* of f) from $X \times Y$ to $\overline{\mathbb{R}}$ by

$$f^-(x,y) = \liminf_{y' \to y} f(x,y') \qquad \textit{and} \qquad f^+(x,y) = \limsup_{x' \to x} f(x',y).$$

Then the functions f^{+-}, f^{-+} given by

$$f^{+-}(x,y) = \liminf_{y'\to y} \limsup_{x'\to x} f(x',y')$$
$$f^{-+}(x,y) = \limsup_{x'\to x} \liminf_{y'\to y} f(x',y')$$

are topologically closed (hence, marginally closed). Observe that f is topologically closed if and only if $f^{+-} \leq f \leq f^{-+}$. Moeover, if f is topologically closed, then all functions $g : X \times Y \to \overline{\mathbb{R}}$ such that $f^{+-} \leq g \leq f^{-+}$, are topologically closed.◇

4. Example. Let $f : X \times Y \to \overline{\mathbb{R}}$ be a quasi concave-convex function. Define the function $f^{(+-)}$, $f^{(-+)} : X \times Y \to \overline{\mathbb{R}}$ by

$$f^{(+-)}(x,y) = \sup_{V \in \mathcal{N}(y)} \limsup_{x'\to x} \inf_{y' \in V} f(x',y')$$
$$f^{(-+)}(x,y) = \inf_{U \in \mathcal{N}(x)} \liminf_{y'\to y} \sup_{x' \in U} f(x',y'),$$

where $\mathcal{N}(x)$ and $\mathcal{N}(y)$ are the filters of neighbourhood of x and y, respectively. It was proved in Greco [3] that the functions $f^{(+-)}$ and $f^{(-+)}$ have saddle value, whenever either X or Y is a compact subset of a locally convex topological vector space. Observe that the functions f such that either $f^{(+-)} = f$ or $f^{(-+)} = f$ are both quasi concave-convex and marginally closed [3]. More generally, a function f is marginally closed if and only if $f^{(+-)} \leq f \leq f^{(-+)}$ [2]. ◇

A *multifunction* Γ from X to Y, denoted by $X \twoheadrightarrow Y$, is a subset of $X \times Y$; its *inverse* Γ^- from Y to X is defined by $\Gamma^- = \{(y,x) \in Y \times X : (x,y) \in \Gamma\}$; the sets $\Gamma x := \{y \in Y : (x,y) \in \Gamma\}$ are called the *values* of Γ; the sets $\Gamma^- y := \{x \in X : (x,y) \in \Gamma\}$ are called the *fibers* of Γ.

The multifunction Γ is said to be *lower semicontinuous*, if, for every open set V in Y, the set $\{x \in X : V \cap \Gamma x \neq \emptyset\}$ is open in X or, equivalently,

$$\cap_{y \in V} \Gamma^* y \quad \textit{is closed in } X.$$

Therefore, the indicator function ψ_Γ is marginally u.s.c. on X if and only if Γ is lower semicontinuous or, equivalently, if and only if its dual multifunction Γ^* has marginally closed values. Obviously, multifunctions with open fibers are lower semicontinuous.

Remark that the definition of lower semicontinuous multifunctions does not require neither the closedness of the values nor their non-emptiness.

Moreover, observe that a function $f : X \times Y \to \overline{\mathbb{R}}$ is marginally u.s.c. on X if and only if, for every real number r, the multifunction $\Gamma_r := \{(x, y) \in X \times Y : f(x, y) < r\}$ from X to Y is lower semicontinuous; dually, f is marginally l.s.c. if and only if, for every real number r, the multifunction $\Gamma_r := \{(y, x) \in Y \times X : f(x, y) > r\}$ from Y to X is lower semicontinuous. In terms of epigraph and hypograph, the function f is marginally u.s.c. if and only if the hypograph

$$\text{hypo}(f) : Y \twoheadrightarrow X \times \mathbb{R} \qquad \textit{defined by}$$
$$\text{hypo}(f) := \{(y, x, r) \in Y \times X \times \mathbb{R} : f(x, y) \geq r\}$$

is lower semicontinuous multifunction; dually, f is marginally l.s.c. if and only if the epigraph

$$\text{epi}(f) : X \twoheadrightarrow Y \times \mathbb{R} \qquad \textit{defined by}$$
$$\text{epi}(f) := \{(x, y, r) \in X \times Y \times \mathbb{R} : f(x, y) \leq r\}$$

is lower semicontinuous multifunction.

As usual, a function f is said to be *quasi concave on* X (resp. *quasi convex on* Y), if for every real number $r \in \mathbb{R}$ and $y_0 \in Y$ (resp. $x_0 \in X$), the set $\{x \in X : f(x, y_0) \geq r\}$ (resp. $\{y \in Y : f(x_0, y) \leq r\}$) is convex.

Theorem A (Minimax Inequality for Marginally Semicontinuous Functions). *Let X, Y be finite-dimensional convex sets such that either X or Y is compact. Let $f, g : X \times Y \to \overline{\mathbb{R}}$ be functions with $f \leq g$. If f is both marginally l.s.c. on Y and quasi concave on X and if g is both marginally u.s.c. on X and quasi convex on Y, then $\inf_Y \sup_X f \leq \sup_X \inf_Y g$.*

The hypothesis of finite-dimensionality of the convex sets is not a *sine cura*: It is not known whether in Theorem A it is possible to drop this hypothesis.

Moreover, Theorem A is not harmless. Now, we will give *four "equivalent" reformulations of Theorem A*, in terms of intersection property, of coincidence property, of fixed point property and of KKM property, respectively.

5. Proposition (Intersection Property for Multifunctions with Margin-ally Closed Values). *Let X, Y be finite-dimensional convex* *sets such that either X or Y is compact. Let $\Delta : X \twoheadrightarrow Y$ be a multifunction such that*

$$\cap_{x \in U} \Delta x \quad \text{is closed for every open subset } U \text{ of } X \tag{7}$$

$$\Delta \tilde{x} \subset \Delta x_0 \cup \Delta x_1 \quad \text{for every } x_0, x_1 \in X \text{ and } \tilde{x} \in [x_0, x_1]. \tag{8}$$

Then $\cap_{x \in X} \Delta x \neq \emptyset$ if and only if there is a lower semicontinuous multifunction $\Omega : X \twoheadrightarrow Y$ with non-empty convex values such that $\Omega \subset \Delta$.

Proof. Define $f := \psi_\Delta$ and $g := \psi_\Omega$ and apply Theorem A to f and g. $\diamond$

6. Proposition (Coincidence Property for Lower Semicontinuous Multifunctions). *Let X, Y be finite-dimensional convex sets such that either X or Y is compact. If $\Omega : X \twoheadrightarrow Y$ and $\Gamma : Y \twoheadrightarrow X$ are lower semicontinuous multifunctions with non-empty convex values (not necessarily closed !), then $\Omega \cap \Gamma^- \neq \emptyset$.*

Proof. Define $\Delta := \Gamma^*$ and apply Proposition 5 to the multifunctions Ω and Δ. $\diamond$

7. Proposition (Fixed Point Property for Lower Semicontinuous Multifunctions). *Let X be a finite-dimensional compact convex set. If $\Omega : X \twoheadrightarrow X$ is a lower semicontinuous multifunction with non-empty convex values (not necessarily closed !), then there exists x such that $x \in \Omega x$.*

Proof. Define $\Gamma : X \to X$ by $\Gamma x := \{x\}$ and apply Proposition 6 to the multifunctions Ω and Γ. $\diamond$

Let $[A]$ denote the convex hull of A.

8. Proposition (KKM Property for Multifunctions with with Margin-ally Closed Values). *Let X be a finite-dimensional compact convex set and let $\Delta : X \twoheadrightarrow X$ be a multifunction. Then $\cap_{x \in X} \Delta x \neq \emptyset$, if Δ verifies such that*

$$\cap_{x \in U} \Delta x \quad \text{is closed for every open subset } U \text{ of } X \tag{9}$$

$$[D] \subset \cup_{x \in D} \Delta x \quad \text{for every finite subset } D \subset X. \tag{10}$$

Proof. Property (9) means that Δ^* is lower semicontinuous; hence the multifunction $\Omega : X \twoheadrightarrow X$ defined, for every $x \in X$, by $\Omega x := [\Delta^* x]$ is lower semicontinuous. Arguing by contradiction, assume that $\cap_{x \in X} \Delta x = \emptyset$; it means that the values of Ω are non-empty. Using Proposition 7, we have that there is $\tilde{x}$ such that $\tilde{x} \in \Omega \tilde{x}$. Therefore, from the definition of Ω it follows that there is a finite set D such that:

$$D \subset \Delta^* \tilde{x} \qquad e \qquad \tilde{x} \in [D]. \tag{11}$$

But $D \subset \Delta^* \tilde{x}$ amounts to $\tilde{x} \notin \cup_{x \in D} \Delta x$, so that (11) contradicts (9). ◇

Proposition 6 yields the coincidence property due to Fan [6]: "*If either X or Y is compact, then $\Omega \cap \Gamma^- \neq \emptyset$, whenever $\Omega : X \twoheadrightarrow Y$ and $\Gamma : Y \twoheadrightarrow X$ have open fibers and non-empty convex values*", because multifunctions with open fibers are lower semicontinuous. Using Theorem A, in Corollary 11 we will show the Liu Minimax Inequality which, as well-known, is the functional counterpart of this coincidence property of Fan.

Proposition 7 includes the celebrated Brouwer Fixed Point Theorem and the fixed point theorem due to Fan [7]: "*There is x with $x \in \Omega x$, if X is compact and $\Omega : X \twoheadrightarrow X$ has open fibers and non-empty convex values*". The well-known functional counterpart of this Fan Fixed Point Property is the famous Fan Minimax Inequality which we will show in Corollary 10.

Proposition 8 yields the classical KKM principle (see Lassonde [9-10] and Dugundji-Granas [11]): "*$\cap_{x \in X} \Delta x \neq \emptyset$, if X is compact and the multifunction $\Delta : X \twoheadrightarrow X$ has closed values and verifies (10)*".

Unfortunately our proof of Theorem A which is based on the finite-dimensional Michael Selection Theorem, requires the Brouwer Fixed Point Theorem.

Proof of the Minimax Inequality. Assume X compact. Arguing by contradiction, suppose that there exists a real number r such that $\inf_Y \sup_X f > r > \sup_X \inf_Y g$. This implies that the multifunctions $\Omega : X \twoheadrightarrow Y$ and $\Delta : Y \twoheadrightarrow X$, defined by

$$\Omega x := \{y \in Y : g(x, y) < r\} \quad \text{and} \quad \Delta y := \{x \in X : f(x, y) > r\},$$

have non-empty values. Moreover, the values of Ω and Δ are convex, because g and f are quasi convex on Y and quasi concave on X, respectively. Since $\{x \in X : V \cap \Omega x \neq \emptyset\} = \{x \in X : \inf_{y \in V} g(x, y) < r\}$ and $\{y \in Y : U \cap \Delta y \neq \emptyset\} = \{y \in Y : \sup_{x \in U} f(x, y) > r\}$, we have that the multifunctions Ω and Δ are lower semicontinuous, because g and f are marginally u.s.c and marginally l.s.c., respectively. Hence, since the values of Ω and Δ are finite-dimensional, by Michael Selection Theorem [4, thm. 3.1"', pag. 368], there are continuous selections ω and δ of Ω and Δ, respectively. Their composition $\delta \circ \omega$ is a continuous function from the finite-dimensional compact convex set X to itself. Hence, from Brouwer Fixed Point Theorem, there are $y_0 \in Y$ and $x_0 \in X$ such that $\omega(x_0) = y_0$ and $\delta(y_0) = x_0$. Therefore, being ω and δ selections of Ω and Δ respectively, we have $g(x_0, y_0) < r < f(x_0, y_0)$, contradicting the inequality $f \leq g$. ◇

If $f = g$, from Theorem A we have:

9. Corollary (Minimax Theorem [1]**).** *Let X, Y be finite-dimensional convex sets such that either X or Y is compact. Then every quasi concave-convex, marginally closed function on $X \times Y$ has saddle value.*

10. Corollary (Fan Minimax Inequality [7]**).** *Let X be a compact convex set (not necessarily finite-dimensional). Let $f : X \times X \to \overline{\mathbb{R}}$ be a function which is quasi concave in the first variable and l.s.c. in the second one. Then* $\min_{y \in X} \sup_{x \in X} f(x,y) \leq \sup_{x \in X} f(x,x)$.

Proof. Let $\mathcal{P}_f(X)$ be the family of the non-empty finite subsets of X; for every $D \in \mathcal{P}_f(X)$, let $[D]$ denote the convex hull of D. Define $g : [D] \times [D] \to \overline{\mathbb{R}}$ by $g(x,y) = +\infty$, if $x \neq y$, and $= \sup_{x \in X} f(x,x)$ otherwise. The function g is quasi convex in the second variable, is marginally u.s.c and, moreover, $f \leq g$ on $[D]$. Hence, by Theorem A, we have:

$$\inf_{y \in X} \sup_{x \in [D]} f(x,y) \leq \inf_{y \in [D]} \sup_{x \in [D]} f(x,y) \leq$$

$$\leq \sup_{x \in [D]} \inf_{y \in [D]} g(x,y) = \sup_{x \in X} f(x,x).$$

On the other hand, the compactness of X and the l.s.c. of f imply the inequality

$$\inf_{y \in X} \sup_{x \in X} f(x,y) = \sup_{D \in \mathcal{P}_f(X)} \inf_{y \in X} \sup_{x \in [D]} f(x,y);$$

this completes the proof. ◇

11. Corollary (Liu Minimax Inequality [8]**).** *Let X, Y be convex sets (not necessarily finite-dimensional) such that either X or Y is compact. Let $f, g : X \times Y \to \overline{\mathbb{R}}$ be functions with $f \leq g$. If f is both quasi concave on X and l.s.c. on Y and if g is both u.s.c. on X and quasi convex on Y, then* $\inf_Y \sup_X f \leq \sup_X \inf_Y g$.

Proof. Assume Y compact. Clearly the corollary is an immediate consequence of Theorem A, if X and Y are finite-dimensional. Hence, if $[D]$ and $[F]$ are the convex hulls of finite non-empty subsets $D \subset X$ and $F \subset Y$, we have:

$$\inf_{[F]} \sup_{[D]} f \leq \sup_{[D]} \inf_{[F]} g. \tag{12}$$

Let $\mathcal{P}_f(X)$ and $\mathcal{P}_f(Y)$be the families of the non-empty finite subsets of X and of Y, respectively. From the compactness of $[D]$ and the u.s.c. of g we have

$$\inf_{F \in \mathcal{P}_f(Y)} \sup_{[D]} \inf_{[F]} g = \sup_{[D]} \inf_{Y} g\,.$$

Hence by (12) we obtain:

$$\inf_{Y} \sup_{[D]} f \leq \sup_{[D]} \inf_{Y} g. \tag{13}$$

Now, the compactness of Y and the l.s.c. of f imply:

$$\inf_{Y} \sup_{X} f = \sup_{D \in \mathcal{P}_f(X)} \inf_{Y} \sup_{[D]} f\,;$$

hence, using (13), we have the required inequality. ◇

12. Corollary (Minimax Theorem for Topologically Closed Functions [3]). *Let X, Y be convex sets such that either X or Y is compact. Every quasi concave-convex topologically closed function $f : X \times Y \to \overline{\mathbb{R}}$ has saddle value.*

Proof. From Liu Minimax Inequality it follows that

$$\inf_Y \sup_X f^- \leq \sup_X \inf_Y f^+.$$

On the other hand it is clear that

$$\sup_X f = \sup_X f^- \quad \textit{and} \quad \inf_Y f = \inf_Y f^+,$$

because f is topologically closed. Hence, $\inf_Y \sup_X f \leq \sup_X \inf_Y f$; that is, f has saddle value. ◇

In the settings of quasi-concave-convex functions, Liu Minimax Inequality is "equivalent" to minimax theorem for topologically closed functions. More precisely, let $\mathcal{QCC}$ denote the family of all quasi concave-convex functions on $X \times Y$; set

$$\begin{aligned}
\mathcal{U}^{qcc} &:= \{f \in \mathcal{QCC} : f \text{ is u.s.c. on } X\} \\
\mathcal{L}^{qcc} &:= \{f \in \mathcal{QCC} : f \text{ is l.s.c. on } Y\} \\
\mathcal{T}^{qcc} &:= \{f \in \mathcal{QCC} : f \text{ is topologically closed }\} \\
\mathcal{T}_q^{qcc} &:= \{f \in \mathcal{QCC} : f \text{ is quasi-topologically closed }\} \\
\mathcal{T}_g^{qcc} &:= \{f \in \mathcal{QCC} : \inf_Y f^+ = \inf_Y f \text{ and } \sup_X f^- = \sup_X f\};
\end{aligned}$$

then we have the following equivalences.

13. Corollary. *Let X and Y be convex sets such that either X or Y is compact. Then the following properties are equivalent:*

$$\inf_Y \sup_X f \leq \sup_X \inf_Y g \qquad \forall f \in \mathcal{L}^{qcc}, \forall g \in \mathcal{U}^{qcc} \text{ such that } f \leq g \tag{14}$$

$$\inf_Y \sup_X f^- \leq \sup_X \inf_Y f^+ \qquad \forall f \in \mathcal{QCC} \tag{15}$$

$$\inf_Y \sup_X f = \sup_X \inf_Y f \qquad \forall f \in \mathcal{T}^{qcc} \tag{16}$$

$$\inf_Y \sup_X f^{+-} = \sup_X \inf_Y f^{+-} \qquad \forall f \in \mathcal{QCC} \tag{17}$$

$$\inf_Y \sup_X f^{-+} = \sup_X \inf_Y f^{-+} \qquad \forall f \in \mathcal{QCC} \tag{18}$$

$$\inf_Y \sup_X f = \sup_X \inf_Y f \qquad \forall f \in \mathcal{T}_q^{qcc} \tag{19}$$

$$\inf_Y \sup_X f = \sup_X \inf_Y f \qquad \forall f \in \mathcal{T}_g^{qcc} \diamond \tag{20}$$

Every topologically closed function is marginally closed. But there are margin-ally closed functions which are not topologically closed; for example, the function ψ_{Ω_1} in Example 1. In spite of this fact, we have the following consequence of theorem A.

Theorem B. *Let X, Y be finite-dimensional convex sets such that X or Y is locally compact. Then every quasi concave-convex, marginally closed function on $X \times Y$ is topologically closed.*

Proof. Assume X locally compact. Let $h : X \times Y \to \overline{\mathbb{R}}$ be a quasi-concave-convex marginally closed function. Choose convex open non-empty subsets $U \subset X$ and $V \subset Y$; and suppose $\overline{U}$ compact. Define the functions $f, g : \overline{U} \times V \to \overline{\mathbb{R}}$ by

$$f(x,y) := \begin{cases} h(x,y) & \text{if } (x,y) \in U \times V \\ -\infty & \text{otherwise} \end{cases} \qquad \text{and} \quad g(x,y) := h(x,y)$$

Observe that $\inf_V \sup_{\overline{U}} f = \inf_V \sup_U h$.
Apply Theorem A to have $\inf_V \sup_{\overline{U}} f \leq \sup_{\overline{U}} \inf_V h$. Therefore we obtain

$$\inf_V \sup_U h \leq \sup_{\overline{U}} \inf_V h. \tag{21}$$

Now, using (21), we will prove the inequality $h \leq h^{-+}$; analogously, the reader can show the other inequality $h^{+-} \leq h$ which remains to prove, for having the required topological closedness of h. Choose the points $x \in X$ and $y \in Y$. Let $\mathcal{N}(y)$ denote the system of convex open nighbourhoods V of y in Y. By (21) we have $\sup_{V \in \mathcal{N}(y)} \inf_V \sup_U h \leq \sup_{V \in \mathcal{N}(y)} \sup_{\overline{U}} \inf_V h$; so that, being h marginally l.s.c. on Y, we obtain $\sup_{x' \in U} h(x', y) \leq \sup_{x' \in \overline{U}} h^-(x', y)$; hence

$$h(x,y) \leq \sup_{x' \in \overline{U}} h^-(x', y) \tag{22}$$

for every U belonging to the family $\mathcal{B}(x)$ of the convex, relatively compact, open neighbourhoods of x in X. Since X is locally compact, the family $\{\overline{U} : U \in \mathcal{B}(x)\}$ is a base of the system of neighbourhoods of x; therefore (22) implies $h(x,y) \leq \inf_{U \in \mathcal{B}(x)} \sup_{x' \in \overline{U}} h^-(x', y) = h^{-+}(x, y)$. ◇

References

1. R.C. Bassanezi and G.H. Greco, *A minimax theorem for marginally u.s.c./l.s.c. functions,* Topol. Methods Nonlinear Anal. **5** (1995), 249-253.
2. G.H. Greco, *Théorème des minimax locaux et fonctions topologiquement fermées,* Partial Differential Equations and the Calculus of Variations. Essays in Honor of Ennio De Giorgi, F. Colombini, A. Marino, L. Modica, S. Spagnolo (eds.), Birkhäuser, Boston, vol. II, 1989, 589-631.

3. G.H. Greco, *Minimax theorems and saddling transformations,* J. Math. Anal. Appl. **147** (1990), 180-197.
4. E. Michael, *Continuous selections I,* Ann. Math. **63** (1956), 361-382.
5. M. Sion, *On general minimax theorems,* Pacific J. Math. **8** (1958), 171-176.
6. K. Fan, *Applications of a theorem concerning sets with convex sections,* Math. Ann. **163** (1966), 189-203.
7. K. Fan, *A minimax inequality and applications,* Inequalities III, O. Shisha (ed), Academic Press, New York, 1972.
8. F.C. Liu, *A note on the von Neumann-Sion minimax principle,* Bull. Inst. Math. Acad. Sinica **6** (1978), 517-524.
9. M. Lassonde, *On the use of KKM multifunctions in fixed point theory and related topics,* J. Math. Anal. Appl. **97** (1983), 151-201.
10. M. Lassonde, *Sur le principe KKM,* C. R. Acad. Sci. Paris, Série I, **310** (1990), 573-576.
11. J. Dugundji and A. Granas, *Fixed Point theory I,* Monografie Matematyczne 61, Varsaw, 1982.

ON VARIATIONAL MINIMAX PROBLEMS UNDER RELAXED COERCIVITY ASSUMPTIONS

JOACHIM GWINNER
Institute of Mathematics
Department of Aerospace Engineering
Universität der Bundeswehr München
D-85577 Neubiberg, Germany
e-mail: `111bgwin@kommsrv.rz.unibw-muenchen.de`

Abstract. In this contribution we continue our research on inf-sup theorems on noncompact sets and investigate convex variational minimax problems in Hilbert spaces under relaxed coercivity assumptions. In addition to the issue of existence we are concerned with well–posedness in the sense of Hadamard.

First we deal with variational minimax problems, where in the equivalent variational inequality formulation, the associated bilinear form satisfies a Gårding inequality, only. In this case, we give new results on boundedness of the solution set and on uniqueness.

Then we turn to variational minimax problems which arise from the Lagrangian associated to an operator equality constraint, in addition to unilateral constraints. Here we extend the classical saddle point theory due to Brezzi including unilateral constraints. Only in a special situation, the standard decomposition works and well–posedness can be easily derived, whereas in the general case by a novel decomposition procedure we can establish solvability invoking the Hahn-Banach Theorem.

1. Introduction

In this contribution we continue our research on inf-sup theorems on non-compact sets [6], [9] extending duality theory of convex optimization. Motivated by unilateral boundary value problems arising in mathematical physics, we study the following variational problem:

$$\text{Find} \quad [u, v] \in K \times Y \quad \text{such that}$$

$$(P) \quad \begin{cases} a(u, x-u) + b(x-u, v) \geq f(x-u), & \forall\, x \in K\,, \\ -b(u, y) + c(v, y) = g(y), & \forall\, y \in Y\,. \end{cases}$$

B. Ricceri and S. Simons (eds.), Minimax Theory and Applications, 53–69.

Here K is a convex closed (generally not compact) nonvoid subset in a Hilbert space X, Y an additional Hilbert space; moreover $a \in \mathcal{L}(X \times X, \mathbf{R})$, $b \in \mathcal{L}(X \times Y, \mathbf{R})$, $c \in \mathcal{L}(Y \times Y, \mathbf{R})$ are continuous bilinear forms and the right hand sides $f \in X' = \mathcal{L}(X, \mathbf{R})$ and $g \in Y'$ are given. Throughout we assume $a \geq 0$, i.e. nonnegativity of the associated quadratic form, and also $c \geq 0$.

We call this problem (P) a " variational minimax problem", since under additional symmetry assumptions (see the subsequent section 2) the sought pair $[u, v]$ is a saddle point of a related convex-concave potential function J on $K \times Y$. Here we are interested in the noncoercive situation; that is, we do neither assume $J(x, y_0) \to \infty$ as $\|x\| \to \infty$ and $J(x_0, y) \to -\infty$ as $\|y\| \to \infty$ for some $x_0 \in K$, $y_0 \in Y$, nor we assume X-ellipticity of a $\Big($that is $(\exists\, \alpha > 0)(\forall\, x \in X)\, \alpha\|x\|^2 \leq a(x, x)\Big)$ and Y-ellipticity of c. In addition to the issue of existence we are concerned with well–posedness in the sense of Hadamard; for a recent study of well–posedness in the sense of Tikhonov we refer to [15] and the references therein.

In the first part of our analysis we instead assume that the bilinear form

$$\beta\left([u, v], [x, y]\right) = a(u, x) + c(v, y) + b(x, v) - b(u, y)$$

satisfies a Gårding inequality: $\exists\, \gamma > 0$, $G : Z = X \times Y \to Z'$ compact such that

$$\gamma\|z\|^2 \leq \beta(z, z) + < Gz, z >, \quad \forall\, z \in Z\,.$$

To motivate the study of such a semicoercive variational inequality, in section 3 we sketch the harmonic Signorini problem and its reformulation as a variational minimax problem by boundary integral operator methods. We point out that the considered unilateral problem is only a scalar model of the Signorini-Neumann contact problems in linear elastostatics and that a similar boundary integral approach leads to such a variational minimax problem, too (see [8]).

It is well known by the pioneering work of Fichera and Stampacchia that in this semicoercive situation, extra conditions on the data are needed to conclude solvability. Moreover, (see e.g. [12] chapters I and VII) these problems are generally ill–posed lacking uniqueness.

Here in section 4, for existence we apply the general theory of [2] and adapt their "compatibility condition" to our setting. We also show that under a slightly sharpened "compatibility condition" the solution set is bounded and under stronger hypotheses uniqueness of the solution of the variational minimax problem.

In the second part of our analysis (section 5) we deal with variational minimax problems which arise from the Lagrangian associated to an operator equality constraint, in addition to unilateral constraints. This leads to variational problems of the form (P) where $c = 0$. Thus we extend the classical saddle point theory due to Brezzi [4] including unilateral constraints.

To treat this more delicate situation, we introduce the operator B associated to the bilinear form b and its kernel $\mathcal{N}(B) =: X_1$. As in the classical saddle point theory, where only equality constraints, but no inequality constraints are considered, we assume that $B : X_1^\perp \to Y'$ is an isomorphism and that a is X_1 - elliptic, only. This gives rise to a block decomposition of the variational problem.

At first in section 5 we make precise this standard block decomposition used in the classical saddle point theory. We illustrate this block decomposition by a simple finite dimensional example taken from the german text book of Hackbusch [10], which demonstrates unique solvability of the associated augmented linear system for any right hand side, although the matrix A associated to the bilinear form a has not full rank. Then by means of this block decomposition we can treat a special case of the convex set K, where K splits into two components and the second component is an appropriate subspace. In this special case we can easily prove well–posedness results. Finally we treat the general case of K, not necessarily a subspace. Here by a novel decomposition procedure we can establish solvability of (P). However, since we invoke a Hahn-Banach separation argument, we need as an additional topological hypothesis that int $K \neq \emptyset$.

2. Some Preliminary Remarks

In this preliminary section we exhibit the relation between convex-concave saddle point problems and the variational minimax problems studied in this paper. Furthermore we shortly review the standard existence theory in the coercive case.

Let K, respectively L be a nonvoid closed convex subset of a Hilbert space X, respectively of a Hilbert space Y. Let $a \in \mathcal{L}(X \times X, \mathbb{R}), b \in \mathcal{L}(X \times Y, \mathbb{R}), c \in \mathcal{L}(Y \times Y, \mathbb{R})$ continuous bilinear forms and the continuous linear forms $f \in X' = \mathcal{L}(X, \mathbb{R})$ and $g \in Y'$ be given. Then one can define the function

$$J(x,y) = \frac{1}{2}a(x,x) + b(x,y) - \frac{1}{2}c(y,y) - f(x) + g(y)$$

and one is interested in the solution of the following *saddle point problem:* Find $[u, v] \in K \times L$ such that

$$J(u,y) \leq J(u,v) \leq J(x,v) \,, \quad \forall [x,y] \in K \times L \,. \tag{2.1}$$

If in addition a is *symmetric*, that is,

$$a(x,x') = a^*(x,x') := a(x',x)\ , \quad \forall\, [x,x'] \in X \times X$$

and $a \geq 0$ (*nonnegativity* of the quadratic form), that is

$$a(x,x) \geq 0 \qquad \forall\, x \in X,$$

and if in addition likewise $c = c^*, c \geq 0$, then J is convex-concave. In this situation, (2.1) is equivalent, using the Gateaux derivative, to

$$J'_x([u,v]; x-u) \geq 0,\ -J'_y([u,v]; y-v) \geq 0 \qquad \forall\, [x,y] \in K \times L.$$

Since an easy computation gives

$$\begin{aligned} J'_x([u,v]; x') &= a(u,x') + b(x',v) - f(x'), \\ J'_y([u,v]; y') &= b(u,y') - c(v,y') + g(y'), \end{aligned}$$

the saddle point problem (2.1) is in the convex situation finally equivalent to the following problem: Find $[u,v] \in K \times L$ such that

$$\begin{cases} a(u,x-u) + b(x-u,v) & \geq f(x-u)\ , \quad \forall\, x \in K; \\ -b(u,y-v) + c(v,y-v) & \geq g(y-v)\ , \quad \forall\, y \in L. \end{cases}$$

Clearly if the convex subset L coincides with the entire space Y, then the second variational inequality above is equivalent to the variational equality

$$-b(u,y) + c(v,y) = g(y)\ , \quad \forall y \in Y$$

and we arrive at a variational minimax problem.

Concerning existence, standard theory (see e.g. [1, Théorème 4.2, Corollaire 4.2], [3, Corollary 3.8]) presumes that for some $[x_0, y_0] \in K \times L$, the functions $y \mapsto -J(x_0, y), x \mapsto J(x, y_0)$ are weakly inf-compact or coercive in the sense that

$$[J(x,y_0) - J(x_0,y)] \to \infty \quad \text{for } x \in K\,, y \in L \text{ and } \|x\| + \|y\| \to \infty\,.$$

There is another approach to existence. Introducing the bilinear form

$$\beta([x,y],[x',y']) = a(x,x') + c(y,y') + b(x',y) - b(x,y') \tag{2.2}$$

we find the variational minimax problem (P) equivalent to the single variational inequality: Find $[u,v] \in K \times L$ such that

$$\beta([u,v],[x,y] - [u,v]) \geq f(x-u) + g(y-v)\ , \quad \forall\, [x,y] \in K \times Y\,.$$

Standard existence theory (see e.g. [13, chap. 2]) presumes that β is $X \times Y$-elliptic, what is equivalent that a is X-elliptic $\Big($that is $(\exists\ \alpha > 0)(\forall\ x \in X)\ \alpha\|x\|^2 \leq a(x,x)\Big)$ and c is Y-elliptic.

As it is shown however in the next section, to tackle an interesting class of unilateral elliptic boundary value problems we have to abandon these strong assumptions of coercivity and ellipticity of the bilinear forms.

3. A Unilateral Boundary Value Problem and its Variational Minimax Formulation

In this section we sketch as a simple unilateral elliptic boundary value problem the harmonic Signorini problem and present an equivalent variational minimax formulation, obtained by a boundary integral approach. Thus we provide the reader with some motivation for the study of the abstract variational minimax problems to follow in the subsequent sections.

Here to keep the matter short and as simple as possible, we only consider the simplest elliptic differential operator, the Laplace operator. The harmonic Signorini problem considered can be looked upon as the scalar model of the friction-free contact problem in elastostatics, where by a boundary integral approach the latter problem transforms also to a variational minimax problem, see [8]. Thus more involved elliptic p.d.e., especially of higher order or vector p.d.e. systems can be similarly treated.

3.1. THE HARMONIC SIGNORINI PROBLEM

Let Ω be a bounded Lipschitz [14] domain in $\mathbb{R}^2$ or $\mathbb{R}^3$. Consider the Poisson equation,

$$-\Delta u = f \text{ in } \Omega,$$

complemented by the boundary conditions

$$\begin{aligned} u = {} & g \text{ on } \Gamma_D, \\ \frac{\partial u}{\partial n} = {} & h \text{ on } \Gamma_N, \\ u \leq {} & g, \frac{\partial u}{\partial n} \leq h, (u-g)\left(\frac{\partial u}{\partial n} - h\right) = 0 \text{ on } \Gamma_S, \end{aligned}$$

where the boundary $\partial\Omega = \Gamma = \overline{\Gamma}_D \cup \overline{\Gamma}_N \cup \overline{\Gamma}_S$, and the boundary parts Γ_D, Γ_N, and Γ_S are mutually disjoint and relatively open. Further we eventually have $\Gamma_D = \emptyset$ (no Dirichlet condition) or $\Gamma_N = \emptyset$ (no Neumann condition), but we are interested in the Signorini conditions on Γ_S, so meas $\Gamma_S > 0$.

Here the boundary value problem is described in its general form with the given inhomogeneous data f, g, and h. By superposition using the solution of a special linear associated boundary value problem, one can get rid two of these three inhomogenities. So without any restriction of generality, we can assume $f = 0$ (and also $g = 0$ or $h = 0$) in what follows. As the standard device to obtain a variational formulation use the Gauss divergence theorem, here

$$0 = \int_\Omega -\Delta u(v-u)\,dx = \int_\Omega \nabla u \cdot \nabla(v-u)\,dx - \int_\Gamma \frac{\partial u}{\partial n}(v-u)\,d\gamma .$$

Further for $v|\Gamma_D = g|\Gamma_D, v|\Gamma_S \leq g|\Gamma_S$, we have

$$\begin{aligned}\int_\Gamma (v-u)\frac{\partial u}{\partial n}\,d\gamma &= \int_{\Gamma_N}(v-u)h\,d\gamma + \int_{\Gamma_S}(v-u)h\,d\gamma \\ &\quad + \int_{\Gamma_S}(v-u)\left(\frac{\partial u}{\partial n} - h\right)d\gamma ,\end{aligned}$$

where the latter boundary integral is nonnegative, since

$$\int_{\Gamma_S}(v-u)\left(\frac{\partial u}{\partial n} - h\right)d\gamma = \int_{\Gamma_S}(v-g)\left(\frac{\partial u}{\partial n} - h\right)d\gamma .$$

To write the resulting variational formulation precisely, we introduce the standard Sobolev space $H^1(\Omega)$ [14] and the convex closed subset

$$K = \{v \in H^1(\Omega) : v|\Gamma_D = g|\Gamma_D, v|\Gamma_S \leq g|\Gamma_S\},$$

which in case of $g = 0$ simplifies to a cone. Thus we arrive at the following variational problem: Find $u \in K$ such that

$$\int_\Omega \nabla u \cdot (\nabla v - \nabla u)\,dx - \int_{\Gamma_N \cup \Gamma_S} h(v-u) \geq 0, \ \forall\, v \in K.$$

This is the standard domain variational inequality, which in view of the symmetry of the bilinear form

$$a(u,v) = \int_\Omega \nabla u \cdot \nabla v\,dx$$

is obviously equivalent to a convex minimization problem. But there are other variational formulations that give rise to variational minimax problems.

3.2. A BOUNDARY VARIATIONAL INEQUALITY – AN EXAMPLE OF A VARIATIONAL MINIMAX PROBLEM

A fundamental solution of the harmonic p.d.e. in $\mathbb{R}^2$ is given by

$$\mathcal{F}(x,y) := -\frac{1}{2\pi}\log|x-y|\,,$$

where

$$|x| = \sqrt{x_1^2 + x_2^2}\,.$$

Then one introduces the bilinear forms $a \in \mathcal{L}(H^{-1/2}(\Gamma) \times H^{-1/2}(\Gamma), \mathbb{R}), b \in \mathcal{L}(H^{-1/2}(\Gamma) \times H^{1/2}(\Gamma), \mathbb{R})$ by

$$\begin{aligned} a(\varphi,\psi) &= \int_\Gamma\int_\Gamma \mathcal{F}(x,y)\,\varphi(y)\,\psi(x)d\gamma(y)d\gamma(x) \\ b(\varphi,v) &= \int_\Gamma\int_\Gamma \Big(\frac{\partial}{\partial n_x}\mathcal{F}(x,y)\Big)\,\varphi(y)\,v(x)d\gamma(y)d\gamma(x) + \frac{1}{2}\int_\Gamma \varphi\,v d\gamma\,, \end{aligned}$$

the linear form

$$l(v) = \int_{\Gamma_N \cup \Gamma_S} hv d\gamma\,,$$

and the convex closed subset

$$K := \{v \in H^{1/2}(\Gamma) : v = g \text{ on } \Gamma_D\,,\ v \le g \text{ on } \Gamma_S\}\,.$$

Using Green's formula and the well-known jump relations for the single and double layer potential one can show (see [11]) that the harmonic Signorini problem of subsection 3.1 is equivalent to the following variational minimax problem: Find $[u,\varphi] \in K \times H^{-1/2}(\Gamma)$ such that

$$(\pi)\quad \begin{cases} a\Big(\dfrac{du}{ds}, \dfrac{dv}{ds} - \dfrac{du}{ds}\Big) + b(\varphi, v-u) \ge l(v-u)\,, & \forall v \in K\,; \\ a(\psi,\varphi) - b(\psi,u) = 0\,, & \forall \psi \in H^{-1/2}(\Gamma)\,; \end{cases}$$

Note that (π) is equivalent to

$$B([u,\varphi]\,,\,[v,\psi] - [u,\varphi]) \ge l(v-u)\,,\quad \forall [v,\psi] \in K \times H^{-1/2}(\Gamma)$$

with the bilinear form

$$B([u,\varphi],[v,\psi]) := a\Big(\frac{du}{ds}\,,\,\frac{dv}{ds}\Big) + a(\psi,\varphi) + b(\varphi,v) - b(\psi,u)\,.$$

However, the bilinear form B is not symmetric, and more importantly, B is not Z - elliptic in the space

$$Z := H^{1/2}(\Gamma) \times H^{-1/2}(\Gamma) .$$

Instead one can show (see [7, Lemmata 2.1,2,2])

LEMMA. The bilinear form $B(\cdot,\cdot)$ is bounded, moreover satisfies a Gårding inequality in $Z \times Z$.

4. The Semicoercive Case

In this section we treat the semicoercive case, where the bilinear forms a and c in (P) are not coercive, but satisfy only a *Gårding inequality*, that is, there exist constants $\gamma_a > 0, \gamma_c > 0$, compact operators $G_a : X \to X', G_c : Y \to Y'$ such that

$$\begin{aligned} \gamma_a \|x\|^2 + \langle G_a x, x\rangle &\leq a(x,x) , \quad \forall\, x \in X ; \\ \gamma_c \|y\|^2 + \langle G_c y, y\rangle &\leq c(y,y) , \quad \forall\, y \in Y . \end{aligned}$$

Here we not only derive an existence result from the general theory of Baiocchi et al. [2], but also give some new results towards well-posedness.

An important tool for the analysis is the *asymptotic cone* of K or using Rockafellar's terminology, the *recession cone* of K, given by

$$K^\infty := \bigcap_{t>0} t(K - x_0) , \qquad x_0 \in K.$$

Note that by the closedness of K, the definition is independent of the chosen x_0. Immediately $\mathbb{R}_+ K^\infty \subseteq K^\infty$ and by convexity of $K, K^\infty + K^\infty \subseteq K^\infty$, that is, K^∞ is always a *cone*; moreover if K is a cone, then $K^\infty = K$.

Now let $M := K \times Y \subseteq Z := X \times Y$, $\lambda := [f,g] \in Z'$ and the bilinear form

$$\begin{aligned} \beta\Big([x,y],[x',y']\Big) := \;& a(x,x') + b(x',y) - b(x,y') + c(y,y'), \\ & \forall\, [x,y],[x',y'] \in X \times Y. \end{aligned}$$

Then our variational minimax problem (P) is immediately equivalent to the single variational inequality: Find $w \in M$ such that

$$\beta(w, z - w) \geq \lambda(z - w) , \quad \forall\, z \in M . \tag{4.1}$$

Here β decomposes in its symmetric part β_{symm} and its antisymmetric part β_{anti},

$$\beta = \beta_{\text{symm}} + \beta_{\text{anti}},$$

$$\beta_{\text{symm}}(z,z') = \tfrac{1}{2}\Big[\beta(z,z') + \beta(z',z)\Big],$$

where in our context above

$$\beta_{\text{symm}}([x,y],[x',y']) = a(x,x') + c(y,y')$$

provided that the bilinear forms a and c are symmetric.
Further let us introduce the kernel of β,

$$\mathcal{N}(\beta) = \{z \in Z : \beta(z,z) = 0\}.$$

Note that

$$\mathcal{N}(\beta) = \mathcal{N}(\beta_{\text{symm}}) = \{z \in Z : \beta_{\text{symm}}(z,\cdot) \equiv 0\}$$

is a subspace because of the Cauchy-Schwarz inequality.

By Minty's Lemma (see e.g. [13]), the variational inequality (4.1) is equivalent to

$$w \in M\,, \qquad \beta(z, w - z) \leq \lambda(w - z)\,, \quad \forall\, z \in M\,.$$

Thus the solution set $\mathcal{S}(\beta, M, \lambda)$ of (4.1) is always convex and closed, possibly empty. Solvability, however, requires additional "compatibility conditions" on the data. Following [2], the *resolvent cone* is the set

$$\mathcal{C} = \mathcal{C}(\beta, M, \lambda) = \{z' \in M^\infty : \sup_{z \in M} \beta(z,z') \leq \lambda(z')\},$$

which is seen (see [2, Lemma 2.1]) to be the asymptotic cone of $\mathcal{S}(\beta, M, \lambda)$, provided $\mathcal{S}(\beta, M, \lambda) \neq \emptyset$. Moreover $\mathcal{C} \cap \mathcal{N}(\beta) = \mathcal{C}$ holds.
Note that by construction $\beta \geq 0$ and β satisfies a Gårding inequality in $Z = X \times Y$. Therefore Theorem 3.1 of [2] applies and guarantees $\mathcal{S}(\beta, M, \lambda) \neq \emptyset$ under the assumption that $\mathcal{C}(\beta, M, \lambda)$ is a subspace. $\mathcal{C}(\beta, M, \lambda)$ being a cone, this assumption reduces to the implication

$$z' \in M^\infty \cap \mathcal{N}(\beta),\, \sup_{z \in M} \beta(z,z') \leq \lambda(z') \Rightarrow -z' \in M^\infty,\, \inf_{z \in M} \beta(z,z') \geq \lambda(z')\,.$$

Note that the slightly stronger assumption $\mathcal{C}(\beta, M, \lambda) = \{0\}$ (which is a more easily verifiable condition) does not only guarantee that $\mathcal{S}(\beta, M, \lambda) \neq \emptyset$, but also the boundedness of the solution set, as shown in the following.

PROPOSITION 1. The condition $\mathcal{C}(\beta, M, \lambda) = \{0\}$, that is

$$z' \in M^\infty \cap \mathcal{N}(\beta)\,,\ \sup_{z \in M} \beta(z,z') \leq \lambda(z') \Rightarrow z' = 0$$

is equivalent to the boundedness of the solution set $\mathcal{S}(\beta, M, \lambda)$.

Proof: Since $\mathcal{S} + \mathcal{C} = \mathcal{S}$, we only have to show that $\mathcal{C} = \{0\}$ implies that $\mathcal{S}$ is bounded. Argue by contradiction and suppose there exist $w_n \in \mathcal{S}(\beta, M, \lambda)$, $\|w_n\| \to \infty \, (n \in \mathbb{N}_0)$. Then eventually for some subsequence

$$z_n = \frac{w_n - w_0}{\|w_n - w_0\|}, \quad z_n \rightharpoonup \hat{z} \text{ (weak convergence)}.$$

Since $\|w_n - w_0\| \to \infty$, for any $t > 0$ there exists $n_0 = n_0(t) \in \mathbb{N}$ such that for all $n \geq n_0$

$$w_0 + tz_n = w_0 + \frac{t}{\|w_n - w_0\|}(w_n - w_0) \in M,$$

hence $\hat{z} \in M^\infty$. Further since $w_n \in \mathcal{S}(\beta, M, \lambda)$, it follows

$$0 \leq \beta(w_n - w_0, w_n - w_0) \leq \lambda(w_n - w_0) - \beta(w_0, w_n - w_0),$$

hence by division by $\|w_n - w_0\|^2$ in the limit $z \in \mathcal{N}(\beta)$. Moreover by Minty's Lemma,

$$\beta(z, w_n - w_0 - z) + \beta(z, w_0) \leq \lambda(w_n - w_0 - z) + \lambda(w_0),$$

hence by division by $\|w_n - w_0\|$ in the limit

$$\beta(z, \hat{z}) \leq \lambda(\hat{z}), \; \forall \, z \in M.$$

Thus $\hat{z} \in \mathcal{C}(\beta, M, \lambda)$ and $\hat{z} = 0$ by assumption. On the other hand by Gårding's inequality

$$\gamma_\beta \|w_n - w_0\|^2 + \langle G_\beta(w_n - w_0), w_n - w_0 \rangle \leq \beta(w_n - w_0, w_n - w_0),$$

where $\gamma_\beta > 0, G_\beta : Z \to Z'$ compact. Hence for some appropriate subsequence $G_\beta z_n \to G_\beta \hat{z} = 0$ and by division by $\|w_n - w_0\|^2$, it follows

$$\gamma_\beta + 0 \leq \beta(\hat{z}, \hat{z}) = 0,$$

a contradiction concluding the proof.

It is well known (see e.g. [12] chapters I and VII) that elliptic variational inequalities lacking coercivity generally lead to ill-posed variational problems. In particular, uniqueness fails in general.

Here we tackle this issue of uniqueness in the context of our specially structured variational minimax problems. In accordance with the setting of section 3.2, we let $\lambda = [f, g] = [l, 0]$ and in view of the discussion in section 3.1 we let K be here a cone. Then the assertion $\mathcal{C}(\beta, M, \lambda) = \{0\}$ is equivalent to the implication

$$\left.\begin{array}{ll} x' \in K \cap \mathcal{N}(a)\,, y' \in \mathcal{N}(c)\,, l(x') \geq 0 & \\ b(x', y) + c(y, y') = 0\,, & \forall\, y \in Y \\ a(x, x') - b(x, y') \leq 0\,, & \forall\, x \in K \end{array}\right\} \Rightarrow x' = 0\,,\, y' = 0\,.$$

In this situation we have the following uniqueness result.

PROPOSITION 2. Let $a = a^*$ and $c = c^*$. Suppose in addition that $\mathcal{N}(a) \cap (K - K) \subseteq K \cup -K$ and the operator $C : Y \to Y'$ via $c(y, y') = \langle Cy, y' \rangle$ is injective. Then $\mathcal{C}(\beta, M, \lambda) = \{0\}$ implies uniqueness.

Proof: Let $[u, v]$, $[u^*, v^*]$ be two solutions of (P). Then the choice $z = [u^*, v^*]$, resp. $[u, v]$ in (4.1) and adding up gives

$$\begin{aligned} 0 \ &\leq a(u^* - u, u - u^*) + c(v^* - v, v - v^*) \\ &= \beta\Big([u^*, v^*] - [u, v], [u, v] - [u^*, v^*]\Big) \leq 0\,. \end{aligned}$$

By the symmetry of a and c, we obtain $a(u - u^*, \cdot) \equiv 0, c(v - v^*, \cdot) \equiv 0$. Moreover by the variational equality in (P)

$$-b(u - u^*, y) + c(v - v^*, y) = -b(u - u^*, y) = 0\,, \quad \forall y \in Y\,.$$

Hence

$$0 = a(u, u^* - u) \geq l(u^* - u)$$

and by the Minty Lemma

$$0 = a(u, u^* - u) \leq l(u^* - u),$$

thus

$$l(u^* - u) = 0\,.$$

By hypothesis $v^* - v \in \mathcal{N}(C) = \{0\}$ and $u^* - u \in K \cup -K$. Therefore $[u^* - u, 0] \in \mathcal{C}(\beta, M, \lambda) \cup -\mathcal{C}(\beta, M, \lambda)$ and the conclusion follows.

Remarks. This uniqueness result applies to the unilateral harmonic problem of section 2. There $\mathcal{N}(a)$ consists of the subspace of the constant functions, thus $\mathbf{R} \cong \mathcal{N}(a) \subseteq K \cup -K$. With the domain formulation, set $Y = \{0\}, c := 0$ and the assumption on C is trivially satisfied. For the boundary integral formulation, c stems from the single layer potential and the operator C is known to be injective for the case of the Laplace equation (in $\mathbf{R}^2$ with the boundary Γ appropriately scaled such that the "transfinite diameter" or "logarithmic capacity" of Γ is different from 1; see e.g. the concluding

Remarks in [5]). – Similar sufficient conditions for uniqueness that however presume the knowledge of the solution were recently given in [16].

5. Lagrangian Minimax Problems

In this section we study variational minimax problems which arise from the Lagrangian associated to an operator equality constraint, in addition to unilateral constraints. This leads to variational problems of the form (P) where $c = 0$.

To treat this more delicate situation, we consider the operator B associated to the bilinear form b and its kernel $\mathcal{N}(B) =: X_1$. As in the classical saddle point theory due to Brezzi [4], where only equality constraints, but no inequality constraints are considered, we assume that $B : X_1^{\perp} \to Y'$ is an isomorphism and that a is X_1 - elliptic, only. This gives rise to a block decomposition of the variational problem.

At first in this section we make precise this standard block decomposition used in the classical saddle point theory. We illustrate this block decomposition by a simple finite dimensional example taken from the german text book of Hackbusch [10], which demonstrates unique solvability of the associated augmented linear system for any right hand side, although the matrix A associated to the bilinear form a has not full rank. Then by means of this block decomposition we can treat a special case of the convex set K, where K splits into two components and the second component is an appropriate subspace. In this special case we prove well–posedness in the sense of Hadamard. Finally we treat the general case of K, not necessarily a subspace. Here by another novel decomposition procedure we can establish solvability of (P). However, since we invoke a Hahn-Banach separation argument, we need as an additional topological hypothesis that int $K \neq \emptyset$.

5.1. A DECOMPOSITION PROCEDURE

First of all let us introduce the operators $A \in \mathcal{L}(X, X'), B \in \mathcal{L}(X, Y')$ and the adjoint operator B^* by

$$\begin{aligned} \langle Au, x\rangle_{X'\times X} &= a(u,x) && (u \in X, x \in X), \\ \langle Bu, y\rangle_{Y'\times Y} &= b(u,y) = \langle u, B^*y\rangle_{X\times X'} && (u \in X, y \in Y) \end{aligned}$$

and the kernel of B by

$$X_1 := \mathcal{N}(B) = \{u \in X : Bu = 0\} = \{u \in X : b(u,y) = 0 \ (\forall\, y \in Y)\}.$$

This gives rise to the orthogonal decomposition

$$X = X_1 \oplus X_2 \,,$$

where $X_2 = X_1^{\perp}$, and leads to $K = K_1 \oplus K_2$, where $K_i := K \cap X_i\ (i = 1,2)$. This decomposition of X yields also

$$X' = X_1' \oplus X_2' .$$

Indeed, for any $x' \in X'$ consider the restrictions $x_i' := x'|X_i\ (i = 1,2)$ and extend these restrictions by zero; then $x_1'|X_2 = 0$ and $x_2'|X_1 = 0$, what characterizes x_1' and x_2'.

By the decompositions of X and of X' we obtain the following block structure of A:

$$A = \begin{bmatrix} A_{11} & A_{12} \\ A_{21} & A_{22} \end{bmatrix},$$

where for $i,j = 1,2$ we have $A_{ij} \in \mathcal{L}(X_j, X_i')$. With the right hand side $f = f_1 + f_2, f_i \in X_i'\ (i = 1,2)$, our variational minimax problem (P) with $c = 0$ reads now: Find $[u_1, u_2, v] \in K_1 \times K_2 \times Y$ such that for any $x_1 \in K_1$, any $x_2 \in K_2$ there holds

$$\begin{cases} \langle A_{11}u_1, x_1 - u_1\rangle + \langle A_{12}u_2, x_1 - u_1\rangle & \geq \langle f_1, x_1 - u_1\rangle , \\ \langle A_{21}u_1, x_2 - u_2\rangle + \langle A_{22}u_2, x_2 - u_2\rangle + \langle B^*v, x_2 - u_2\rangle & \geq \langle f_2, x_2 - u_2\rangle , \\ \qquad - Bu_2 & = g\,. \end{cases}$$

Note that $B^* = \begin{bmatrix} 0 \\ B^* \end{bmatrix} : Y \to \begin{bmatrix} 0 \\ X_2' \end{bmatrix}$ is an isomorphism by the assumption on B. Hence there exist the inverse operators $B^{-1} \in \mathcal{L}(Y', X_2)$ and$(B^*)^{-1} = (B^{-1})^* \in \mathcal{L}(X_2', Y)$, shortly written B^{-*} in the following.

5.2. A SIMPLE EXAMPLE

For illustration we copy the following simple finite dimensional example from the german textbook of Hackbusch ([10]).
Let $K = X = \mathbb{R}^2, Y = \mathbb{R}$ and

$$A = \begin{pmatrix} 1 & -1 \\ -1 & 1 \end{pmatrix}, B = (1 \quad 1).$$

Then clearly the rank of the matrix A is $1 < 2$. Nevertheless, the associated linear system

$$\begin{cases} \begin{pmatrix} 1 & -1 \\ -1 & 1 \end{pmatrix}\begin{pmatrix} u_1 \\ u_2 \end{pmatrix} + \begin{pmatrix} 1 \\ 1 \end{pmatrix} v & = \begin{pmatrix} f_1 \\ f_2 \end{pmatrix} \\ -u_1 \quad -u_2 & = \quad g \end{cases}$$

is uniquely solvable for any right hand side $[f_1, f_2, g]$. Indeed, the extended coefficient matrix

$$\begin{pmatrix} 1 & -1 & 1 & f_1 \\ -1 & 1 & 1 & f_2 \\ -1 & -1 & 0 & g \end{pmatrix}$$

transforms by the Gauss elimination procedure to

$$\begin{pmatrix} 1 & -1 & 1 & f_1 \\ 0 & 1 & -\frac{1}{2} & -\frac{1}{2}(f_1+g) \\ 0 & 0 & 1 & \frac{1}{2}(f_1+f_2) \end{pmatrix},$$

what is uniquely solvable by back substitution.
Here we find

$$\begin{aligned} X_1 &= \mathcal{N}(B) = \left\{[x,y] \in \mathbf{R}^2 : x+y=0\right\} = \{[x,-x] : x \in \mathbf{R}\}, \\ X_2 &= \mathbf{R}\begin{pmatrix} 1 \\ 1 \end{pmatrix} = \left\{[y,y] : y \in \mathbf{R}\right\}, \\ X_1' &= X_1,\ X_2' = X_2 \end{aligned}$$

and

$$A = \begin{bmatrix} 2 & 0 \\ 0 & 0 \end{bmatrix}$$

with $A_{11} \neq 0$.

5.3. A SPECIAL CASE

Let us now treat the case $K = K_1 \times X_2$, that is, $K_2 = X_2$ is a subspace. In this case we can easily show the well-posedness of the variational minimax problem (P) with $c = 0$ under the hypotheses stated above in the beginning of the section, namely under the hypothesis that A_{11} is X_1 - elliptic and $B : X_2 \to Y'$ is an isomorphism.

In the case considered, the variational minimax problem (P) (with $c = 0$) reads: Find $[u_1, u_2, v] \in K_1 \times X_2 \times Y$ such that

$$\begin{cases} \langle A_{11}u_1 & +A_{12}u_2, x_1 - u_1\rangle & & \geq \langle f_1, x_1 - u_1\rangle,\ \forall\, x_1 \in K_1, \\ A_{21}u_1 & +A_{22}u_2 & +B^*v & = f_2, \\ & -Bu_2 & & = g. \end{cases}$$

Therefore for existence, solve for u_2 in the last equation. Then insert this u_2 in the variational inequality above what leads to a variational inequality in the unknown u_1 alone, what in view of X_1-ellipticity is solvable. Finally plug in u_1 and u_2 in the middle equation and solve for v.

To show well-posedness, let $[u_1^I, u_2^I, v^I] \in K_1 \times X_2 \times Y$ be a solution to the data f_1^I, f_2^I, g^I; respectively $[u_1^{II}, u_2^{II}, f^{II}] \in K_1 \times X_2 \times Y$ be a solution to the data $f_1^{II}, f_2^{II}, g^{II}$. We follow the same order in the reasoning as above. Thus we first estimate from the last equation

$$\|u_2^I - u_2^{II}\| \leq \|B^{-1}\|\, \|g^I - g^{II}\|.$$

Then denoting the X_1-ellipticity constant of A_{11} by $\alpha > 0$ we obtain from the variational inequalities for u_1^I, resp. for u_1^{II} with the choice $x_1 = u_1^{II}$, resp. $x_1 = u_1^I$ by addition that

$$\begin{aligned} \alpha\|u_1^I - u_1^{II}\|^2 &\le \langle A_{11}(u_1^I - u_1^{II}), u_1^I - u_1^{II}\rangle \\ \le \langle f_1^I - f_1^{II}, u_1^{II} - u_1^I\rangle &- \langle A_{12}(u_2^I - u_2^{II}), u_1^{II} - u_1^I\rangle \,, \end{aligned}$$

hence

$$\|u_1^I - u_1^{II}\| \le \frac{1}{\alpha}\left\{\|f_1^I - f_1^{II}\| + \|A_{12}\|\,\|B^{-1}\|\,\|g^I - g^{II}\|\right\}.$$

Finally the middle equation gives the estimate

$$\begin{aligned} &\|v^I - v^{II}\| \\ &\le \|B^{-*}\|\,\left\{\|f_2^I - f_2^{II}\| + \|A_{21}\|\,\|u_1^I - u_1^{II}\| + \|A_{22}\|\,\|u_2^I - u_2^{II}\|\right\} \\ &\le \|B^{-*}\|\,\left\{\|f_2^I - f_2^{II}\| + \tfrac{\|A_{21}\|}{\alpha}\left[\|f_1^I - f_1^{II}\| + \|A_{12}\|\,\|B^{-1}\|\,\|g^I - g^{II}\|\right]\right. \\ &\qquad \left. + \|A_{22}\|\,\|B^{-1}\|\,\|g^I - g^{II}\|\right\}. \end{aligned}$$

5.4. THE GENERAL CASE

Now we deal with general Lagrangian minimax problems (P) (with $c = 0$). Here we shortly write $W =: \mathcal{N}(B)$. In addition to W- ellipticity of the bilinear form a and the assumption that the operator $B : W^\perp \to Y'$ is an isomorphism we have to impose further assumptions, which, on the other hand, are trivially satisfied in the classical saddle point theory without unilateral constraints. In particular we need the regularity asumption that K has nonvoid interior. Then by a novel approach we can establish solvability of (P). More precisely we have the following result.

THEOREM. Suppose in addition that int $K \neq \emptyset$ and $-g \in B[K]$. Then there exists a solution to (P).

The *Proof* runs in two steps.
First step: Solution of a simplified variational inequality.
By surjectivity of B, there exists $u_\perp \in W^\perp$ such that $-Bu_\perp = g$. Let $K_0 = K - u_\perp$, $u_0 = u - u_\perp$. Then (P) is equivalent to finding $[u_0, v] \in K_0 \times Y$ such that for all $x_0 \in K_0$, respectively for all $y \in Y$

$$\begin{cases} a(u_0, x_0 - u_0) + b(x_0 - u_0, v) & \ge f(x_0 - u_0) - a(u_\perp, x_0 - u_0)\,, \\ b(u_1, y) & = 0\,, \end{cases}$$

where the latter variational equality simply means $u_1 \in W$. Consider in particular any $x_0 \in K_0 \cap W$, where this latter set is nonvoid in view of the assumption $-g \in B[K]$. Then $x_0 - u_0 \in W$ or $b(x_0 - u_0, y) = 0$ for all $y \in Y$. Thus the variational inequality above simplifies to the problem of finding $u_0 \in K_0 \cap W$ such that

$$a(u_0, x_0 - u_0) \geq f(x_0 - u_0) - a(u_\perp, x_0 - u_0)\,,\ \forall\, x_0 \in K_0 \cap W.$$

Since $K_0 \cap W$ is closed and convex and since a is W-elliptic, there exists a (unique) solution $u_0 \in K_0 \cap W$. Hence $u_0 + u_\perp \in K$ and $b(u_0, y) = 0$ for all $y \in Y$.

Second step: Construction of an appropriate linear form in $W^\perp$.
By the solution procedure above, we are left with the problem of finding $v \in Y$ such that

$$\begin{array}{rcl} \langle x_0 - u_0, B^* v\rangle = b(x_0 - u_0, v) & \geq & \\ f(x_0 - u_0) - a(u_\perp + u_0, x_0 - u_0) & =: & \lambda(x_0 - u_0)\,,\ \forall\, x_0 \in K_0\,. \end{array}$$

Note that by construction, $\lambda(x_0 - u_0) \leq 0$ for any $x_0 \in K_0 \cap W$, and by assumption, $B^* : Y \to W^\perp$ is an isomorphism. Thus it remains to find some $\lambda_0 \in W^\perp$ such that $\lambda_0(\tilde{x}) \geq \lambda(\tilde{x})$ for any $\tilde{x} \in K - u_\perp - u_0$, where $\lambda \in X'$ is given and satisfies $\lambda(\tilde{x}) \leq 0$ for any $\tilde{x} \in K - u_\perp - u_0 \cap W$.
To demonstrate the existence of the desired linear form λ, we apply the Hahn-Banach theorem. To this end, introduce the following convex sets

$$\begin{array}{l} R := \{[x.r] \in X \times \mathbf{R} : r < \lambda(x)\,, x \in K - u_\perp - u_0\} \\ S := \{[w, 0] : w \in W\} \end{array}$$

Then $R \cap S = \emptyset$, for otherwise, for some $\tilde{x} \in W \cap K - u_\perp - u_0, r = 0$ would result in $0 < \lambda(\tilde{x})$, a contradiction. By assumption, int $R \neq \emptyset$. Therefore the separation theorem yields some $[\tilde{\lambda}, \tilde{\rho}] \in X' \times \mathbf{R}\backslash\{0\}$ and some $\gamma \in \mathbf{R}$ such that

$$\tilde{\lambda}(w) \leq \gamma < \tilde{\lambda}(x) - \tilde{\rho} r,\ \forall\, w \in W; [x, r] \in R.$$

Keeping w, x fixed, letting $r \to -\infty$ shows $\tilde{\rho} \geq 0$.
We claim that $\tilde{\rho} > 0$. Assume not. Then in virtue of int $K \neq \emptyset$, there exists some zero neighborhood $O = -O$ such that for any $o \in O$ we have $\tilde{\lambda}(x + o) > \gamma$, hence $\tilde{\lambda} = 0$ contradicting $[\tilde{\lambda}, \tilde{\rho}] \neq 0$.
Without any restriction of generality we can assume $\tilde{\rho} = 1$. Since W is a subspace, $0 \leq \gamma$ and $\tilde{\lambda}(w) = 0$ for any $w \in W$ follows. Thus $\tilde{\lambda} \in W^\perp$. Moreover keeping $\tilde{x} \in K - u_\perp - u_0$ fixed, the limit process $r \to \lambda(\tilde{x})$ shows that

$$\tilde{\lambda}(\tilde{x}) - \lambda(\tilde{x}) \geq \gamma \geq 0,\ \forall\, \tilde{x} \in K - u_\perp - u_0\,.$$

Thus $\tilde{\lambda}$ is the desired linear form concluding the proof.

References

1. A. Auslender, *Problèmes de Minimax via l'Analyse Convexe et les Inégalités Variationnelles: Théorie et Algorithmes,* Lect. Notes Econ. Math. Syst. 77 Springer-Verlag, Berlin, 1972.
2. C. Baiocchi, F. Gastaldi and F. Tomarelli, *Some existence results on non–coercive variational inequalities,* Ann. Scuola Norm. Sup. Pisa cl. Sci., IV, **13** (1986), 617-659.
3. V. Barbu and Th. Precupanu, *Convexity and Optimization in Banach Spaces,* Editura Academiei, Bucuresti and D. Reidel, Dordrecht, 1986.
4. F. Brezzi, *On the existence, uniqueness and approximation of saddle-point problems arising from Lagrangian multipliers,* RAIRO Anal. Numér. **8** (1974), 129-151.
5. M. Costabel, *Boundary integral operators on Lipschitz domains: Elementary results,* SIAM J. Math. Anal. **19** (1988), 613-626.
6. J. Gwinner and J.-C. Pomerol, *On weak* closedness, coerciveness, and inf-sup theorems,* Arch. Math. **52** (1989), 159-167.
7. J. Gwinner, *Boundary element convergence for unilateral harmonic problems,* in C. Bandle, J. Bemelmans, M. Chipot, M. Grüter, and J. Saint Jean Paulin (eds.), Progress in partial differential equations: calculus of variations,applications, Pitman Res. Notes Math., 267, Longman and J. Wiley, 1992, 200-213.
8. J. Gwinner and E.P. Stephan, *A boundary element procedure for contact problems in plane linear elastostatics,* RAIRO Modél. Math. Anal. Numér. **27** (1993), 457-480.
9. J. Gwinner and W. Oettli, *Theorems of the alternative and duality for inf-sup problems,* Math. Oper. Res. **19** (1994), 238-256.
10. W. Hackbusch, *Theorie und Numerik elliptischer Differentialgleichungen,* Teubner, Stuttgart, 1986.
11. H. Han, *A direct boundary element method for Signorini problems,* Math. Computation **55** (1990), 115-128.
12. A. Kaplan and R. Tichatschke, *Stable Methods for Ill–Posed Variational Problems,* Akademie Verlag, Berlin, 1994.
13. D. Kinderlehrer and G. Stampacchia, *An Introduction to Variational Inequalities and Their Applications,* Academic Press, New York, 1984.
14. J. Nečas, *Les méthodes directes en théorie des équations élliptiques,* Academia, Prague and Masson, Paris, 1967.
15. F. Patrone, *Well-posedness for Nash equilibria and related topics,* in R. Lucchetti and J. Revalski (eds.), Recent Developments in Well-posed Variational Problems, Kluwer, Dordrecht, 1995, 211-227.
16. W. Spann, *Error estimates for the approximation of semicoercive variational inequalities,* Numer. Math. **69** (1994), 103-116.

A TOPOLOGICAL INVESTIGATION OF THE FINITE INTERSECTION PROPERTY

CHARLES D. HORVATH
Département de Mathématiques
Université de Perpignan
66860 Perpignan cedex, France
e-mail: horvath@univ-perp.fr

Abstract. When does a finite family of sets $\{R_0,\cdots,R_n\}$ have a nonempty intersection? From the topology of the unions formed from these sets we obtain sufficient conditions. Homotopy or homology properties will guarantee that the intersection is not empty.

1. Introduction

Often the solution of a problem requires that one shows that some family of sets has the finite intersection property. In minimax theory the famous result of Knaster-Kuratowski-Mazurkiewicz is the main ingredient and its infinite dimensional extension by Ky Fan has proved itself to be an elegant and powerful tool. Even though after the work of König, [18], and Kindler, [17], we know how to prove minimax theorems using only connectedness, there are still some inequalities, like Ky Fan inequality, that can not be obtained that way, simply because they are just another form of Brouwer's fixed point theorem. We will have little to say about connectedness and intersection theorems. The condition is too weak to give us what we want. No intersection theorem generalizing Fan's theorem has been obtained, and we would venture to say could be obtained outside of dimension one or maybe two, using only connectedness, even though some authors have claimed otherwise.

Theorem 1 (KKM) *Let $R_i \subseteq \Delta_n$, $i = 0, \cdots, n$ a family of subsets of the n-dimensional standard simplex Δ_n. Assume that each R_i is closed or that each R_i is open, and that for each subset $\{e_{i_0}, \ldots, e_{i_k}\}$ oj the vertices of the*

B. Ricceri and S. Simons (eds.), Minimax Theory and Applications, 71–90.

simplex the face spaned by these vertices is contained in $\cup_{j=0}^{j=k} R_{i_j}$. *Then the set* $\cap_{i=0}^{i=n} R_i$ *is not empty.*

In the original result all the sets were supposed to be closed. That they also can be open was shown by Lassonde, [19].

To prove theorem 10 we will need the following result, which can be derived from KKM.

Theorem 2 (Alexandroff-Pasynkoff) *Let* $R_i \subseteq \Delta_n$, $i = 0, \cdots, n$ *a family of subsets of the n-dimensional standard simplex* Δ_n. *Assume that each* R_i *is closed or that each* R_i *is open, and that for each* i *the face* Δ_i *spaned by the set of vertices* $\{e_j : j \neq i\}$ *is contained in* R_i. *Then the set* $\cap_{i=0}^{i=n} R_i$ *is not empty.*

Theorem 3 (Fan) *Let* E *be a topological vector space,* $X \subseteq E$ *a convex subset, and* $\Omega : X \longrightarrow E$ *a correspondence such that:*
(a) either all the values of Ω *are closed or all the values are open and, for each* $x \in X$, $x \in \Omega(x)$;
(b) for each $x_1, x_2 \in X$ *and for each* $y \in [x_1, x_2]$, $\Omega(y) \subseteq \Omega(x_1) \cup \Omega(x_2)$.

Then the family $\{\Omega(x) : x \in X\}$ *has the finite intersection property.*

Condition (b) is equivalent to: (b') for any $y \in E$ the set $X \setminus \Omega^{-1}(y)$ is convex.

Conditions (a) and (b) together imply that $co(A) \subseteq \cup_{a \in A} \Omega(a)$ for any nonempty finite subset $A \subseteq X$, Ω is a KKM correspondence. The important point here is that condition (b) involves the geometric structure of the space X, we must have a convex hull operator, while condition (b') is closer to a topological property amenable to different possible generalizations.

But first let us see that to the correspondence $\Omega : X \longrightarrow X$ is associated a hull operator, larger than the convex hull operator but intrinsically defined from Ω. We assume that $x \in \Omega(x)$ for each $x \in X$ and, that (b), or equivalently (b'), holds.

To a finite nonempty subset $A \subseteq X$ we associate the set

$$\Delta_\Omega(A) = \{y \in X : \Omega(y) \subseteq \cup_{a \in A} \Omega(a)\}$$

Obviously $A \subseteq \Delta_\Omega(A)$ and $\Delta_\Omega(A) \subseteq \Delta_\Omega(B)$ if $A \subseteq B$. Notice that these two properties are purely set theoretical.

Now, let us see that $\Delta_\Omega(A)$ is convex.

If $y_1, y_2 \in \Delta_\Omega(A)$ then $\Omega(y_1) \cup \Omega(y_2) \subseteq \cup_{a \in A} \Omega(a)$.
From (b) we have $\Omega(y) \subseteq \Omega(y_1) \cup \Omega(y_2)$ if $y \in [y_1, y_2]$, and therefore $[y_1, y_2] \subseteq \Delta_\Omega(A)$.

Finally, we notice that $\Delta_\Omega(A) \subseteq \cup_{a\in A}\Omega(a)$.

If $y \in \Delta_\Omega(A)$ then $\Omega(y) \subseteq \cup_{a\in A}\Omega(a)$, but $y \in \Omega(y)$ by hypothesis, consequently, $y \in \Omega(y) \subseteq \cup_{a\in A}\Omega(a)$. The argument here is also set theoretical.

In conclusion we have an operator $A \mapsto \Delta_\Omega(A)$ whose properties are fixed by set theoretical properties of the correspondence Ω and with one geometric, topological, property: $\Delta_\Omega(A)$ is convex.

That $\Delta_\Omega(A)$ is convex can been see differently. Observe that $y \in \Delta_\Omega(A)$ if and only if $A \cap \Omega^{-1}(x) = \emptyset$ implies $y \in X \setminus \Omega^{-1}(x)$, in other words

$$\Delta_\Omega(A) = \cap\{X \setminus \Omega^{-1}(x) : A \cap \Omega^{-1}(x) = \emptyset\}$$

The strategy we will follow to generalize Fan's theorem is now clear. A correspondence $\Omega : X \longrightarrow X$ on a topological space yields a hull operator $A \mapsto \Delta_\Omega(A)$. If the sets $\Delta_\Omega(A)$ have nice enough topological properties we will be able to show that the family $\{\Omega(x) : x \in X\}$ has the finite intersection property. The topological properties in question are mainly the subject matter of this paper.

There are other well known intersection theorems that have proved useful inconnection with minimax or systems of inequalities. For example, Helly's theorem was used by Rockafellar, [22], and Klee's theorem was used by Flåm-Greco, [12].

Theorem 4 (Helly) *A family of convex subsets of* $\mathbb{R}^n$ *has the finite intersection property if and only if any subfamily of at most* $n+1$ *sets has a nonempty intersection.*

Theorem 5 (Klee) *A family of* $n+1$ *closed convex sets in a topological vector space has a nonempty intersection if any* n *members of the family meet and if the union of the* $n+1$ *sets is convex.*

Convexity is a key ingredient of all these results, or so it seems. In the search for minimax theorems without convexity many authors have been led to topological intersection theorems. Some are genuine generalizations of Klee's and Helly's theorems, some generalize KKM, but one should say that the proofs rely on KKM itself.

Often these generalized convexities have been introduced in an ad hoc fashion in order to prove some minimax theorem, but one might ask if there are spaces with enough geometric structure to carry a form of convexity to which the previous results could be generalized.

So, these are the two themes of this paper, the finite intersection property for families of non convex sets and topological spaces with a natural convexity structure.

There will be few complete proofs, otherwise the paper would grow to unmanageable size. The list of references is also far from being complete, one could most probably list a few hundred items.

2. The Finite Intersection Property

The first topological notion that comes to mind is connectedness. On a one dimensional simplex, an interval, we do not have to go further than the definition of connectedness to prove KKM. But in higher dimensions connectedness is not good enough. Consider the following example:

The space X is the boundary of a two dimensional simplex, whose vertices we call a, b and c.

Let $F_a = [a, b]$, $F_b = [b, c]$ and $F_c = [a, c]$.
Then, $a \in F_a, \cdots$, $[a, b] \subseteq F_a \cup F_b \cdots$, and $[a, b] \cup [b, c] \cup [a, c] = X$, but $[a, b] \cap [b, c] \cap [a, c] = \emptyset$.

So, connectedness is good enough to establish that a family of sets has the binary intersection property, (any two sets of the family meet), but does not allow the passage to the finite intersection property. One has to impose higher order connectedness properties to have the n-ary intersection property, or conditions that reduce the finite intersection property to the binary intersection property, as in Kindler, [17], or König, [18] for example. But the conditions are so stringent that there is little hope of encompassing KKM.

Definition 1 *A topological space C is n-connected if any continuous function*
$f : \partial\Delta_{n+1} \longrightarrow C$ from the boundary of the $n+1$-dimensional simplex into C has a continuous extension $g : \Delta_{n+1} \longrightarrow C$. It is $\mathbf{C}^\infty$ if it is n-connected for any $n \geq 0$.

Our n-connected sets are also called n-aspherical, $n - C$ sets. Sets which are $k - C$ for each $0 \leq k \leq n$ are C^n sets.

Obviously, a -1-connected set is a non empty set, and a 0-connected set is a pathconnected set. A convex subset of a topological vector space is n-connected for any $n \geq 0$. More generally any contractible space is n-connected for any n, but the implication can not be reversed.

Let $\mathcal{F}(\Delta_n)$ be the set of all faces of the standard simplex Δ_n and Δ_n^0 the set of its vertices.

The following result was obtained by Horvath and Lassonde, (1) of theorem 2 in [16].

Theorem 6 *In a topological space, let $\{R_i : i \in \Delta_n^0\}$ be a family of sets, all closed or all open. If there is a family of sets $\{C_\sigma : \sigma \in \mathcal{F}(\Delta_n)\}$ such*

that:
(a) $C_\sigma \subseteq C_\tau$ *for every* $\sigma, \tau \in \mathcal{F}(\Delta_n)$ *such that* $\sigma \subseteq \tau$;
(b) C_σ *is* $(\dim \sigma - 1)$*-connected for every* $\sigma \in \mathcal{F}(\Delta_n)$;
(c) $C_\sigma \subseteq \cup\{R_i : i \in \sigma^0\}$ *for every* $\sigma \in \mathcal{F}(\Delta_n)$.

Then $\cap\{R_i : i \in \Delta_n^0\} \neq \emptyset$.

From theorem 6 one can then derive the following result, which appears in Horvath and Lassonde, [16].

Theorem 7 *Let n sets in a topological space be given such that all the sets are closed or all are open, and either of the following properties is satisfied:*
(a) For each $k \leq n$ *the union of k of the sets is* $(k-2)$*-connected; or*
(b) The union of n sets is $(n-2)$*-connected and for each* $k \leq n-1$ *the intersection of every k of the sets is nonempty and* $(n-k-2)$*-connected.*

Then the sets have a nonempty intersection.

Next is a topological version of Fan's theorem for n-connected sets.

Theorem 8 *Let X be a* $\mathbf{C}^\infty$ *topological space and* $\Omega : X \longrightarrow X$ *a correspondence such that:*
(a) Either all the values of Ω *are closed or all the values are open and, for each* $x \in X$, $x \in \Omega(x)$;
(b) For each nonempty subset $B \subseteq X$ *the set* $\bigcap_{x\in B} X \setminus \Omega^{-1}(x)$ *is* $\mathbf{C}^\infty$.

Then the family $\{\Omega(x) : x \in X\}$ *has the finite intersection property.*

Proof. For each nonempty finite subset $A \subseteq X$ let

$$\Delta_\Omega(A) = \cap\{X \setminus \Omega^{-1}(x) : A \cap \Omega^{-1}(x) = \emptyset\}$$

We know that $\Delta_\Omega(A) \neq \emptyset$, $\Delta_\Omega(A) \subseteq \Delta_\Omega(B)$ if $A \subseteq B$ and $\Delta_\Omega(A) \subseteq \cup_{a\in A}\Omega(a)$.

Now fix $A \subseteq X$ and enumerate its elements, say $A = \{a_0, \ldots, a_n\}$. To a face σ of the n-dimensional simplex Δ_n associate the set $C_\sigma = \Delta_\Omega(\{a_i : i \in \sigma^0\})$ The conclusion now follows from theorem 6. □

It is well known that from Fan's intersection theorem we obtain the Fan-Browder fixed point theorem. Similarly, from the previous intersection theorem we obtain a fixed point theorem.

Theorem 9 *Let X be a compact* $\mathbf{C}^\infty$ *topological space and,* $F : X \longrightarrow X$ *a correspondence such that:*
(a) For each $x \in X$ *the set* $F^{-1}(x)$ *is open;*
(b) For each $x \in X$ *the set* $F(x)$ *is not empty;*
(c) For each nonempty subset $B \subseteq X$ *the set* $\bigcap_{x\in B} F(x)$ *is* $\mathbf{C}^\infty$.

Then there is a point $x^* \in X$ *such that* $x^* \in F(x^*)$.

Proof. Let $\Omega(x) = X \setminus F^{-1}(x)$. Then $\Omega(x)$ is compact and $\bigcap_{x \in X} \Omega(x) = \emptyset$. Therefore the family $\{\Omega(x) : x \in X\}$ does not have the finite intersection property. Since each set $\bigcap_{x \in B} X \setminus \Omega^{-1}(x)$ is $\mathbf{C}^\infty$ there must be a point $x^* \in X$ such that $x^* \in X \setminus \Omega(x^*)$. □

Theorem 6 extends the result of Knaster-Kuratowski-Mazurkiewicz while theorem 7 is an extension of Klee's theorem. This can bee seen by taking $C_\sigma = \sigma$ in theorem 6 and for theorem 7 one has to notice that the union of a family of convex sets whose intersection is not empty is starshaped, therefore contractible, and consequently n-connected for any n. The compactness assumption in theorem 9 can be relaxed in different ways.

For $n = 2$ theorems 6 and 7 require the sets to be pathconnected. But connectedness is good enough to prove that two closed, or two open, sets have a nonempty intersection. So our assumptions seem too strong. We will give now another version of our intersection results with conditions weaker than pathconnectedness.

Definition 2 *A subspace C of a topological space X is $n-$**UV** in X if, for any neighborhood V of C in X there is a neighborhood U of C contained in V such that any continuous function $f : \partial\Delta_{n+1} \longrightarrow U$ has a continuous extension $g : \Delta_{n+1} \longrightarrow V$. Such a pair of open sets will be called an $n-$**UV** pair for C. $\mathbf{UV^n}$ means $k -$ **UV** for each $k \leq n$.*

The closure of the graph of the function $x \mapsto \sin \frac{1}{x}$, $0 < x \leq 1$ is $n-$**UV** in $\mathbb{R}^2$ for any n, but is not pathconnected.

In a locally compact, locally connected metric space, a compact subspace is $0 -$ **UV** if and only if it is connected. In a locally compact $\mathbf{LC^{n-1}}$ space, (a space is $\mathbf{LC^m}$ if the singletons are $\mathbf{UV^m}$), a compact n-connected subspace is also $n -$ **UV**, lemma 5 in [1]. Being an $n -$ **UV** set is not a property of the space C itself but a property of the embedding of C in X. For **ANR** the $n -$ **UV** property does not depend on the embedding, and a compact **ANR** subspace of the Hilbert cube has property $n -$ **UV** if and only if it is n-connected, see Armentrout, [1] or Daverman [7] chapter 3.

To keep things simple we will say that a topological space P is a polyhedron if it is a subspace of $\mathbb{R}^n$ for some $n \in \mathbb{N}$, and if there is a finite family of simplices $\Sigma(P)$ of $\mathbb{R}^n$ such that $\sigma \cap \tau \in \Sigma(P)$ if $\sigma, \tau \in \Sigma(P)$ and $\sigma \cap \tau \neq \emptyset$, and $P = \cup\{\sigma : \sigma \in \Sigma(P)\}$.

We put $\dim P = \max\{\dim \sigma : \sigma \in \Sigma(P)\}$.

Lemma 1 *Let X be a topological space, P a finite polyhedron, $\Sigma(P)$ the set of simplices of P and $\{F_\sigma : \sigma \in \Sigma(P)\}$ a family of nonempty subsets of*

X such that:
(a) F_σ *is* $(\dim \sigma - 1) - \mathbf{UV}$ *if* $\dim \sigma \geq 1$;
=20 *(b) if* $\sigma \subseteq \tau$ *then* $F_\sigma \subseteq F_\tau$.

Then, for any family of open sets $\{W_\sigma : \sigma \in \Sigma(P)\}$ *such that* $F_\sigma \subseteq W_\sigma$ *for each* $\sigma \in \Sigma(P)$, *there exists a continuous function* $f : P \longrightarrow X$ *such that for each* $\sigma \in \Sigma(P)$, $f(\sigma) \subseteq W_\sigma$.

Proof. Let $\Sigma^{(k)}(P) = \{\sigma \in \Sigma(P) : \dim \sigma = k\}$ and $P^n = \cup\{\sigma : \sigma \in \Sigma^{(k)}(P), 0 \leq k \leq n\}$.

For $\sigma \in \Sigma(P)$ let $\Sigma_{(k)}(\sigma) = \{\sigma' \in \Sigma(P) : \dim \sigma' = k, \sigma \subseteq \sigma'\}$, and for each $p \in P^0$ choose a point $f_0(p)$ in F_p.

The function $f_0 : P^0 \longrightarrow X$ is continuous since the topology of P^0 is discrete.

For each $\sigma \in \Sigma^{(m)}(P)$, where $m = \dim P$, let (U_σ^m, V_σ^m) be a $(\dim \sigma - 1) - \mathbf{UV}$ pair for F_σ such that $V_\sigma^m \subseteq W_\sigma$.

For $\sigma \in \Sigma^{(m-1)}(P)$ let

$$V_\sigma^{m-1} = W_\sigma \cap \bigcap \{U_{\sigma'}^m : \sigma' \in \Sigma_{(m)}(\sigma)\}$$

We can now pick an open set U_σ^{m-1} such that $(U_\sigma^{m-1}, V_\sigma^{m-1})$ be a $(\dim \sigma - 1) - \mathbf{UV}$ pair for F_σ.

If $\sigma \in \Sigma^{(m-1)}(P)$ and $\Sigma_{(m)}(\sigma) = \emptyset$ let $(U_\sigma^{m-1}, V_\sigma^{m-1})$be an arbitrary $(\dim \sigma - 1) - \mathbf{UV}$ pair for F_σ such that $V_\sigma^{m-1} \subseteq W_\sigma$.

We can iterate this procedure to obtain a set of pairs of open sets

$$\{(U_\sigma, V_\sigma) : \sigma \in \Sigma(P)\}$$

such that

1. (U_σ, V_σ) is a $(\dim \sigma - 1) - \mathbf{UV}$ pair for F_σ such that $V_\sigma \subseteq W_\sigma$.
2. If $\sigma \subseteq \sigma'$ and $\dim \sigma' = 1 + \dim \sigma$ then $V_\sigma \subseteq U_{\sigma'}$.

For each $\sigma \in \Sigma^{(1)}(P)$, we have a continous function

$$f_{0|\partial\sigma} : \partial\sigma \longrightarrow \bigcup_{p \in \partial\sigma} F_p$$

From the inclusions

$$\bigcup_{p \in \partial\sigma} F_p \subseteq F_\sigma \subseteq U_\sigma$$

we know that the function $f_{0|\partial\sigma} : \partial\sigma \longrightarrow U_\sigma$ has a continuous extension

$f_{1,\sigma} : \sigma \longrightarrow V_\sigma$.

For $\sigma, \sigma' \in \Sigma^{(1)}(P)$ we have

$$f_{1,\sigma|\sigma\cap\sigma'} = f_{0|\sigma\cap\sigma'} = f_{1,\sigma'|\sigma\cap\sigma'}$$

We therefore have a continuous function
$f_1 : P^1 \longrightarrow X$ such that $f_{1|P^0} = f_0$ and $f_1(\sigma) \subseteq V_\sigma$ for each $\sigma \in \Sigma^{(1)}(P)$.

Now for each $\sigma \in \Sigma^{(2)}(P)$, we have a continuous function

$$f_{1|\partial\sigma} : \partial\sigma \longrightarrow \bigcup\{V_{\sigma'} : \sigma' \subseteq \sigma, \dim\sigma' = 1\}$$

But, from $\bigcup\{V_{\sigma'} : \sigma' \subseteq \sigma, \dim\sigma' = 1\} \subseteq U_\sigma$, we get a continuous function
$f_{1|\partial\sigma} : \partial\sigma \longrightarrow U_\sigma$ which can be extended to a continuous function $f_{2,\sigma} : \sigma \longrightarrow V_\sigma$.

One can see that if $\sigma_1, \sigma_2 \in \Sigma^{(2)}(P)$ and $\sigma_1 \cap \sigma_2 \neq \emptyset$ then the functions $f_{2,\sigma_1|\sigma_1\cap\sigma_2}$ and $f_{2,\sigma_2|\sigma_1\cap\sigma_2}$ coincide on $\sigma_1 \cap \sigma_2$.

We have obtained a continuous function $f_2 : P^2 \longrightarrow X$ such that $f_2(\sigma) \subseteq V_\sigma$ for each $\sigma \in \Sigma^{(2)}(P)$.

After a finite number of iterations of the same construction we get a continuous function $f : P \longrightarrow X$ having the required property.□

From the lemma we derive the following Klee like intersection theorem.

Theorem 10 *Let X be a topological space and $\{R_i : i = 0, \cdots, n\}$ a family of nonempty subspaces of X all of which are closed, in which case X is assumed to be normal, or all of which are open. If one of the sets of hypotheses (A) or (B) below is verified then $\bigcap_{i=0}^{i=n} R_i \neq \emptyset$.*
(A)
For each nonempty subset J of $\{0, \cdots, n\}$ the set $\bigcup_{j\in J} R_j$ is (card(J)-2)- **UV**.
(B)
(a) For each $j \in \{0, \cdots, n\}$ the set $\bigcap_{i\neq j} R_i$ is not empty;
*(b) If J is a nonempty subset of $\{0, \cdots, n\}$ of cardinality at most n then the set $\bigcap_{j\in J} R_j$ is (n-1-card(J))-***UV***;*
*(c) The set $\bigcup_{i=0}^{i=n} R_i$ is (n-1)-***UV***.*

Proof. First we assume that X is normal and that all the R_i are closed subspaces.

Assume that (A) holds.

For each face σ of the n-dimensional simplex Δ_n let $F_\sigma = \bigcup_{i\in\sigma^0} R_i$ where σ^0 is the set of vertices of σ. The set F_σ is $(\dim\sigma - 1) - \mathbf{UV}$.

For each $i \in \{0,\cdots,n\}$ let A_i be an open neighborhood of R_i and let $W_\sigma = \bigcup_{i\in\sigma^0} A_i$.

By lemma 1 there exists a continuous function $f : \Delta_n \longrightarrow X$ such that $f(\sigma) \subseteq \bigcup_{i\in\sigma^0} A_i$ for each face σ of Δ_n.

Therefore for each face σ of Δ_n we have $\sigma \subseteq f^{-1}(\bigcup_{i\in\sigma^0} A_i)$.
Invoking the KKM theorem for open sets we conclude that $\bigcap_{i=0}^{i=n} f^{-1}(A_i) \neq \emptyset$ and consequently that $\bigcap_{i=0}^{i=n} A_i \neq \emptyset$.

The sets A_i being arbitrary open neighborhoods of the closed sets R_i in the normal space X we have $\bigcap_{i=0}^{i=n} R_i \neq \emptyset$.

Now we assume that (B) holds.

For each proper face σ of Δ_n let $F_\sigma = \bigcup_{i\notin\sigma^0} R_i$ and $F_{\Delta_n} = \bigcap_{i=0}^{i=n} R_i$.

As before let A_i be an open neighborhood of R_i and put $W_\sigma = \bigcap_{i\notin\sigma^0} A_i$.

Now we find a continuous function $f : \Delta_n \longrightarrow X$ such that for each face σ of Δ_n we have $\sigma \subseteq f^{-1}(\bigcap_{i\notin\sigma^0} A_i)$ and therefore for each $i \in \{0,\cdots,n\}$ we have $\sigma_i \subseteq f^{-1}(A_i)$, where σ_i is the $(n-1)$-dimensional face of Δ_n that does not contain the i-th vertex.
By theorem 2 we have $\bigcap_{i=0}^{i=n} f^{-1}(A_i) \neq \emptyset$, and therefore also $\bigcap_{i=0}^{i=n} A_i \neq \emptyset$.
And we conclude as previously that $\bigcap_{i=0}^{i=n} R_i \neq \emptyset$.

To complete the demonstration we deal with the second case, X is arbitrary and all the R_i are open subspaces.

We proceed exactly as previously but since the sets R_i are open we can take $A_i = R_i$ for each $i \in \{0,\cdots,n\}$ and we obtain directly that $\bigcap_{i=0}^{i=n} R_i \neq \emptyset$. □

A subspace S of a topological space X is $\mathbf{UV}^\omega$ in X if it is n-**UV** for each $n \in \mathbb{N}$.

Corollary 1 *Let X and Y be topological spaces, with X normal, and $\Omega : Y \longrightarrow X$ a correspondence such that:*
(a) For each $y \in Y$ the set $\Omega(y)$ is closed and X is normal or, for each $y \in Y$ the set $\Omega(y)$ is open;
*(b) For any nonempty finite subset $A \subseteq Y$ the set $\bigcup_{y\in A} \Omega(y)$ is (card(A)-2)-***UV**.

Then the family $\{\Omega(y) : y \in Y\}$ has the finite intersection property.

The next result is a topological version of Fan's theorem for **UV** sets. It is proved as theorem 8.

Theorem 11 *Let X be a* $\mathbf{UV}^\omega$ *topological space and* $\Omega : X \longrightarrow X$ *a correspondence such that:*
(a) Either all the values of Ω *are closed or all the values are open and, for each* $x \in X$, $x \in \Omega(x)$;
(b) For each nonempty subset $B \subseteq X$ *the set* $\bigcap_{x\in B} X \setminus \Omega^{-1}(x)$ *is* $\mathbf{UV}^\omega$;

Then the family $\{\Omega(x) : x \in X\}$ *has the finite intersection property.*

From theorem 11 one can derive a fixed point theorem exactly as it was done from theorem 8.

Helly himself gave a topological version of his theorem using singular homology. His conditions are on the intersections of the sets, in contrast ours are on the union.

Theorem 12 (Helly) *Let* $\mathcal{F}$ *be a family of open sets in* $\mathbb{R}^n$. *Assume that the following two conditions hold:*
(a) Any subfamily of at most $n+1$ *members has a nonempty intersection,*
(b) The intersection of any subfamily of at most n *members is acyclic.*
Then $\mathcal{F}$ *has the finite intersection property.*

Helly's theorem was generalized to connected topological manifolds without boundary by Debrunner, see theorem 3 in [8] for a proof. Debrunner's result relies on a kind of duality theorem, lemma A_m in [8], which is a Klee like intersection result.

Theorem 13 (Debrunner) *Let* $\{R_0, \cdots, R_n\}$ *be a family of* $n+1$ *open acyclic subsets in a topological space* X *such that the intersection of at most* n *members of the family is nonempty and acyclic. Then*
(a) if $\cap_{i=0}^{i=n} R_i = \emptyset$, *then* $\tilde{H}_q(\cup_{i=0}^{i=n} R_i)$ *is nonzero for* $q = n-1$ *and zero for* $q \neq n-1$. *In particular, if* $\cup_{i=0}^{i=n} R_i$ *is acyclic then* $\cap_{i=0}^{i=n} R_i \neq \emptyset$.
(b) if $\cap_{i=0}^{i=n} R_i \neq \emptyset$, *then* $\tilde{H}_q(\cup_{i=0}^{i=n} R_i)$ *is isomorphic to* $\tilde{H}_{q-n}(\cap_{i=0}^{i=n} R_i)$ *for each integer* q.

The proof is done by induction using the Mayer-Vietoris sequence. $\tilde{H}_q(X), q \geq 0$, are the reduced singular homology groups of the space X. The space X is acyclic if $\tilde{H}_q(X) = 0, q \geq 0$. Convex sets are acyclic, more generally contractible spaces are acyclic.

Dugundji gave two generalizations of Helly's theorem. One for open sets, [11], and another for closed sets, [10].

Theorem 14 (Dugundji) *Let $\mathcal{F}$ be a family of $m \geq n+2$ open sets in $\mathbb{R}^n$. Assume that the following two conditions hold:*
(a) any subfamily of at most $m-1$ members has a nonempty intersection;
(b) any subfamily of k members has a $(m-k-2)$-connected intersection.
Then $\mathcal{F}$ has a nonempty intersection.

Theorem 15 (Dugundji) *Let $\mathcal{F} = \{R_0, \cdots, R_m\}$ be a family of $m+1$ closed sets in a metric space. Assume that the following two conditions hold:*
(a) any subfamily of at most m members has a nonempty intersection;
(b) any subfamily of k members has a $(m-k-1)$-connected intersection;
(c) $\tilde{H}_{m-1}(\cup_{i=1}^{i=m} R_i) = 0$.
Then $\mathcal{F}$ has a nonempty intersection.

3. Topological Spaces with a Convexity Structure

Definition 3 *A preconvexity on a space X is a couple $(\Delta,\ D(\Delta))$ where $D(\Delta)$ is a family of nonempty subsets of X containing all the finite subsets, and Δ is a function from $D(\Delta)$ into the power set $\mathcal{P}(X)$ such that:*
(a) for any $A \in D(\Delta)$ the set $\Delta(A)$ is not empty;
(b) if $A \subseteq B \in D(\Delta)$ then $A \in D(\Delta)$ and $\Delta(A) \subseteq \Delta(B)$.

We have seen that any correspondence $\Omega : X \longrightarrow X$ such that $x \in \Omega(x)$ for each $x \in X$ defines a preconvexity Δ_Ω on X.

An operator for which (b) holds, as well as $A \subseteq \Delta(A)$ and $\Delta(\Delta(A)) = \Delta(A)$ is called a closure operator. If furthermore $\Delta(A) = \cup\{\Delta(B) : B \subseteq A, card(B) < \infty\}$ then Δ is an algebraic closure operator. One can show that any preconvexity generates a smallest algebraic closure operator. We will refer only to preconvexities even if the operator Δ has further properties.

From a given preconvexity one can generate new preconvexities.
Let $(\Delta,\ D(\Delta))$ be a preconvexity on X and $T : X \longrightarrow X$ a function.
(a) $(T^*\Delta)(A) = \Delta(T(A))$ with $D(T^*\Delta) = \{A : T(A) \subseteq D(\Delta)\}$
(b) $(T_*\Delta)(A) = T(\Delta(A))$ with $D(T_*\Delta) = D(\Delta)$
(c) if $r : X \longrightarrow Y$ is an onto function and if $D(\Delta) = \mathcal{P}(X)$, we define a preconvexity on Y by $\Delta_Y(A) = r(\Delta(r^{-1}(A)))$.

The last construction shows that there is a preconvexity structure on any absolute retract Y, since an absolute retract can be embedded in a norm ed space E and there is a retraction $r : E \longrightarrow Y$.

If X is a metric space and if there is enough compatibility between the preconvexity $(\Delta,\ D(\Delta))$ and the topology, then one can show that X is an absolute retract. The first result of this kind is due to Himmelberg, [14].

Some topological spaces have a rich enough algebraic or geometric structure to carry a natural preconvexity. This will be the subject matter of the next paragraphs. But we are not interested in arbitrary preconvexities, we want preconvexities for which Fan's like results hold.

Definition 4 *A preconvexity* $(\Delta,\ D(\Delta))$ *on a topological space* X *is called a KKM preconvexity if for any finite subset* $\{x_0, \cdots, x_n\} \subseteq X$ *and any finite family* $\{R_0, \cdots, R_n\}$ *of closed or open subsets such that* $\Delta(\{x_i : i \in I\}) \subseteq \cup_{i \in I} R_i$ *for any nonempty* $I \subseteq \{0, \cdots, n\}$*, the intersection* $\cap_{i=0}^{i=n} R_i$ *is not empty.*

From the results of the previous section the proof of the following theorem is now obvious.

Theorem 16 *For a preconvexity* $(\Delta,\ D(\Delta))$ *on a topological space to be a KKM preconvexity it is sufficient that for any finite subset* $A \subseteq X$ *the set* $\Delta(A)$ *be* $(card(A) - 2)$*-connected or* $(card(A) - 2)$-**UV** .

3.1. ORDER CONVEXITIES

On any partially ordered set $(X, \preceq)$ one can define a preconvexity by
$\Delta_{\preceq}(A) = \cap\{C : A \subseteq C \in \mathcal{C}(\preceq)\}$, where
$\mathcal{C}(\preceq) = \{C \subseteq X : [x, y] \subseteq C \text{ if } x, y \in C \text{ and } x \preceq y\}$.

If the order is not linear, $\Delta_{\preceq}$ might not have the required topological properties, for example if A is a two elements set in $\mathbb{R}^2$ and if the points are not comparable then $\Delta_{\preceq}(A)$ is a discrete set. Semilattices provide the right kind of structure.

Definition 5 *A partially ordered set* $(X, \preceq)$ *is a sup-semilattice if each nonempty finite set* $A \subseteq X$ *has a least upper bound* $\sup A$*, it is bounded if there is an element* $1 \in X$ *such that* $x \preceq 1$ *for any* $x \in X$*.* $(X, \preceq)$ *is a topological semilattice if* X *is a topological space and the function* $X \times X \longrightarrow X,\ (x, y) \mapsto \sup\{x, y\}$ *is continuous.*

We have given the definition of a sup-semilattice, we could obviously also consider inf-semilattices. When no confusion can arise we will simply use the word semilattice. It is also evident that each nonempty finite set will have a least upper bound if any two elements have a least upperbound.

The key result is the following lemma, Brown, [5]

Lemma 2 (Brown) *If* $(X, \preceq)$ *is a pathconnected bounded topological semilattice then any continuous function* $f : [0, 1]^n \longrightarrow X$ *which is constant on the boundary of the cube* $[0, 1]^n$ *is homotopic to a constant through an homotopy* $h : [0, 1]^n \times [0, 1] \longrightarrow X$ *such that for each* $t \in [0, 1]$ *the function* $h(-, t) : [0, 1]^n \longrightarrow X$ *is constant on the boundary of the cube.*

Proof. Let 1 be the largest element of X. Since X is pathconnected we can assume that $f(\partial([0,1]^n)) = \{1\}$.

Let $h(t_1, \cdots, t_n, s) = \sup\{f(t_1, \cdots, t_n), f(t_1 - st_1, \cdots, t_n - st_n)\}$ □

As a consequence of Brown's lemma we have the following result.

Lemma 3 *A pathconnected bounded topological semilattice is* $\mathbf{C}^\infty$.

Proof. The boundary of Δ_{n+1} is homeomorphic to the n-dimensional euclidean sphere S^n.

S^n is obtained from the $(n+1)$-dimensional cube $[0,1]^{n+1}$ by identification of the boundary to a single point. Therefore a continuous function $f : S^n \longrightarrow X$ can be identified to a continuous function $f' : [0,1]^{n+1} \longrightarrow X$ constant on the boundary. By Brown's lemma f' is homotopic to a constant function with the partial functions $h(-,t)$ constant on the boundary of $[0,1]^{n+1}$.

The function $f : S^n \longrightarrow X$ is therefore homotopic to a constant.

If $H : S^n \times [0,1] \longrightarrow X$ is a continuous function with $H(-,0) = x_0$ and $H(-,1) = f$ then

$$g(u) = \begin{cases} x_0 & \text{if } u = 0 \\ H(\| u \|^{-1} u, \| u \|) & \text{if } u \neq 0 \end{cases}$$

is a continuous extension of f to the $(n+1)$-dimensional ball. □

Theorem 17 *In a topological semilattice* $(X, \preceq)$ *with pathconnected intervals the preconvexity defined on nonempty finite subsets by* $\Delta_{\sup}(A) = \cup_{x \in A}[x, \sup A]$ *is KKM.*

Proof. First, notice that $\cup_{x \in A}[x, \sup A]$ is a bounded semilattice. Furthermore, each interval $[x, \sup A]$ is pathconnected by hypothesis, and therefore $\cup_{x \in A}[x, \sup A]$ is also pathconnected. By lemma 3, $\Delta_{\sup}(A)$ is $\mathbf{C}^\infty$ □

A subset C of a topological semilattice $(X, \preceq)$ is convex if the following two conditions hold:

(i) if $x_1, x_2 \in C$ then $\sup\{x_1, x_2\} \in C$;

(ii) if $x_1, x_2 \in C$ and $x_1 \preceq x_2$ then $[x_1, x_2] \subseteq C$,

or if the following single condition holds:

(iii) if $x_1, x_2 \in C$ then $[x_1, \sup\{x_1, x_2\}] \subseteq C$

Since an arbitrary intersection of convex sets is convex Fan's intersection theorem and Browder-Fan's fixed point theorem take the following form.

Theorem 18 *Let* $(X, \preceq)$ *be a topological semilattice with pathconnected intervals, and* $\Omega : X \longrightarrow X$ *a correspondence such that:*

(a) Either all the values of Ω *are closed or all the values are open, and for each* $x \in X$, $x \in \Omega(x)$;

(b) For each $x_1, x_2 \in X$, $\Omega([x_1, \sup\{x_1, x_2\}]) \subseteq \Omega(x_1) \cup \Omega(x_2)$.

Then the family $\{\Omega(x) : x \in X\}$ has the finite intersection property.

Condition (b) means that for each $x \in X$ the set $X \setminus \Omega^{-1}(x)$ is convex.

Theorem 19 *Let $(X, \preceq)$ be a compact topological semilattice with pathconnected intervals, $F : X \longrightarrow X$ a correspondence such that:*
(a) For each $x \in X$ the set $F^{-1}(x)$ is open;
(b) For each $x \in X$ the set $F(x)$ is not empty and for each $x_1, x_2 \in F(x)$ we have $[x_1, \sup\{x_1, x_2\}] \subseteq F(x)$. =20

Then there is a point $x^* \in X$ such that $x^* \in F(x^*)$.

Examples of semilattices with pathconnected intervals are given by cubes $[0, 1]^n$, more generally by Tychonoff cubes $[0, 1]^S$. One can easily see that there are subsets of $\mathbb{R}^n$ that are not convex, for the linear structure, but are sup-semilattices for the order defined by the positive cone.

3.2. CONVEXITY IN METRIC SPACES.

In a metric space (X, d) define the ball-hull $\Delta_d(A)$ of a bounded subset A of X to be the intersection of the family of closed balls containing it.

Let us say that two closed balls in a metric space are adjacent if the distance between the centers is at most equal to the sum of the radii.

A metric space is said to be hyperconvex if any family of pairwise adjacent closed balls has a nonempty intersection. One can show that a metric space is hyperconvex if the family of closed balls has the binary intersection property and if it is metrically convex (i.e. if $d(x, y) = a + b$, with $a \geq 0$ and $b \geq 0$, then there is a point z such that $d(x, z) = a$ and $d(z, y) = b$). L^∞ spaces are hyperconvex.

Theorem 20 *In any hyperconvex metric space the ball hull defines a KKM preconvexity.*

In a hyperconvex metric space the intersection of a family of closed balls is hyperconvex, Aronszajn - Panitchpakdi, [2], or Baillon, [3], [23] or Horvath [15].
Fan's intersection theorem takes the following form:

Theorem 21 *Let (X, d) be an hyperconvex metric space, and $\Omega : X \longrightarrow X$ a correspondence such that:*
(a) Either all the values of Ω are closed or all the values are open and, for each $x \in X$, $x \in \Omega(x)$;
(b) For each $x \in X$ and for each finite subset $A \subseteq X \setminus \Omega^{-1}(x)$ we have $\Delta_d(A) \subseteq X \setminus \Omega^{-1}(x)$

Then the family $\{\Omega(x) : x \in X\}$ has the finite intersection property.

Proof. As we have seen, conditions (a) and (b) imply that $\Delta_d(A) \subseteq \cup_{a\in A}\Omega(a)$ for each finite subset $A \subseteq X$, and, since Δ_d is a KKM preconvexity the conclusion follows.□

If A is a bounded subset of X then $p \in \Delta_d(A)$ if and only if $d(x,p) \leq \sup_{y\in A} d(x,y)$ for any $x \in X$.

3.3. BETWEENNESS CONVEXITY.

If x, y, z are three points on the unit interval $[0,1] \subseteq \mathbb{R}$, then one of them is between the other two, name it $\mu_0(x,y,z)$. It is not hard to see that $\mu_0(x,y,z) = \min\{\max(x,y), \max(x,z), \max(y,z)\}$. This shows that what matters is the lattice structure. The continuous function $\mu_0 : [0,1]^3 \longrightarrow [0,1]$ yields a continuous function $\mu : ([0,1]^{\mathbb{N}})^3 \longrightarrow [0,1]^{\mathbb{N}}$ given by $\mu(x,y,z)_i = \mu_0(x_i, y_i, z_i)$.

Notice that $\mu(x,x,y) = \mu(x,y,x) = \mu(y,x,x) = x$. A subset $C \subseteq [0,1]^{\mathbb{N}}$ is μ-convex if $\mu(C^3) \subseteq C$.

Theorem 22 *Let X be a closed connected μ-convex subspace of $[0,1]^{\mathbb{N}}$ and $\Omega : X \longrightarrow X$ a correspondence such that:*
(a) Either all the values of Ω are closed or all the values are o pen and, for each $x \in X$, $x \in \Omega(x)$;
(b) For each $x \in X$ the set $\Omega^{-1}(x)$ is open;
(c) For each $x \in X$ the set $X \setminus \Omega^{-1}(x)$ is μ-convex;
(d) For each nonempty subset $B \subseteq X$ the set $\bigcap_{x\in B} X\setminus\Omega^{-1}(x)$ is connected.

Then $\{\Omega(x) : x \in X\}$ has the finite intersection property.

Once we have noticed that (b), (c) and (d) imply that for any nonempty finite subset B of X the set $\bigcap_{x\in B} X \setminus \Omega^{-1}(x)$ is closed and μ-connected, and if we admit that closed connected μ-convex subspaces of $[0,1]^{\mathbb{N}}$ are absolute retracts, this theorem is proved exactly as theorem 11. We could even replace the Hilbert cube by an arbitrary Tychonoff cube and add the hypothesis that the sets $\bigcap_{x\in B} X \setminus \Omega^{-1}(x)$ are G_δ, since they would then be absolute extensors, notice that this implies that they are pathconnected. For these results see Szymanski, [24], characterization 1.5 and theorem 1.14. The openness condition on the sets $\Omega^{-1}(x)$ can be removed, and the G_δ assumption replaced by pathconnectedness, if we use a difficult result of Van Mill and Van de Vel, theorem 2.5 [26].

Definition 6 *A mixer on a metric space X is a continuous function $\mu : X^3 \longrightarrow X$ such that* $\lim \mu(x_n, y_n, z_n) = \lim \mu(x_n, z_n, y_n) = \lim \mu(z_n, x_n, y_n) = x^*$ *if* $\lim x_n = \lim y_n = x^*$.

If the metric space X is compact then any continuous function $\mu : X^3 \longrightarrow X$ such that $\mu(x,x,y) = \mu(x,y,x) = \mu(y,x,x) = x$ is a mixer.

Consequently any compact, metrizable topological lattice has a mixer: $\mu_0(x, y, z) = \min\{\max(x, y), \max(x, z), \max(y, z)\}$.

Theorem 23 (Van Mill-Van de Vell) *A pathconnected metric space on which there is a mixer is* $\mathbf{C}^\infty$.

Now, given a metric space with a mixer, for example the Hilbert cube $[0, 1]^{\mathbb{N}}$, we can define μ-convex sets as we did previously.

Taking into account the results of the previous section and the theorem of Van Mill-Van de Vel, we have the following Fan like theorem.

Theorem 24 *Let X be a pathconnected metric space with a mixer $\mu : X^3 \longrightarrow X$ and $\Omega : X \longrightarrow X$ a correspondence such that:*
(a) For each $x \in X$ the set $\Omega(x)$ is closed and $x \in \Omega(x)$;
(b) For each $x \in X$ the set $X \setminus \Omega^{-1}(x)$ is μ-convex ;
(c) For each nonempty subset $B \subseteq X$ the set $\bigcap_{x \in B} X \setminus \Omega^{-1}(x)$ is pathconnected.

Then $\{\Omega(x) : x \in X\}$ has the finite intersection property.

In case X is a compact metric space and $\mu : X^3 \longrightarrow X$ is a continuous function such that $\mu(x, x, y) = \mu(x, y, x) = \mu(y, x, x) = x$ it is sufficient to assume that X is connected to deduce that it is $\mathbf{C}^\infty$, theorem 1.3 in [25]. With theorem 9 we get the following fixed point theorem.

Theorem 25 *Let X be a compact connected metric space with a continuous function $\mu : X^3 \longrightarrow X$ such that $\mu(x, x, y) = \mu(x, y, x) = \mu(y, x, x) = x$, and $F : X \longrightarrow X$ a correspondence such that:*
(a) For each $x \in X$ the set $F^{-1}(x)$ is open;
(b) For each $x \in X$ the set $F(x)$ is not empty and μ-convex;
(c) For each nonempty subset $B \subseteq X$ the set $\bigcap_{x \in B} F(x)$ is connected.

Then there is a point $x^* \in X$ such that $x^* \in F(x^*)$.

3.4. SUPERCOMPACT SPACES.

A family of sets is binary if any subfamily with the binary intersection property has a nonempty intersection.

A family of subsets $\mathcal{B}$ of a set X is normal if, for any pair $B_1, B_2 \in \mathcal{B}$ such that $B_1 \cap B_2 = \emptyset$, there exists $B'_1, B'_2 \in \mathcal{B}$ such that $B_i \cap B'_i = \emptyset$ and $X = B'_1 \cup B'_2$.

Definition 7 *A topological space is normally supercompact if it has a normal and binary subbase of closed sets.*

Supercompact spaces have been extensively studied by Van Mill and Van de Vel. A connected and metrizable normally supercompact space is an absolute retract. This was originally proved by Van Mill. For this result as well as for the following, and for further references see Horvath [15].

Theorem 26 *Let X be a normally supercompact metrizable space and $\mathcal{B}$ be a normal and binary subbase of closed sets. Let $\Delta_{\mathcal{B}}(A) = \cap\{B : A \subseteq B \in \mathcal{B}\}$.*

If X is connected then $\Delta_{\mathcal{B}}$ is a KKM preconvexity.

We let the reader state the obvious Fan like intersection and fixed point theorems.

There is a close relationship between midpoint convexities, interval convexities, semilattices and topological spaces with a binary subbase.

3.5. INTERVAL CONVEXITY

An interval structure on a space X is given by a function $I : X \times X \longrightarrow \mathcal{P}(X)$ such that :

(i) $\{x, y\} \subseteq I(x, y)$
(ii) $I(x, y) = I(y, x)$
(iii) if $z, w \in I(x, y)$ then $I(z, w) \subseteq I(x, y)$
(iv) $I(x, y) \cap I(x, z) \cap I(y, z) \neq \emptyset$

A subset $C \subseteq X$ is I-convex if $I(x, y) \subseteq C$ for any $(x, y) \in C \times C$.

To obtain a Fan like theorem, which we will not state, one needs one more property, that the topology has a normal closed subbase of I-convex sets, since in that case the space is normally supercompact, [27]. One should notice here that the normality condition is a separation condition which in vector spaces is a consequence of the Stone-Kakutani theorem, two disjoint convex sets can be enlarged to maximal disjoint convex sets.

3.6. DECOMPOSABLE SETS.

Let E be a Banach space, μ a nonatomic probability measure on a measurable space T, and $L^1(T, E)$ the Banach space of μ-measurable functions $f : T \longrightarrow E$.

A subset $C \subseteq L^1(T, E)$ is decomposable if, for any measurable set $A \subseteq T$ and any $f, g \in C$, one has $f\chi_{T\setminus A} + g\chi_A \in C$. An intersection of closed decomposable sets is obviously closed and decomposable. A closed decomposable set is a retract of $L^1(T, E)$, it is therefore an absolute retract, Bressan and Colombo [4].

The closed decomposable hull of a subset $D \subseteq L^1(T, E)$ is therefore a KKM preconvexity on $L^1(T, E)$.

3.7. GEODESIC CONVEXITY.

Geodesic convexity on a riemannian manifold was investigated by Rapcsák, [21] for the purpose of nonlinear optimization. Eventhough it seems to be a natural generalization there might be a scarcity of interesting objects.

Assume that X is a differentiable manifold with a given riemannian structure.

A function $f : X \longrightarrow \mathbb{R}$ is convex if, for any geodesic $\gamma : [0,1] \longrightarrow X$, the function $f \circ \gamma : [0,1] \longrightarrow X$ is convex.

A subset $C \subseteq X$ is totally convex if any geodesic with both endpoints in C is entirely in C. The sublevel sets of a convex function are totally convex.

Unfortunately, on a connected riemannian manifold of finite volume, a convex function is constant. And, in a compact connected riemannian manifold there is only one nonempty totally convex subset: the manifold itself.

Even without the compactness condition, or the finiteness condition, the existence of nontrivial convex functions imposes severe restrictions: if there exists a strictly convex function $f : X \longrightarrow X$ all of whose closed sublevel sets are compact then X is diffeomorphic to $\mathbb{R}^n$. For all these questions one can look at Walter [29] and the bibliography therein.

4. Conclusion

We have given examples of spaces with a rich enough geometric structure to yield KKM like and Fan like intersection theorems. But a lot has not been said, at least as far as minimax results are concerned. In each case, what are the convex, or the quasiconvex functions? In semilattices one can see that increasing functions will be convex, but with maybe the most natural convexity of all, the geodesic convexity, there might very well be only trivial convex or even quasiconvex functions. This does not prevent us from stating abstract intersection theorems, but they might not have much interest. Also, what do convex sets look like? And, if we want Fan like theorems, what do compact convex sets look like? If they are trivial, the whole space, or singletons, as with geodesic convexity or decomposable sets, a Fan like result is pointless. Much effort has gone into replacing convexity by topological conditions, the minimal topological apparatus needed to establish the classical minimax theorems was soon clarified thanks to the work, among others, of König and Kindler. But, altogether, it seems that the topological approach has given a very large number of abstract formulations of classical results and few new results outside of a purely postulational framework. We hope that the examples presented here will lead to interesting results in a framework where some natural convexity structure is available.

Finally, let us mention the book by M. Van de Vel on topological convexities, [28], and the paper by Coppel, [6] where one can find a characterization of abstract convex sets that can be embedded as convex subsets of linear spaces.

References

1. S.Armentrout, *UV properties of Compact Sets,* Trans. Amer. Math. Soc. **143** (1969), 487-498.
2. N.Aronszajn and P. Panitchpakdi, *Extensions of Uniformly Continuous Transformations and Hyperconvex Metric Spaces,* Pac. J.Math. **6** (1956), 405-439.
3. J.B. Baillon, *Nonexpansive Mappings and Hyperconvex Spaces,* Contemp. Math. **72** (1988), 11-19.
4. A. Bressan and G. Colombo, *Extensions and Selections of Maps with Decomposable Values,* Stud. Math. **90** (1988), 69-86.
5. D.R. Brown, *Topological Semilattices on the Two Cell,* Pac. J. Math. **15** no.1 (1965), 35-46.
6. W.A. Coppel, *Axioms for Convexity,* Bull. Austral. Math. Soc. **47** (1993), 179-197.
7. R. Daverman, *Decomposition of Manifolds,* Academic Press, 1986.
8. H.E. Debrunner, *Helly Type Theorem Derived from Basic Singular Homology,* Amer. Math. Monthly **77** (1970), 375-380.
9. J. Dugundji, *Modified Vietoris Theorems for Homotopy,* Fund. Math. **LXVI** (1970), 223-235.
10. J. Dugundji, *Maps Into Nerves of Closed Coverings,* Ann. Sc. Norm. Sup. di Pisa bf XXI, Fasc.II, (1967), 121-136.
11. J. Dugundji, *A Duality Property of Nerves,* Fund. Math., **LIX** (1966), 213-219.
12. S. D. Flåm and G. H. Greco, *Minimax and Intersection Theorems,* Fixed Point Theory and Applications, M. A. Théra and J. B. Baillon (editors), Longman Scientific and Technical (1990), 123- 140.
13. E. Helly, *Über Mengen Konvexer Körper mit Gemeinschaftlichen Punkten,* J. Jber. Deutsch. Math. Verein. **32** (1923) 175-176.
14. C. J. Himmelberg, *Some Theorems on Equiconnected and Locally Equiconnected Spaces,* Trans. Amer. Math. Soc. **115** (1965), 43-53.
15. C. Horvath, *Extension and Selection Theorems in Topological Spaces,* Annales de la Faculté des Sciences de Toulouse **2** (1993), 253-269.
16. C. Horvath and M. Lassonde, *Intersection of Sets with n-Connected Unions,* Proc. Amer. Math. Soc. **125** (1997), 1209-1214.
17. J. Kindler, *Topological Intersection Theorems,* Proc. Amer. Math. Soc. **117** (1993), 1003-1011.
18. H. König, *A General Minimax Theorem Based on Connectedness,* Arch. Math. **59** (1992), 55-64.
19. M. Lassonde, *Sur le Principe KKM,* C.R. Acad. Sci. Paris, Série I Math. **310** (1990), 573-576.
20. L. Pasicki, *Retracts in Metric Spaces,* Proc. Amer. Math. Soc. **78** (1980), 595-600.
21. T. Rapcsák, *Geodesic Convexity in Nonlinear Optimization,* J. Optim. Theory Appl. **69** (1991), 169-183.
22. T. Rockafellar, *Helly's Theorem and Minima of Convex Functions,* Duke Math. J. **32** (1965), 381-397;
23. R. Sine, *Hyperconvexity and Nonexpansive Multifunctions,* Trans. Amer. Math. Soc. **315** Nb.2 (1989), 755-767.

24. A. Szymanski, *Q-Binary Spaces,* Top. Appl. **43** (1992), 117-130.
25. J. van Mill and M. van de Vel, *On an Internal Property of Absolute Retracts,* Topology Proc. 4 (1979), 193-200.
26. J. van Mill and M. van de Vel, *On an Internal Property of Absolute Retracts, II,* Top. Appl. **13** (1982), 59-68.
27. J. van Mill and E. Wattel, *An External Characterization of Spaces which Admit Binary Normal Subbases,* Amer. J. Math. **100** (1978), 987-994.
28. M. Van de Vel, *Topological Convexity,* North Holland, 1993.
29. R. Walter, *Konvexität in riemannschen Mannigfaltigkeiten,* Jber. d. Dt. Math. Verein. **83** (1981), 1-31.

MINIMAX RESULTS AND RANDOMIZATION FOR CERTAIN STOCHASTIC GAMES

ALBRECHT IRLE
Mathematisches Seminar
University of Kiel
Ludewig-Meyn-Str. 4
D-24098 Kiel, Germany
e-mail: irle@math.uni-kiel.de

Abstract. Randomization and minimax results are considered for the following type of stochastic two-person zero-sum games: two players sequentially observe a stochastic process. Player I selects a stopping time and player II the distribution of the process. The pay-off is given by the expected value of the stopped process. We discuss in detail the different ways of introducing randomized strategies. Minimax results are given for stochastic processes with discrete and with continuous time parameter.

1. Introduction

Let $(\Omega, \mathcal{A})$ be a measurable space and $(\mathcal{A}_s)_{s \in S}$ an increasing sequence of sub-σ-algebras of $\mathcal{A}$ with $\mathcal{A}_\infty = \sigma(\bigcup_{s \in S} \mathcal{A}_s)$ where $S = \bar{N} = N \cup \{\infty\}$ with $N = \{1, 2, \ldots\}$, the discrete time case, or $S = [0, \infty]$, the continuos time case.

A stopping time T is a mapping $T : \Omega \to S$ with

$$\{\omega : T(\omega) \leq t\} \in \mathcal{A}_t \quad \text{for all} \quad t \in S.$$

We denote by

$$\mathcal{S} \text{ the set of all stopping times .}$$

These notions may be interpreted as follows: S is the set of possible time points when to stop some process evolving in time. $\mathcal{A}_s$ describes the information available at time s. A stopping time is a strategy for stopping,

B. Ricceri and S. Simons (eds.), Minimax Theory and Applications, 91–103.

which is non-anticipative in the sense that the event to stop not later than at time s belongs to $\mathcal{A}_s$.

Let furthermore $\mathcal{M}$ be a set of probability measures on $(\Omega, \mathcal{A})$ and $(Z_s)_{s \in S}$ a stochastic process with values in $[-\infty, +\infty]$ such that each Z_s is $\mathcal{A}_\infty$-measurable and $Z_\infty = \lim Z_n$.

We assume for each $P \in \mathcal{M}$

(A 1) $\quad E_P \sup_s Z_s^+ < \infty$ for each $P \in \mathcal{M}$,

or

(A 1') $\quad E_P \sup_s Z_s^- < \infty$ for each $P \in \mathcal{M}$.

Here $E_P = \int dP$ denotes expectation under P and the above assumption ensures that the integrals which will arise in the following are well-defined.

A two-person zero-sum game of the following type will be called a game of stopping:

Player I picks a stopping time $T \in \mathcal{S}$, player II picks a distribution $P \in \mathcal{M}$ and the reward for player I, hence loss for player II, is given by

$$A(T, P) = E_P Z_T = \int Z_{T(\omega)}(\omega)\, P(d\omega) .$$

Special games of this type have e.g. been treated by Gödde (1991), Hill and Krengel (1991).

Minimax theorems of game theory, yielding validity of inf sup = sup inf under suitable assumptions on the set of strategies and the reward, usually rely on some generalized conditions of convexity, resp. concavity for these items. In games of stopping, $\mathcal{S}$ and the mapping $T \to E_P Z_T$ in general do not fulfill such conditions which is reflected by the fact that typically $\sup_{\mathcal{S}} \inf_{\mathcal{M}} A(T, P) < \inf_{\mathcal{M}} \sup_{\mathcal{S}} A(T, P)$ although $P \to E_P Z_T$ is a linear mapping and $\mathcal{M}$ may usually be assumed a convex subset of the set of all probability measures.

As a common remedy of game theory, player I turns to randomized strategies. There are different ways to introduce randomized stopping times which will be discussed in the next section 2.

Sections 3 and 4 give an overview on results for the discrete time case. Section 3 gives a suitable compact embedding for the set of randomized stopping times which shows via Choquet representation theory the equivalence of different notions of randomization. Section 4 gives minimax results under different sets of assumptions.

In Section 5, the continuous time case is treated. Using results from discrete time and a suitable approximation technique, a new minimax result is obtained.

2. Randomization of Stopping Times

The first two types of randomization can be introduced for general time sets S.

Mixed stopping times.
Let $\sigma(\mathcal{S})$ be a σ-algebra on $\mathcal{S}$ such that $\{T\} \in \sigma(\mathcal{S})$ for all $T \in \mathcal{S}$ and $T \to A(T,P)$ is measurable for all $P \in \mathcal{M}$. A probability measure τ on $(\mathcal{S}, \sigma(\mathcal{S}))$ is called a mixed stopping time, with reward given by

$$A(\tau, P) = \int_{\mathcal{S}} A(T,P)\,\tau(dT)\,.$$

Denoting the set of all mixed stopping times by $\tilde{\mathcal{S}}$, we call

$$(\tilde{\mathcal{S}}, \mathcal{M}, A) \quad \text{game of mixed stopping} \ .$$

Randomized stopping times.
Extend Ω, $\mathcal{A}$, $\mathcal{A}_s$ and any P to

$$\Omega^* = \Omega \times [0,1],\ \mathcal{A}^* = \mathcal{A} \otimes \mathcal{B},\ \mathcal{A}_s^* = \mathcal{A}_s \otimes \mathcal{B},\ P^* = P \otimes \lambda,$$

where $\mathcal{B}$ denotes the Borel-σ-algebra of $[0,1]$ and λ Lebesgue measure on $[0,1]$. A stopping time T^* with respect to these enlarged measurable space and filtration is called a randomized stopping time with reward given by

$$A(T^*, P) = E_{P^*}\, Z_T = \int Z_{T(\omega,x)}(\omega)\, P^*(d\omega, dx)\,.$$

Denoting the set of all mixed stopping times by $\mathcal{S}^*$, we call

$$(\mathcal{S}^*, \mathcal{M}, A) \quad \text{game of randomized stopping} \ .$$

To see how these two different notions are connected we look at a mixed stopping time τ which forms a discrete probability measure,

$$\tau = \sum_n c_n \delta_{T_n}$$

with δ denoting Dirac measure. Define intervals $I_n = [\sum_{i=1}^{n-1} c_i, \sum_{i=1}^{n} c_i)$ and a randomized stopping time T^* by

$$T^* = \sum_n T_n 1_{I_n}.$$

Then we have

$$A(\tau, P) = A(T^*, P) \text{ for all } P.$$

Randomization in the discrete time case.

Let us now consider the discrete time case where we can give the following intuitive meaning to the concept of randomization. We shall express in mathematical terms that at each time point we may decide either to proceed with one more observation or to stop. This decision may be based on observations obtained so far and on an additional random experiment at our disposal like throwing a coin or using a random number generator. We define

$$F = \{a = (a_j)_{j \in \bar{N}} : 0 \le a_j \le 1\,,\, a_\infty = 1\,,\, a_j = 1 \ \Rightarrow \ a_i = 1 \text{ for } i > j\},$$

$$G = \{b = (b_j)_{j \in \bar{N}} : 0 \le b_j \le 1\,,\, \sum_{j \in N} b_j = 1\}\,.$$

Note that

$\varphi : F \to G\,,\ \varphi(a)_j = a_j \prod_{1<j} (1 - a_i)\,,$

$\Psi : G \to F\,,\ \Psi(b)_j = b_j \,/ \sum_{i \ge j} b_i$ with $0/0 = 1\,,$

fulfill $\varphi \circ \Psi$ = identity (G), $\Psi \circ \varphi$ = identity (F), whence yield a $1-1$ correspondence between F and G.

A randomized stopping time in F-representation is given by a mapping $f = (f_n)_{n \in \bar{N}} : \Omega \to F$ such that each f_n is $\mathcal{A}_n$-measurable, with reward

$$A(f, P) = \sum_n \int Z_n\, f_n \sum_{i<n} (1 - f_i)\, dP\,.$$

Denoting the set of all randomized stopping times in F-representation by $\mathcal{F}$, we call $(\mathcal{F}, \mathcal{M}, A)$ game of randomized stopping in F-representation.

A randomized stopping time in G-representation is given by a mapping $g = (g_n)_{n \in \bar{N}} : \Omega \to G$ such that each g_n is $\mathcal{A}_n$-measurable, with reward

$$A(g, P) = \sum_n \int Z_n\, g_n\, dP\,.$$

Denoting the set of all randomized stopping times in G-representation by $\mathcal{G}$, we call

$$(\mathcal{G}, \mathcal{M}, A) \text{ game of randomized stopping in } G\text{-representation}.$$

For a randomized stopping time in F-representation, f_n is to be interpreted as conditional probability that stopping occurs at time n, given the observations in $\mathcal{A}_n$ and given that stopping has not occured before that time. In G-representation, g_n gives the conditional probability that stopping occurs at time n given the observations in $\mathcal{A}_n$. Note that a stopping time $T \in \mathcal{S}$ has the F-representation $(1_{\{T \leq n\}})_{n \in \bar{N}}$ and the G-representation $(1_{\{T=n\}})_{n \in \bar{N}}$.

Due to the $1-1$ correspondence between F and G and the definition of the reward, it is immediate that $(\mathcal{F}, \mathcal{M}, A)$ and $(\mathcal{G}, \mathcal{M}, A)$ are equivalent from the point of view of game theory. We shall now show that this equivalence also extends to $(\mathcal{S}^*, \mathcal{M}, A)$.

Let $T^* \in \mathcal{S}^*$. We want to find a corresponding $g \in \mathcal{G}$. Define $\mathcal{A}_n$-measurable g_n by

$$g_n(\omega) = \lambda(\{x : T^*(\omega, x) = n\}).$$

This defines $g \in \mathcal{G}$ such that for all P

$$A(T^*, P) = A(g, P).$$

Conversely, for $g \in \mathcal{G}$ we may define the corresponding T^* by

$$T^*(\omega, x) = n \text{ for } x \in [\sum_{i=1}^{n-1} g_i(\omega), \sum_{i=1}^{n} g_i(\omega)),\ n < \infty,$$

$$T^*(\omega, x) = \infty \text{ for } x \in [\sum_{n<\infty} g_n(\omega), 1].$$

We have thus seen that the different games of randomized stopping are equivalent. The question of equivalence for the game of mixed stopping will be treated in the following section.

3. Compact Embedding and Equivalence of Randomization

It will now be convenient to use the G-representation for games of randomized stopping, since $\mathcal{G}$ is a convex set and the mapping $g \to A(g, P) = \sum_n \int Z_n g_n \, dP$ is linear.

The following assumption (A2) will allow us to use tools from functional analysis for our treatment of games of stopping.

(A2) We assume that $\mathcal{M}|\mathcal{A}_\infty = \{P|\mathcal{A}_\infty : P \in \mathcal{M}\}$ is dominated by some probability measure Q on $(\Omega, \mathcal{A}_\infty)$ so that we have densities

$$f_P = dP|\mathcal{A}_\infty / dQ|\mathcal{A}_\infty \quad \text{for each} \quad P \in \mathcal{M}.$$

Without loss of generality we suppose

$$\{A \in \mathcal{A}_\infty : Q(A) = 0\} \subset \mathcal{A}_n \quad \text{for each} \quad n,$$

and furthermore, that Q is equivalent to $\mathcal{M}|\mathcal{A}_\infty$, i.e. $P(A) = 0$ for all $P \in \mathcal{M}$ iff $Q(A) = 0$ for $A \in \mathcal{A}_\infty$.

Let $c(\bar{N})$ be the space of continuous functions on $\bar{N}$, i.e. the space of bounded sequences $(c_n)_{n \in \bar{N}}$ with $c_\infty = \lim c_n$. Then the dual space is given by

$$c(\bar{N})^* = \ell_1(\bar{N}),$$

i.e. the space of all sequences $(\ell_n)_{n \in \bar{N}}$ with $\sum_{n \in \bar{N}} |\ell_n| < \infty$.

Furthermore we consider

$$L_1(c) = L_1(\Omega, \mathcal{A}_\infty, Q\,;\, c(\bar{N})),$$

the Banach space of integrable functions with values in $c(\bar{N})$ where $\|f\|_1 = \int \sup_{n \in \bar{N}} |f_n| dQ$ for $f = (f_n)_{n \in \bar{N}} \in L_1(c)$. The dual space is given by

$$L_1(c)^* = L_\infty(\ell_1) = L_\infty(\Omega, \mathcal{A}_\infty, Q\,;\, \ell_1(\bar{N})),$$

the Banach space of essentially bounded functions with values in $\ell_1(\bar{N})$, where $\|h\|_\infty = \text{ess sup}\,(\sum_{n \in \bar{N}} |h_n|)$ for $h = (h_n)_{n \in \bar{N}} \in L_\infty(\ell_1)$ and ess sup denotes the essential supremum with respect to the probability measure Q; see e.g. Neveu (1975), p. 201.

We shall consider the w^*-topology on $L_\infty(\ell_1)$ induced by this duality which is the weakest topology on $L_\infty(\ell_1)$ rendering the linear functionals $L_\infty(\ell_1) \ni h \to \sum_{n \in \bar{N}} \int f_n h_n \, dQ$ for each $f \in L_1(c)$ continuous.

For any $g \in \mathcal{G}$, $P \in \mathcal{M}$ we now have

$$A(g, P) = \sum_{n \in \bar{N}} \int Z_n \, g_n \, f_P \, dQ$$

and $g = g'$ Q-a.s. implies $A(g,P) = A(g',P)$ for all $P \in \mathcal{M}$.
Hence we may consider $\mathcal{G}$ as a subset of $L_\infty(\ell_1)$,

$$\mathcal{G} = \{g = (g_n)_{n\in\bar{N}} \in L_\infty(\ell_1) : 0 \le g_n \le 1,\ g_n\ \mathcal{A}_n\text{—measurable},\ \sum_{n\in\bar{N}} g_n = 1\}.$$

Here inequalities, resp. equalities are considered Q-a.s. and we shall not distinguish between equivalence classes of functions and their representants in the following.

Lemma 1 *$\mathcal{G}$ is w^*-compact in $L_\infty(\ell_1)$.*

To obtain equivalence of games of mixed stopping and games of randomized stopping we investigate

$$\text{ext}(\mathcal{G}) = \{g \in \mathcal{G} : g \text{ extremal point of } \mathcal{G}\},$$

i.e. the set of those elements of $\mathcal{G}$ which cannot be written as non-trivial convex combination of elements of $\mathcal{G}$.

Lemma 2 *$ext(\mathcal{G}) = \{g \in \mathcal{G} : g_n \in \{0,1\}$ for all $n \in \bar{N}\}$.*

For a proof of Lemma 1 and Lemma 2 we refer to Irle (1995).

Lemma 2 shows that the extremal points of $\mathcal{G}$ are the stopping times $T \in \mathcal{G}$ in their G-representation: For any $T \in \mathcal{G}$, the sequence $(1_{\{T=n\}})_{n\in\bar{N}}$ obviously defines an element of ext$(\mathcal{G})$. Conversely, any $g \in \text{ext}(\mathcal{G})$ is given by $g = (1_{A_n})_{n\in\bar{N}}$, $A_n \in \mathcal{A}_n$, and g is the G-representation of the stopping time T given by $\{T = n\} = A_n$. Note that $\sum_{n\in\bar{N}} 1_{A_n} = 1$ implies $A_n \cap A_m = \emptyset$ for $n \ne m$, $\bigcup_n A_n = \Omega$.

In general, the set of extremal points need not be a Borel set. To avoid this difficulty we assume that $L_1(\Omega, \mathcal{A}_\infty, Q)$ is separable which implies that $L_1(c)$ is separable. Then as a topological space with its w^*-topology, $\mathcal{G}$ is metrizable which implies that ext$(\mathcal{G})$ is a Borel set in $\mathcal{G}$, see Phelps (1966, sect. 1). Identifying $\mathcal{G}$ with ext$(\mathcal{G})$ we choose $\sigma(\mathcal{G})$ as the σ-algebra of Borel sets, and let $\tilde{S}$ be the set of all probability measures on $\mathcal{G}$ with respect to this σ-algebra.

Theorem 1 *Assume (A1), (A2) and that $L_1(\Omega, \mathcal{A}_\infty, Q)$ is separable.*

(i) For any $\tau \in \tilde{S}$ there exists $g \in \mathcal{G}$ such that

$$A(\tau, P) = A(g, P) \text{ for all } P \in \mathcal{M}.$$

(ii) For any $g \in \mathcal{G}$ there exists $\tau \in \tilde{S}$ such that

$$A(\tau, P) = A(g, P) \text{ for all } P \in \mathcal{M}.$$

For a proof of (i) we use $\mathcal{G}$-valued integration to define

$$g = \int_{\mathcal{G}} id \, d\tau,$$

for (ii) Choquet's representation theorem is applied, see Irle (1995) for details.

4. Minimax Results in Discrete Time

We assume $S = \bar{N}$.

The compact embedding of $\mathcal{G}$ into $L_\infty(\ell_1)$ as given in Lemma 1 immediately yields a minimax result, since $A(g, P)$ is linear in both arguments.

Theorem 2 *Assume (A1), (A2) and that $\mathcal{M}|\mathcal{A}_\infty$ is convex. Then the game of randomized stopping has a value, i.e.*

$$\inf_{\mathcal{M}} \sup_{\mathcal{G}} A(g, P) = \sup_{\mathcal{G}} \inf_{\mathcal{M}} A(g, P).$$

Whereas the compact embedding of Section 3 gives a satisfactory treatment for games of stopping under assumption (A2), i.e. for dominated sets of probability measures, hence in particular for problems with countable sample space, they do not show how to treat such games when (A2) is not fulfilled.
The following example from Irle (1995) shows that, even in the case $S = \{1, 2\}$ and for convex $\mathcal{M}$, without (A2) we do not have a value in general.

Example.

Let $(\Omega, \mathcal{A})$ be the unit interval $[0, 1]$ with its Borel σ-algebra. The time set is $\{1, 2\}$. We define

$$\mathcal{A}_1 = \{A \subset \Omega : A \text{ countable or } A^c \text{ countable}\}, \ \mathcal{A}_2 = \mathcal{A}.$$

Choose $B \in \mathcal{A}$ such that B and B^c are uncountable and set

$$Z_1 = 1_B, \ Z_2 = 1 - 1_B.$$

Let furthermore

$$\mathcal{M} = \{P : P \text{ probability measure on } (\Omega, \mathcal{A}) \text{ with countable support } C(P)\}.$$

It then may be shown that this game

$$(\mathcal{G}, \mathcal{M}, A) \quad \text{with } A(g, P) = \int (Z_1 \, g_1 + Z_2 \, g_2) \, dP$$

does not have a value.

Looking for minimax theorems without (A2), we now turn to topological assumptions.

(A3) Ω is a topological space, $\mathcal{A}$ a σ-algebra which contains the Baire sets, and Z_s is continuous and bounded from below for all $s < \infty$. Furthermore for any $s < \infty$, probability measure P on $(\Omega, \mathcal{A})$, $\varepsilon > 0$ and $\mathcal{A}_s$-measurable $h : \Omega \to [0,1]$, there exists a continuous $\mathcal{A}_s$-measurable $\hat{h} = \hat{h}(P, s, \varepsilon) : \Omega \to [0,1]$ such that $P(h \neq \hat{h}) \leq \varepsilon$.

Denote by $M(\Omega)$ the set of all finite measures on $(\Omega, \mathcal{A})$, by $C(\Omega)$ the set of all real-valued functions on Ω which are bounded and continuous.

The w^*-topology on $M(\Omega)$ is defined as the weakest topology on $M(\Omega)$ such that the functionals $P \to \int h \, dP$, $h \in C(\Omega)$, are continuous.

Due to the method of proof we have to replace (A1) by (A1'). Then the following result is given in Irle (1995) for $S = \bar{N}$.

Theorem 3 *Assume (A1'), (A3) and that* $\mathcal{M}$ *is* w^**-compact and convex. Then the game of randomized stopping has a value, i.e.*

$$\inf_{\mathcal{M}} \sup_{\mathcal{G}} A(g, P) = \sup_{\mathcal{G}} \inf_{\mathcal{M}} A(g, P) \, .$$

For compact metric Ω, $C(\Omega)^* = M(\Omega)$ hence the set of all probability measures on $(\Omega, \mathcal{A})$ is w^*-compact. As immediate consequence we thus obtain

Corollary 1 *Assume (A1'), (A3). Let* Ω *be compact metric with* σ*-algebra of Borel sets* $\mathcal{A}$*. If* $\mathcal{M}$ *is* w^**-closed, then the game of randomized stopping has a value.*

In particular, if $\mathcal{M}$ is the set of all probability measures, then the game of randomized stopping has a value for compact metric Ω.

5. A Minimax Result in Continuous Time

We now turn to the case $S = [0, \infty]$ and consider the game of randomized stopping

$$(\mathcal{S}^*, \mathcal{M}, A).$$

It is our aim to extend Theorem 3 to this setting of continuous time.

In our treatment we shall use an approximation by discrete problems. For this let

$$S_m = \{k2^{-m} : k \in \bar{N}\} \ , \ m = 1, 2, \dots ,$$

and denote by

$\mathcal{S}_m^*$ the set of all randomized stopping times with values in S_m.

Theorem 4 *Assume (A1'), (A3) and that $\mathcal{M}$ is w^*-compact and convex. Assume that for all ω the mapping $s \to Z_s(\omega)$ is right-continuous. Let*

$$\mathcal{R} \subseteq \mathcal{S}^* \text{ such that } \bigcup_m \mathcal{S}_m^* \subseteq \mathcal{R}.$$

Then the game of randomized stopping with strategy set $\mathcal{R}$ has a value, i.e.

$$\inf_{\mathcal{M}} \sup_{\mathcal{R}} A(g, P) = \sup_{\mathcal{R}} \inf_{\mathcal{M}} A(g, P) \, .$$

Proof. Let $T^* \in \mathcal{S}^*$. Define $T_m^* \in \mathcal{S}_m^*$ by

$$T_m^* = \inf\{k2^{-m} : T \leq k2^{-m}\}.$$

Then from right-continuity and Fatou's lemma we have

$$E_P Z_{T^*} \leq \liminf_m E_P Z_{T_m^*} .$$

This shows

$$\inf_{\mathcal{M}} \sup_{\mathcal{R}} A(T^*, P) = \inf_{\mathcal{M}} \sup_{N} \sup_{\mathcal{S}_m^*} A(T^*, P) = \inf_{\mathcal{M}} \sup_{N} B(m, P)$$

with

$$B(m, P) = \sup_{\mathcal{S}_m^*} A(T^*, P).$$

Let us now look at the game

$$(N, \mathcal{M}, B).$$

Here $B(m, P)$ is increasing in m and konvex in P. We can thus apply a standard minimax result if we can show w^*-lower semicontinuity in P. Since the mappings $P \to A(T^*, P)$ are not w^*-lower semicontinuous for general $T^* \in \mathcal{S}_m^*$, we shall employ a suitable continuos approximation.

Let us fix m. Let $T^* \in \mathcal{S}_m^*$. Then we will use the $\mathcal{G}$-representation g for T^*, see Section 2, such that

$$A(T^*, P) = \sum_{n \in \bar{N}} \int Z_{n2^{-m}}\, g_n \, dP$$

with $\mathcal{A}_{n2^{-m}}$-measurable g_n, $g_n(\omega) = \lambda(\{x : T^*(\omega, x) = n2^{-m}\})$.. Denote by

$$\mathcal{C}_m = \{T^* \in \mathcal{S}_m^*\text{: There exists } k = k(g) < \infty \text{ with } g_n = 0 \text{ for } n > k\,,\ g_n \text{ continuous for } n \le k\}\,.$$

Then for any $T^* \in \mathcal{C}_m$

$$P \to A(T^*, P) = \sup_{j \in N} \int \sum_{n \le k(g)} \min\{Z_{n2^{-m}}, j\}\, g_n \, dP$$

is w^*-lower semicontinuous, since $\sum_{n \le k(g)} \min\{Z_{n2^{-m}}, j\}\, g_n \in C(\Omega)$.

Thus it is sufficient to show that for any $P \in \mathcal{M}$

$$\sup_{\mathcal{C}_m} A(T^*, P) = \sup_{\mathcal{S}_m^*} A(T^*, P)\,.$$

Let $P \in \mathcal{M}$, $\varepsilon > 0$, $T^* \in \mathcal{S}_m^*$ with corresponding $\mathcal{G}$-representation $g \in \mathcal{G}$.

Then using (A1') and Fatou's Lemma

$$\begin{aligned}
&\sum_{n \in \bar{N}} \int Z_{n2^{-m}}\, g_n \, dP \\
&= \sum_{n < \infty} \int Z_{n2^{-m}}\, g_n \, dP + \int Z_\infty (1 - \sum_{n < \infty} g_n)\, dP \\
&\le \lim_{k \to \infty} \sum_{n < k} \int Z_{n2^{-m}}\, g_n \, dP + \limsup_{k \to \infty} \int Z_{k2^{-m}} (1 - \sum_{n < k} g_n)\, dP\,.
\end{aligned}$$

We thus may choose k such that

$$A(T^*, P) \le \sum_{n < k} \int Z_{n2^{-m}}\, g_n \, dP + \int Z_{k2^{-m}} (1 - \sum_{n < k} g_n)\, dP + \varepsilon = A(\hat{g}, P) + \varepsilon$$

where $\hat{g}_n = g_n$ for $n < k$, $\hat{g}_k = 1 - \sum_{n < k} g_n$, $\hat{g}_n = 0$ for $n > k$.

Next, choose $\beta > 0$ such

$$A(\hat{g}, P) \leq \sum_n \int \min\{Z_{n2^{-m}}, \beta\}\, \hat{g}_n \, dP + \varepsilon .$$

We now use the F-representation of $\hat{g}$,

$$\hat{f} = \Psi(\hat{g}), \quad \text{hence} \quad \hat{f}_n = 1 \quad \text{for} \quad n > k .$$

For any $n \leq k$ choose $h_n : \Omega \longrightarrow [0,1]$, $\mathcal{A}_{n2^{-m}}$–measurable and continuous, such that

$$P(\hat{f}_n \neq h_n) \leq \varepsilon/(k^2\gamma)$$

where $\gamma = \sup_{n \leq k} \sup_{\Omega} |\min\{Z_{n2^{-m}}(w), \beta\}| + 1$.

Define $f' \in F$ by

$$f'_n = h_n \quad \text{for } n \leq k , \; f'_n = 1 \quad \text{for } n > k .$$

It follows that

$$g' = \varphi(f') \text{ defines an element of } \mathcal{C}_m$$

with $g'_n = 0$ for $n > k$ and for $n \leq k$

$$P(g'_n \neq \hat{g}_n) \leq \sum_{i \leq n} P(\hat{f}_i \neq h_i) \leq \varepsilon/(k\gamma) .$$

Furthermore

$$A(\hat{g}, P) - A(g', P) \leq \sum_n \int \min\{Z_{n2^{-m}}, \beta\} \, (\hat{g}_n - g'_n) \, dP$$
$$+ \quad \varepsilon \leq \gamma \sum_{n \leq k} P(g'_n \neq \hat{g}_n) \leq 2\varepsilon .$$

This proves

$$\sup_{\mathcal{C}_m} A(T^*, P) = \sup_{\mathcal{S}^*_m} A(T^*, P) .$$

We thus obtain w^*-lower semicontinuity of $B(m, P)$ in P, hence by a standard minimax result

$$\inf_{\mathcal{M}} \sup_{N} \; B(m, P) = \sup_{N} \inf_{\mathcal{M}} \; B(m, P).$$

This implies

$$\begin{aligned}
\inf_{\mathcal{M}} \sup_{\mathcal{R}} A(T^*, P) &= \inf_{\mathcal{M}} \sup_{N} \sup_{\mathcal{S}_m^*} A(T^*, P) \\
&= \inf_{\mathcal{M}} \sup_{N} B(m, P) \\
&= \sup_{N} \inf_{\mathcal{M}} B(m, P) \\
&= \sup_{N} \inf_{\mathcal{M}} \sup_{\mathcal{S}_m^*} A(T^*, P) \\
&= \sup_{N} \sup_{\mathcal{S}_m^*} \inf_{\mathcal{M}} A(T^*, P)
\end{aligned}$$

where we have used the minimax result of Section 4 for discrete time. The result now follows from

$$\sup_{N} \sup_{\mathcal{S}_m^*} \inf_{\mathcal{M}} , A(T^*, P) \leq \sup_{\mathcal{R}} \inf_{\mathcal{M}} A(T^*, P).$$

As an open question we point out the problem to obtain a minimax result in the continuous time case under assumption (A2).

References

1. M. Gödde, *Statistical games against a prophet – proof of a minimax conjecture,* Statist. Papers **32** (1991), 75-81.
2. T.P. Hill and U. Krengel, *Minimax-optimal stop rules and distributions in secretary problems,* Ann. Prob. **19** (1991), 342-353.
3. A. Irle, *Games of stopping with infinite horizon,* Z. Oper. Res. **42** (1995), 345-359.
4. J. Neveu, *Discrete-parameter Martingales,* North-Holland, Amsterdam, 1975.
5. R. Phelps, *Lectures on Choquet's Theorem,* Van Nostrand, Princeton, 1966.

INTERSECTION THEOREMS, MINIMAX THEOREMS AND ABSTRACT CONNECTEDNESS

JÜRGEN KINDLER
Fachbereich Mathematik
Technische Hochschule Darmstadt
Schlossgartenstr. 7
D-64289 Darmstadt, Germany
e-mail: kindler@mathematik.th-darmstadt.de

1. Introduction

Let S be a nonvoid set and 2^S the power set of S. Then every nonvoid subset $\mathcal{P} \subset 2^S$ is called a *paving* in S and $(S, \mathcal{P})$ is a *paved space*. Especially $\mathcal{E}(S)$ is the paving of all nonvoid finite subsets of S.

The paving $\mathcal{P}$ in S is said to have
the *pairwise intersection property* (PIP) iff $P_1 \cap P_2 \neq \emptyset \; \forall P_1, P_2 \in \mathcal{P}$
the *finite intersection property* (FIP) iff $\bigcap\{P : P \in \mathcal{R}\} \neq \emptyset \; \forall \; \mathcal{R} \in \mathcal{E}(\mathcal{P})$, and
the *global intersection property* (GIP) iff $\bigcap\{P : P \in \mathcal{P}\} \neq \emptyset$.

A paving is called *compact* iff every subpaving with FIP has GIP, and it is said to be $\cap_f$*-closed* ($\cap_a$*-closed*) iff it is closed under finite (arbitrary) intersections.

If S is a topological space, then we denote by $\mathcal{F}(S)$, $\mathcal{G}(S)$ and $\mathcal{K}(S)$ the pavings of all closed, open and compact subsets, respectively. Of course, the paving $\mathcal{K}(S) \cap \mathcal{F}(S)$ is always compact, and $\mathcal{F}(S)$ is compact iff S is compact.

Let X, Y be nonvoid sets and Φ a *correspondence* from X to Y, i.e., $\Phi : X \to 2^Y$ with $\Phi(x) \neq \emptyset \quad \forall x \in X$ with *value set* $\text{val}\,\Phi := \{\Phi(x) : x \in X\}$.
A function $f : X \to Y$ is a *selector* for Φ iff $f(x) \in \Phi(x)$, $x \in X$. Of course, val Φ has GIP iff Φ possesses a constant selector. We set
$\Phi^{\cap}(A) := \bigcap_{x \in A} \Phi(x)$, $A \in 2^X$ (with $\Phi^{\cap}(\emptyset) = Y$),

B. Ricceri and S. Simons (eds.), Minimax Theory and Applications, 105–120.

$\Phi^{\cup}(A) := \bigcup_{x\in A}\Phi(x),\ \emptyset \neq A \in 2^X$, and
$\Phi_E(x) := \Phi^{\cap}(E \cup \{x\}),\ x \in X,\ E \subset X$,
and we say that Φ *has property* P *hereditarily* iff every $\Phi_E, E \in \mathcal{E}_\Phi(X) := \{E \in \mathcal{E}(X) \cup \{\emptyset\} :\ \Phi_E(x) \neq \emptyset\ \forall x \in X\}$ has property P.

Remark 1.
a) val Φ has FIP iff it has PIP hereditarily.
b) val Φ has GIP iff it is compact and has FIP .

The present investigations were motivated by the following observation.

Lemma 1 *Let Y be a convex subset of a topological vector space and let $b : [0,1] \times Y \to I\!R$. Suppose that either*

a) every function $b(\cdot, y),\ y \in Y$, is upper semicontinuous and every function $b(t,\cdot),\ t \in [0,1]$, is quasiconvex, or

b) Y is compact, every function $b(\cdot, y),\ y \in Y$, is lower semicontinuous and quasiconcave, and every function $b(t,\cdot),\ t \in [0,1]$, is lower semicontinuous and quasiconvex.

Then, for $\Psi(t) = \{b(t,\cdot) \leq \lambda\},\ t \in [0,1]$, with $\lambda > \sup_{t\in[0,1]}\inf_{y\in Y} b(t,y)$ the image $\Psi^{\cup}([0,1])$ is connected.

Proof. a) follows from Lemma 2 and 7 and Remark 3 b), and b) follows from Lemma 6 below.

This lemma can be used for a simple proof of the following classical minimax theorem. The usc–version is due to Sion [17] and the lsc–version is (a simplified version of) a theorem of Tuy [19, 20]:

Sion–Tuy minimax theorem. *Let X and Y be convex subsets of two topological vector spaces with Y compact and let $a : X \times Y \to I\!R$. Assume that every function $a(\cdot, y),\ y \in Y$, is upper (lower) semicontinuous and quasiconcave, and every function $a(x,\cdot),\ x \in X$, is lower semicontinuous and quasiconvex.*
Then $\min_{y\in Y}\sup_{x\in X} a(x,y) = \sup_{x\in X}\inf_{y\in Y} a(x,y)$.

Sketch of a proof. For $\lambda > \sup_{x\in X}\inf_{y\in Y} a(x,y)$ let $\Phi(x) := \{a(x,\cdot) \leq \lambda\} = \{y \in Y : a(x,y) \leq \lambda\},\ x \in X$. It is sufficient to show that $\bigcap_{x\in X}\Phi(x) \neq \emptyset$. Let $\{x_1, x_2\} \subset X$. Then from the lower semicontinuity of $a(x,\cdot)$ we infer that Φ is closed-valued, and quasiconcavity of $a(\cdot, y)$ implies $\Phi^{\cup}([x_1,x_2]) = \Phi(x_1) \cup \Phi(x_2)$. By Lemma 1, applied to the function $b(t,y) = a(tx_1 + (1-t)x_2, y)$, the set $\Phi^{\cup}([x_1,x_2])$ is connected, and therefore, $\Phi(x_1) \cap \Phi(x_2) \neq \emptyset$. Since all essential properties of Φ are hereditary, the assertion follows with Remark 1. □

Only one year after the publication of Sion's minimax theorem Wu Wen–Tsün [21] observed that connectedness rather than convexity was the crucial property needed in its proof. After it had been observed [6, 7] that some kind of abstract connectedness of the intersections of certain level–sets is *necessary* for the validity of the minimax relation, several notions of abstract connectedness [6, 11, 13, 15, 16] have been studied in recent years.

A second fundamental observation is due to Komiya [12] who pointed out that some semicontinuity conditions can be replaced by the condition of quartercontinuity of a level–set correspondence, a condition which is also *necessary.* In the present paper the topological concept of quartercontinuity is extended to correspondences between arbitrary paved spaces. In combination with the concept of abstract connectedness we generalize a result of Correa et al. [2] which states that a quartercontinuous correspondence between two paved spaces with connected values is connected (cf. Lemma 3 below). This result will be the central tool for the present investigations.

2. Abstract Continuity

Let $\Phi : X \to 2^Y$ be a correspondence. For $B \subset Y$ let
$\Phi^+(B) := \{x \in X : \Phi(x) \subset B\}$ denote the *upper inverse* and
$\Phi^-(B) := \{x \in X : \Phi(x) \cap B \neq \emptyset\}$ the *lower inverse* of Φ.

Let $(X, \mathcal{P})$ and $(Y, \mathcal{Q})$ be paved spaces. We say that
Φ is $\mathcal{P} - \mathcal{Q}$*–semicontinuous* iff $\Phi^-(\mathcal{Q}) \subset \mathcal{P}$ (i.e., $\Phi^-(Q) \in \mathcal{P}\ \forall Q \in \mathcal{Q}$)
Φ is $\mathcal{P} - \mathcal{Q}$*–quartercontinuous* iff $\forall Q \in \mathcal{Q}\ \exists\ P \in \mathcal{P} : \Phi^+(Q) \subset P \subset \Phi^-(Q)$.

Remark 2. val Φ has PIP (FIP) iff Φ is (hereditarily) $\mathcal{P} - \mathcal{Q}$–quartercontinuous for all pavings $\mathcal{P}$ in X with $\{\emptyset, X\} \in \mathcal{P}$ and all pavings $\mathcal{Q}$ in Y.

Remark 3. Let $(X, \mathcal{P})$ and $(Y, \mathcal{Q})$ be paved spaces.

a) If $\mathcal{Q}$ is $\cap_f$–closed, then every $\mathcal{Q}$–valued $\mathcal{P} - \mathcal{Q}$–semicontinuous correspondence $\Phi : X \to 2^Y$ is hereditarily $\mathcal{P} - \mathcal{Q}$–semicontinuous.
b) Let $\Phi, \Psi : X \to 2^Y$ be correspondences. If Ψ is $\mathcal{P} - \mathcal{Q}$–quartercontinuous and $\Psi \subset \Phi$, i.e., $\Psi(x) \subset \Phi(x)\ \forall x \in X$, then Φ is also $\mathcal{P} - \mathcal{Q}$–quartercontinuous.

Proof. a) $\Phi_E^-(Q) = \Phi^-(\Phi^\cap(E) \cap Q) \in \mathcal{P}$, since $\Phi^\cap(E) \cap Q \in \mathcal{Q}$.
b) This follows from $\Phi^+(Q) \subset \Psi^+(Q) \subset \Psi^-(Q) \subset \Phi^-(Q)$. □

Example 1. Let X and Y be topological spaces and $\Phi : X \to 2^Y$ a correspondence. Then

a) Φ is upper semicontinuous (i.e., for all $x \in X$ and for every open $G \supset \Phi(x)$ there exists a neighborhood V of x with $\Phi^{\cup}(V) \subset G$) iff Φ is $\mathcal{F}(X) - \mathcal{F}(Y)$–semicontinuous.

b) Φ is lower semicontinuous (i.e., for all $x \in X$ and for every open G with $\Phi(x) \cap G \neq \emptyset$ there exists a neighborhood $V$of x with $\Phi(v) \cap G \neq \emptyset$ for all $v \in V$) iff Φ is $\mathcal{G}(X) - \mathcal{G}(Y)$–semicontinuous.

c) Φ is quartercontinuous (i.e., for all $x \in X$ and for every open $G \supset \Phi(x)$ there exists a neighborhood V of x with $\Phi(v) \cap G \neq \emptyset$ for all $v \in V$) iff Φ is $\mathcal{F}(X) - \mathcal{F}(Y)$–quartercontinuous iff Φ is $\mathcal{G}(X) - \mathcal{G}(Y)$–quartercontinuous.

Proof. a) and b) are obvious.
c) **1.** Let Φ be quartercontinuous. Then for $G \in \mathcal{G}(Y)$ and for all $x \in \Phi^+(G)$ there exists an open neighborhood V_x of x with $V_x \subset \Phi^-(G)$. For $V = \bigcup_{x \in \Phi^+(G)} V_x$ we have $\Phi^+(G) \subset V \subset \Phi^-(G)$, i.e., Φ is $\mathcal{G}(X) - \mathcal{G}(Y)$–quartercontinuous.
2. Conversely, let Φ be $\mathcal{G}(X) - \mathcal{G}(Y)$–quartercontinuous, and let $G \in \mathcal{G}(Y)$ and $x \in \Phi^+(G)$. Then for $V \in \mathcal{G}(X)$ with $\Phi^+(G) \subset V \subset \Phi^-(G)$ we have $x \in V$ and $\Phi(v) \cap G \neq \emptyset \; \forall v \in V$.
3. The second equivalence follows from the identity $\Phi^+(Y - B) = X - \Phi^-(B)$, $B \subset Y$. □

The notion of quartercontinuity was introduced by Komiya [12] and, under the name of semicontinuity, by Correa et al. [2]. Their main result, stated below, will be generalized in the following to a more abstract setting. More general topological connectedness theorems have also been obtained by Del Prete and Di Iorio [3].

Lemma 2 ([2]). *Let X and Y be topological spaces. Then every quartercontinuous correspondence $\Phi : X \to 2^Y$ with connected values is connected (i.e., $\Phi^{\cup}(C)$ is connected for every connected $C \subset X$).*

3. Abstract Connectedness

Let $(S, \mathcal{P})$ be a paved space. Then a subset $T \subset S$ is called *connected (for $\mathcal{P}$)* (König [11]) iff for all $A_1, A_2 \in \mathcal{P}$ the relations

$$T \subset A_1 \cup A_2,\; T \cap A_1 \neq \emptyset,\; T \cap A_2 \neq \emptyset, \text{ and } T \cap A_1 \cap A_2 = \emptyset$$

cannot hold simultaneously. If $\mathcal{K}$ is another paving in S , then $\mathcal{K}$ is called *connected for $\mathcal{P}$* iff every $T \in \mathcal{K}$ is connected for $\mathcal{P}$. Finally, $\mathcal{P}$ is called *connected* iff $\mathcal{P}$ is connected for $\mathcal{P}$.

Example 2. Let S be a topological space and $\emptyset \neq T \subset S$. Then
T is connected iff T is connected for $\mathcal{F}(S)$ iff T is connected for $\mathcal{G}(S)$.

Let $(X, \mathcal{P})$ and $(Y, \mathcal{Q})$ be paved spaces and $\Phi : X \to 2^Y$ a correspondence. Then $\mathcal{Q}_\Phi = \{Q \in \mathcal{Q} : \Phi^+(Q) = \Phi^-(Q)\}$ is the paving of *Φ-invertible Q's.*

Remark 4. Every $\mathcal{P} - \mathcal{Q}-$quartercontinuous correspondence Φ is $\mathcal{P} - \mathcal{Q}_\Phi-$semicontinuous.

We now present an abstract version of Lemma 2 above.

Lemma 3 *Let $(X, \mathcal{P})$ and $(Y, \mathcal{Q})$ be paved spaces and $\Phi : X \to 2^Y$ a correspondence such that val Φ is connected for $\mathcal{Q}$.*
a) If X is connected for $\mathcal{P}$ and if Φ is $\mathcal{P} - \mathcal{Q}_\Phi-$semicontinuous, then $\Phi^\cup(X)$ is connected for $\mathcal{Q}$.
b) If Φ is $\mathcal{P} - \mathcal{Q}$-quartercontinuous and if $C \subset X$ is connected for $\mathcal{P}$, then $\Phi^\cup(C)$ is connected for $\mathcal{Q}$.

Proof. a) Let $\{Q_1, Q_2\} \subset \mathcal{Q}$ such that

$$\Phi^\cup(X) \subset Q_1 \cup Q_2 \text{ and } \Phi^\cup(X) \cap Q_1 \cap Q_2 = \emptyset.$$

Since val Φ is connected for $\mathcal{Q}$, we obtain $\Phi^+(Q_i) = \Phi^-(Q_i) =: P_i$, $i \in \{1, 2\}$, and

$$P_1 \cup P_2 = X, \;\; P_1 \cap P_2 = \emptyset.$$

From the $\mathcal{P} - \mathcal{Q}_\Phi-$semicontinuity of Φ we infer that $P_i \in \mathcal{P}$, $i \in \{1, 2\}$. Since X is connected for $\mathcal{P}$, it follows that $P_1 = \emptyset$, say, which leads to $\Phi^\cup(X) \cap Q_1 = \emptyset$.
b) Apply a) to the restriction $\Psi = \Phi_{|C}$ which is $\mathcal{P} \cap C - \mathcal{Q}$-quartercontinuous and therefore $\mathcal{P} \cap C - \mathcal{Q}_\Psi-$ semicontinuous. Observe that C is connected for $\mathcal{P} \cap C$. □

An *(abstract) interval space* is a triplet $(S, \mathcal{P}, \langle \cdot, \cdot \rangle)$ where $(S, \mathcal{P})$ is a paved space and $\langle \cdot, \cdot \rangle : S \times S \to 2^S$ is a correspondence with $\langle s, t \rangle \supset \{s, t\}$ for all $s, t \in S$ such that every "interval" $\langle s, t \rangle$ is connected for $\mathcal{P}$.

$(\mathcal{P}, \langle \cdot, \cdot \rangle)$ is called an *interval structure* for S and $\langle \cdot, \cdot \rangle$ is an *interval function* for $(S, \mathcal{P})$. A subset $T \subset S$ is *convex* iff $\{s, t\} \subset T \;\Rightarrow\; \langle s, t \rangle \subset T$. A function $f : S \to I\!R$ is called *quasiconcave* iff every set $\{f \geq \alpha\} := \{s \in S : f(s) \geq \alpha\}$, $\alpha \in I\!R$, is convex which is equivalent to $f(s) \geq f(s_1) \wedge f(s_2)$ for all $s \in \langle s_1, s_2 \rangle$, $(s_1, s_2) \in S \times S$.

Remark 5. Let $(S, \mathcal{P}, \langle \cdot, \cdot \rangle)$ be an interval space and $\mathcal{C}$ the paving of all convex subsets of S. Then $\mathcal{C}$ is an *alignment* (i.e., $\{\emptyset, S\} \subset \mathcal{C}$, $\mathcal{C}$ is $\cap_a-$stable and $\bigcup_{A \in \mathcal{A}} A \in \mathcal{C}$ for every chain $\mathcal{A} \subset \mathcal{C}$), and every $C \in \mathcal{C}$ is connected for $\mathcal{P}$.

Topological interval spaces, where S is a topological space and $\mathcal{P} = \mathcal{F}(S)$, were introduced by Stachó [18].

4. Intersection Theorems

We shall now study the problem whether a family of nonvoid subsets C_x, $x \in X$, of a set Y has a common point or, equivalently, the correspondence $\Phi : X \to 2^Y$ with $\Phi(x) = C_x$, $x \in X$, has a constant selector.

4.1. ABSTRACT INTERSECTION THEOREMS

Theorem 1 *A correspondence $\Phi : X \to 2^Y$ has a constant selector iff there exists an interval structure $(\mathcal{P}, \langle\cdot,\cdot\rangle)$ on X and a compact paving $\mathcal{Q}$ in Y such that*

(i) Φ is $\mathcal{Q}$–valued,
(ii) $\Phi(x) \subset \Phi(x_1) \cup \Phi(x_2)$ for all $x \in \langle x_1, x_2\rangle, (x_1, x_2) \in X \times X$,
(iii) every set $\Phi^{\cap}(A)$, $A \in \mathcal{E}(X)$, is connected for $\mathcal{Q}$, and
(iv) Φ is hereditarily $\mathcal{P} - \mathcal{Q}$–quartercontinuous.

Proof. Let $\{x_1, x_2\} \subset X$ and $E \in \mathcal{E}_\Phi(X)$. Then (ii) implies $C := \Phi_E^{\cup}(\langle x_1, x_2\rangle) \subset \Phi(x_1) \cup \Phi(x_2)$. Of course, $C \cap \Phi(x_i) \supset \Phi_E(x_i) \neq \emptyset$, $i \in \{1, 2\}$. From (iii) and (iv), together with Lemma 3, we conclude that C is connected for $\mathcal{Q}$, and with (i) and (ii) we arrive at $\emptyset \neq C \cap \Phi(x_1) \cap \Phi(x_2) = \Phi_E(x_1) \cap \Phi_E(x_2)$. Hence, val Φ has PIP hereditarily. But val $\Phi \subset \mathcal{Q}$ is compact, so by Remark 1 Φ has a constant selector.
Conversely, if Φ has a constant selector, then for $\mathcal{P} = \{\emptyset, X\}$, $\langle x_1, x_2\rangle = \{x_1, x_2\}$, and $\mathcal{Q} =$ val Φ conditions (i) – (iv) are satisfied. □

Remark 6. An inspection of our proof shows that Theorem 1 remains valid when condition (iv) is replaced by the weaker condition

(iv)* for every pair $(x_1, x_2) \in X \times X$ and every $E \in \mathcal{E}_\Phi(X)$ the restriction of Φ_E on $\langle x_1, x_2\rangle$ is $\mathcal{P} \cap \langle x_1, x_2\rangle - \mathcal{Q}$–quartercontinuous.

A topological version of the following corollary has been proved in [8]:

Corollary 1 *Let $(X, \mathcal{P})$ and $(Y, \mathcal{Q})$ be paved spaces and $\Phi : X \to 2^Y$ a correspondence such that*

(i) every set $\Phi^+(\Phi(x_1) \cup \Phi(x_2))$, $(x_1, x_2) \in X \times X$, is connected for $\mathcal{P}$,
(ii) $\mathcal{Q}$ is $\cap_f$–closed, connected and compact,
(iii) Φ is $\mathcal{Q}$–valued, and
(iv) Φ is hereditarily $\mathcal{P} - \mathcal{Q}$–quartercontinuous.

Then Φ has a constant selector.

Proof. Set $\langle x_1, x_2\rangle := \Phi^+(\Phi(x_1) \cup \Phi(x_2))$. Then condition (i) implies that $(\mathcal{P}, \langle\cdot,\cdot\rangle)$ is an interval structure for X. Moreover, we have $x \in \langle x_1, x_2\rangle \Leftrightarrow \Phi(x) \subset \Phi(x_1) \cup \Phi(x_2)$, and conditions (ii) and (iii) yield that $\Phi^{\cap}(A) \in \mathcal{Q}$ is connected for $\mathcal{Q}$ for all $A \in \mathcal{E}(X)$. □

If $\Phi : X \to 2^Y$ is a correspondence, then $\Phi^* : Y \to 2^X$, $\Phi^*(y) := \{x \in X : y \notin \Phi(x)\}$, $y \in Y$, is its *dual.*

Remark 7.

a) Φ^* is a correspondence iff $\Phi^\cap(X) = \emptyset$.
b) If $(\mathcal{P}, \langle\cdot,\cdot\rangle)$ is an interval structure for X, then Φ^* is convex–valued iff $\Phi(x) \subset \Phi(x_1) \cup \Phi(x_2)$ for all $x \in \langle x_1, x_2\rangle, (x_1, x_2) \in X \times X$.

Lemma 4 *Let $\mathcal{P}$ be a $\cap_a$–closed paving in X such that Φ^* is $\mathcal{P}$–valued. Then Φ is hereditarily $\mathcal{P} - 2^Y$–quartercontinuous.*

Proof. This follows from $\Phi_E^+(B) = \bigcap\{\Phi^*(y) : y \in \Phi^\cap(E) - B\} \in \mathcal{P}, B \subset Y$, $E \in \mathcal{E}_\Phi(X)$. □

Corollary 2 ([7]). *Let $\Phi : X \to 2^Y$ be a correspondence such that* val Φ *is compact and the pavings $\mathcal{P} := \{\Phi^{*\cap}(B) : B \subset Y\}$ and $\mathcal{Q} := \{\Phi^\cap(A) : A \in \mathcal{E}(X)\}$ are connected. Then Φ has a constant selector.*

Proof. By Lemma 4, Φ is hereditarily $\mathcal{P} - \mathcal{Q}$–quartercontinuous. Furthermore, we have $\Phi^+(\Phi(x_1) \cup \Phi(x_2)) = \bigcap\{\Phi^*(y) : y \in Y - \Phi(x_1) \cup \Phi(x_2)\} \in \mathcal{P}$. Of course, $\mathcal{Q}$ is compact if (and only if) val Φ is compact. Hence, Corollary 1 applies. □

Corollary 3 *Let $(X, \mathcal{P}, \langle\cdot,\cdot\rangle)$ be an interval space, $(Y, \mathcal{Q})$ a paved space with $\cap_f$–closed, connected and compact $\mathcal{Q}$ and $\Phi : X \to 2^Y$ a $\mathcal{Q}$–valued correspondence with convex–valued dual Φ^*. If moreover $\Phi^{*\cap}(Q) \in \mathcal{P}$ for all $Q \in \mathcal{Q}$, then Φ has a constant selector.*

Proof. A subset of X is connected for $\mathcal{P}$ iff it is connected for $\mathcal{P}^C := \{X - P : P \in \mathcal{P}\}$. In particular, $(X, \mathcal{P}^C, \langle\cdot,\cdot\rangle)$ is an interval space. Furthermore, Φ is hereditarily $\mathcal{P}^C - \mathcal{Q}$– quartercontinuous, since $\Phi_E^-(Q) = X - \Phi^{*\cap}(\Phi^\cap(E) \cap Q) \in \mathcal{P}^C$, $Q \in \mathcal{Q}$. Hence, by Remark 7 b), Theorem 1 applies. □

4.2. CORRESPONDENCES WITH CLOSED GRAPH

The first part of the following observation is well–known at least for compact spaces Y [10]:

Lemma 5 *Let X and Y be topological spaces. Then every correspondence $\Phi : X \to 2^Y$ with closed graph is $\mathcal{F}(X)$–$\mathcal{K}(Y)$–semicontinuous. If, moreover, Φ is compact–valued, then it is hereditarily $\mathcal{F}(X) - \mathcal{K}(Y) \cap \mathcal{F}(Y)$-semi-continuous.*

Proof. **1.** Let $F \in \mathcal{K}(Y)$, $A := \Phi^-(F)$, and let (x_n) be a net in A with $x_n \to x \in X$. We have to show that $x \in A$:
Since $x_n \in A$ there exist $y_n \in \Phi(x_n) \cap F$, and since F is compact, there

exists a subnet $y_m \to y \in F$. Therefore, $y \in \Phi(x) \cap F$ as Gr Φ is closed, and we obtain $x \in A$.
2. From Step 1 we infer that Φ is $\mathcal{F}(X) - \mathcal{K}(Y)$–semicontinuous. In particular, Φ is $\mathcal{F}(X) - \mathcal{K}(Y) \cap \mathcal{F}(Y)$–semicontinuous. But Φ is $\mathcal{K}(Y) \cap \mathcal{F}(Y)$–valued and $\mathcal{K}(Y) \cap \mathcal{F}(Y)$ is $\cap_f$–closed. Thus, by Remark 3 a), Φ is hereditarily $\mathcal{F}(X) - \mathcal{K}(Y) \cap \mathcal{F}(Y)$–semicontinuous. □

Predecessors of the following corollary are due to Joó [5], Stachó [18], Simons [15] and Kindler [8].

Corollary 4 *Let $(X, \mathcal{F}(X), \langle\cdot,\cdot\rangle)$ be a topological interval space, Y a topological space and $\Phi : X \to 2^Y$ a correspondence such that*

(i) every nonvoid set $\Phi^\cap(A)$, $A \in \mathcal{E}(X)$, is compact and connected for $\mathcal{F}(Y) \cap \mathcal{K}(Y)$,
(ii) Φ^* is convex–valued, and
(iii) Gr $\Phi \cap (\langle x_1, x_2\rangle \times Y)$ is closed in $\langle x_1, x_2\rangle \times Y$ $\forall x_1, x_2 \in X$.

Then Φ has a constant selector.

Proof. Let $x_1, x_2 \in X$ and $E \in \mathcal{E}_\Phi(X)$. Set $\mathcal{P} = \mathcal{F}(X)$ and $\mathcal{Q} = \mathcal{F}(Y) \cap \mathcal{K}(Y)$. Let Ψ denote the restriction of Φ_E on $\langle x_1, x_2\rangle$. Then by (iii) Gr Ψ is closed. Therefore, by Lemma 5 and Remark 7 b), Theorem 1 in the version of Remark 6 applies. □

Lemma 6 (cf. [1], Proposition 11.14) *Let $X \subset I\!R^n$ be a polytope, Y a topological space and $\Phi : X \to 2^Y$ a correspondence such that Φ has closed values and Φ^* has open convex values. Then* Gr Φ *is closed. If, in addition, Φ has compact connected values, then $\Phi^\cup(X)$ is connected for $\mathcal{K}(Y) \cap \mathcal{F}(Y)$.*

Proof. For $(\overline{x}, \overline{y}) \in X \times Y -$ Gr Φ choose a nonvoid finite subset $A \subset X$ such that the convex hull $[A]$ is a neighborhood of $\overline{x}$ with $[A] \subset \Phi^*(\overline{y})$. Since Φ^* is convex–valued, we have $\Phi^\cup([A]) = \Phi^\cup(A)$. Hence $W := Y - \Phi^\cup([A])$ is open, and $[A] \subset \Phi^*(\overline{y})$ implies $\overline{y} \in W$. Therefore, $[A] \times W$ is a neighborhood of $(\overline{x}, \overline{y})$ with $[A] \times W \subset X \times Y -$ Gr Φ, i.e., Gr Φ is closed.
The second part follows with Lemma 3 and 5. □

Corollary 5 *Let X and Y be topological spaces, and let $\Phi : X \to 2^Y$ be a correspondence such that Φ^* has open values and every nonvoid set $\Phi^\cap(A)$, $A \in \mathcal{E}(X)$, is closed, compact and connected. Assume moreover that for all $x_1, x_2 \in X$ with $x_1 \neq x_2$ there exists a continuous map $h : [0,1] \to X$ with $h(0) = x_1$, $h(1) = x_2$ and*

$$\Phi(h(t)) \subset \Phi(h(t_1)) \cup \Phi(h(t_2)) \quad \textit{for all} \quad 0 \le t_1 \le t \le t_2 \le 1.$$

Then Φ has a constant selector.

Proof. For a fixed pair x_1, x_2 choose h as above and define $\Psi(t) := \Phi(h(t))$, $t \in [0,1]$. Then Ψ^* has open convex values. By Lemma 6, Gr Ψ is closed, hence by Corollary 4 we have $\Phi(x_1) \cap \Phi(x_2) = \Psi(0) \cap \Psi(1) \neq \emptyset$. The assertion follows with Remark 1 since all assumptions on Φ are hereditary. □

Example 3. Let $X \subset I\!R^n$ be convex, Y a topological space and $\Phi : X \to 2^Y$ a correspondence such that Φ^* has open convex values and every nonvoid set $\Phi^\cap(A)$, $A \in \mathcal{E}(X)$, is closed, compact and connected. Then Φ has a constant selector.

Proof. Apply Corollary 5 with $h(t) = (1-t)x_1 + tx_2$ and observe Remark 7 b). □

5. Minimax Theorems

5.1. ABSTRACT MINIMAX THEOREMS

For a function $a : X \times Y \to I\!R$ we set

$$\mathcal{K}(a) := \left\{ H \in 2^Y - \{\emptyset\} : \sup_{x\in X}\inf_{y\in H} a(x,y) = \sup_{x\in X}\inf_{y\in Y} a(x,y) \right\}$$

and

$$\mathcal{E}(a) := \left\{ A \in \mathcal{E}(X) : \bigcap_{t\in A} \left\{ a(t,\cdot) \leq \sup_{x\in X}\inf_{y\in Y} a(x,y) \right\} \in \mathcal{K}(a) \right\}.$$

A nonvoid subset B of $I\!R$ is called a *border set* for a iff $\lambda > \sup_{x\in X}\inf_{y\in Y} a(x,y)$, $\lambda \in B$, and $\inf B = \sup_{x\in X}\inf_{y\in Y} a(x,y)$.

In our proof of the Sion–Tuy minimax theorem we implicitely used the following simple fact:

Remark 8 (cf. [7], Remark 10). Let X and Y be nonvoid sets and $a : X \times Y \to I\!R$. Then the following are equivalent:

(a) $\min_{y\in Y}\sup_{x\in X} a(x,y) = \sup_{x\in X}\inf_{y\in Y} a(x,y)$.

(b) There exists a compact paving $\mathcal{Q}$ in Y and a border set B for a such that for every $\lambda \in B$ the correspondence $\Phi(x) = \{a(x,\cdot) \leq \lambda\}$, $x \in X$, is $\mathcal{Q}$–valued and has a constant selector.

Hence, our intersection theorems can be used to derive minimax theorems. As an immediate consequence of Theorem 1 and Remark 8 we obtain:

Example 4. Let $(X, \mathcal{P})$ and $(Y, \mathcal{Q})$ be paved spaces with $\mathcal{Q}$ compact, and let $a : X \times Y \to I\!R$ such that

(i) every "interval" $\langle x_1, x_2\rangle := \{x \in X : a(x,\cdot) \geq a(x_1,\cdot) \wedge a(x_2,\cdot)\}$, $(x_1, x_2) \in X \times X$, is connected for $\mathcal{P}$,

and let B be a border set for a such that for every $\lambda \in B$ the correspondence $\Phi : X \to 2^Y, \Phi(x) = \{a(x,\cdot) \leq \lambda\}$ satisfies

(ii) Φ is $\mathcal{Q}$–valued,
(iii) every set $\Phi^\cap(A)$, $A \in \mathcal{E}(X)$, is connected for $\mathcal{Q}$, and
(iv) Φ is hereditarily $\mathcal{P} - \mathcal{Q}$–quartercontinuous.

Then $\min_{y\in Y} \sup_{x\in X} a(x,y) = \sup_{x\in X} \inf_{y\in Y} a(x,y)$.

For a subset C of $X \times Y$ we consider the sections $C_x := \{y \in Y : (x,y) \in C\}$, $x \in X$, and the cosections $C^*_y := \{x \in X : (x,y) \notin C\}$, $y \in Y$.

We shall now present a more flexible version of Example 4, which has the character of a "unifying minimax theorem":

Theorem 2 *Let X and Y be nonvoid sets and $a : X \times Y \to I\!R$. Then the following are equivalent:*

(a) $\min_{y\in Y} \sup_{x\in X} a(x,y) = \sup_{x\in X} \inf_{y\in Y} a(x,y)$.
(b) There exists a compact paving $\mathcal{Q}$ in Y and a border set B for a such that

(i) $\{a(x,\cdot) \leq \lambda\} \in \mathcal{Q}$ for all $x \in X$, $\lambda \in B$,

and for every $\lambda \in B$ and $A \in \mathcal{E}(a) \cup \{\emptyset\}$ there exists a $\mu \in B$, two subsets C and D of $X \times Y$ and an interval structure $(\mathcal{P}, \langle\cdot,\cdot\rangle)$ for X such that

(ii) $\{a < \mu\} \subset C \subset D \subset \{a \leq \lambda\}$,

(iii) every set $\bigcap_{t\in A\cup\{x\}} C_t$, $x \in X$, is connected for $\mathcal{Q}$,

(iv) every cosection D^*_y, $y \in Y$, is convex, and

(v) the correspondence Ξ with $\Xi(x) = \bigcap_{t\in A\cup\{x\}} \{a(t,\cdot) < \mu\}$ is $\mathcal{P} - \mathcal{Q}$–quartercontinuous.

Proof. Let $\alpha := \sup_{x\in X} \inf_{y\in Y} a(x,y)$.
(b) $\Rightarrow$ (a): We show by induction on $k = |A|$ that

$$F_A := \bigcap_{t\in A}\{a(t,\cdot) \leq \alpha\} \in \mathcal{K}(a)\ \forall A \in \mathcal{E}(X) \cup \{\emptyset\} \qquad (*)$$

The case $|A| = 0$, i.e., $F_A = F_\emptyset = Y$ is obvious. Suppose that $F_A \in \mathcal{K}(a)$ for all $A \in \mathcal{E}(X) \cup \{\emptyset\}$ with $|A| = k$. Let $E \in \mathcal{E}(X)$ with $|E| = k+1$. Take $x_1 \in E$ and set $A = E - \{x_1\}$. For $\lambda \in B$ and for $A (\in \mathcal{E}(a) \cup \{\emptyset\})$ choose μ, C, D and $(\mathcal{P}, \langle\cdot,\cdot\rangle)$ as in (b). Then, by Remark 3 b), the correspondence $\Lambda(\supset \Xi)$ with $\Lambda(x) = \bigcap_{t\in A\cup\{x\}} C_t$, $x \in X$, is also $\mathcal{P} - \mathcal{Q}$–quartercontinuous, and val Λ is connected for $\mathcal{Q}$. For arbitrary $x_2 \in X$

it follows from (iv) together with Remark 7 b) that $Z := \Lambda^{\cup}(\langle x_1, x_2\rangle) \subset \bigcup_{x\in\langle x_1,x_2\rangle} C_x \subset \bigcup_{x\in\langle x_1,x_2\rangle} D_x = D_{x_1} \cup D_{x_2} \subset \{a(x_1,\cdot) \le \lambda\} \cup \{a(x_2,\cdot) \le \lambda\}$ and $Z \cap \{a(x_i,\cdot) \le \lambda\} \supset \Lambda(x_i) \ne \emptyset,\ i \in \{1,2\}$. But, by Lemma 3, Z is connected for $\mathcal{Q}$, and with (i) we arrive at

$$\emptyset \ne Z \cap \{a(x_1,\cdot) \le \lambda\} \cap \{a(x_2,\cdot) \le \lambda\} \subset \bigcap_{t\in A\cup\{x_1,x_2\}} \{a(t,\cdot) \le \lambda\}$$

for all $\lambda \in B$, $x_2 \in X$. Since $\mathcal{Q}$ is compact, it follows with (i) that $F_{E\cup\{x\}} \ne \emptyset\ \forall\, x \in X$, which implies $F_E \in \mathcal{K}(a)$.
Finally, from $(*)$ we infer that the family $\{a(x,\cdot) \le \lambda\},\ x \in X,\ \lambda \in B$, of sets in $\mathcal{Q}$ has the finite intersection property. Therefore these sets possess a common point $\hat{y}$ which yields $\sup_{x\in X} a(x,\hat{y}) \le \sup_{x\in X} \inf_{y\in Y} a(x,y)$.
(a) $\Rightarrow$ (b): Take $B = (\alpha,\infty), \mathcal{Q} = \{\{a(x,\cdot) \le \lambda\} : x \in X,\ \lambda \in B\}$, $\mu = \lambda,\ C = D = \{a \le \lambda\},\ \mathcal{P} = \{\emptyset, X\}$ and $\langle\cdot,\cdot\rangle = \{\cdot,\cdot\}$. □

Remark 9. Let $a : X \times Y \to I\!R$ and B a border set for a.

a) Let $\lambda \in B$ and $\Phi(x) = \{a(x,\cdot) \le \lambda\},\ x \in X$. Then $\mathcal{E}(a) \cup \{\emptyset\} \subset \mathcal{E}_\Phi(X)$.

b) The implications (i) $\Rightarrow$ (ii) $\Leftrightarrow$ (iii) hold for the following conditions:

(i) For all $\lambda \in B$ the correspondences Φ with $\Phi(x) = \{a(x,\cdot) \le \lambda\}$, $x \in X$, are hereditarily $\mathcal{P}-\mathcal{Q}-$ quartercontinuous.

(ii) For all $A \in \mathcal{E}(a) \cup \{\emptyset\}$ and for all $\lambda \in B$ the correspondences Ξ with $\Xi(x) = \bigcap_{t\in A\cup\{x\}} \{a(t,\cdot) < \lambda\}$ are $\mathcal{P}-\mathcal{Q}-$ quartercontinuous.

(iii) For all $A \in \mathcal{E}(a) \cup \{\emptyset\}$ and for all $\lambda \in B$ the correspondences Λ with $\Lambda(x) = \bigcap_{t\in A\cup\{x\}} \{a(t,\cdot) \le \lambda\}$ are $\mathcal{P}-\mathcal{Q}-$ quartercontinuous.

This follows from a) together with Remark 3 b).

c) If for $A \in \mathcal{E}(X)$ and $\alpha \in I\!R$ all sets $\bigcap_{t\in A}\{a(t,\cdot) \le \lambda\},\ \lambda > \alpha$, are connected for $\mathcal{Q}$, then also all sets $\bigcap_{t\in A}\{a(t,\cdot) < \lambda\},\ \lambda > \alpha$, are connected for $\mathcal{Q}$ (cf. [11], Consequence 2.4).

d) If $(X, \mathcal{P}, \langle\cdot,\cdot\rangle)$ is an interval space, then for fixed $y \in Y$ and $\alpha \in I\!R$

every set $\{a(\cdot,y) \ge \lambda\},\ \lambda > \alpha$, is convex iff
every set $\{a(\cdot,y) > \lambda\},\ \lambda > \alpha$, is convex.

This follows with Remark 5 from the identities $\{a(\cdot,y) \ge \lambda\} = \bigcap_{n\in I\!N}\{a(\cdot,y) > \lambda - \frac{1}{n}\}$ and $\{a(\cdot,y) > \lambda\} = \bigcup_{n\in I\!N}\{a(\cdot,y) \ge \lambda + \frac{1}{n}\}$.

Our first application of Theorem 2 generalizes Example 4 as well as a topological minimax theorem of Komiya [12] (compare Example 8 below):

Example 5. Let $(X, \mathcal{P}, \langle\cdot,\cdot\rangle)$ be an interval space and $(Y, \mathcal{Q})$ a paved space with compact $\mathcal{Q}$, let $a : X \times Y \to I\!R$, and let B be a border set for a such that for every $\lambda \in B$

(i) every set $\{a(\cdot, y) > \lambda\}$, $y \in Y$, is convex,
(ii) $\{a(x, \cdot) \le \lambda\} \in \mathcal{Q}$, for all $x \in X$,
(iii) every set $\bigcap_{x \in A}\{a(x, \cdot) \le \lambda\}$, $A \in \mathcal{E}(X)$, is connected for $\mathcal{Q}$, and
(iv) for all sets $F := \bigcap_{x \in A}\{a(x, \cdot) \le \alpha\}, A \in \mathcal{E}(X) \cup \{\emptyset\}$, with $F \in \mathcal{K}(a)$ and $\alpha = \sup_{x \in X} \inf_{y \in Y} a(x, y)$ the correspondence Λ with $\Lambda(x) = \{a(x, \cdot) \le \lambda\} \cap F$ is $\mathcal{P} - \mathcal{Q}-$ quartercontinuous.

Then $\min_{y \in Y} \sup_{x \in X} a(x, y) = \sup_{x \in X} \inf_{y \in Y} a(x, y)$.

Proof. Apply Theorem 2 with $C = D = \{a \le \lambda\}$ together with Remarks 3 b) and 9 b). □

Lemma 7 *Let $a : X \times Y \to I\!R$ and let $\mathcal{P}$ be a $\cap_a-$closed paving in X. Let H be a nonvoid subset of Y and $\lambda > \sup_{x \in X} \inf_{y \in H} a(x, y)$ such that $\{a(\cdot, y) \ge \lambda\} \in \mathcal{P}$ for all $y \in H$. Then $\Xi(x) := \{a(x, \cdot) < \lambda\} \cap H, x \in X$, is a $\mathcal{P} - 2^Y-$ quartercontinuous correspondence.*

Proof. This follows from $\Xi^+(Q) = \bigcap_{y \in H - Q}\{a(\cdot, y) \ge \lambda\} \in \mathcal{P},\ Q \in 2^Y$. □

Corollary 6 *Let $(X, \mathcal{P})$ and $(Y, \mathcal{Q})$ be paved spaces with $\cap_a-$closed $\mathcal{P}$ and compact $\mathcal{Q}$. Let $a : X \times Y \to I\!R$ and B a border set for a such that*

(i) $\{a(x, \cdot) \le \lambda\} \in \mathcal{Q}$ for all $x \in X$, $\lambda \in B$, and
(ii) $\{a(\cdot, y) \ge \lambda\} \in \mathcal{P}$ for all $y \in Y$, $\lambda \in B$.

Suppose that for every $\lambda \in B$ and for every $A \in \mathcal{E}(a) \cup \{\emptyset\}$ there exists a $\mu \in B$, subsets C and D of $X \times Y$ with $\{a < \mu\} \subset C \subset D \subset \{a \le \lambda\}$ and an interval function $\langle \cdot, \cdot \rangle$ for $(X, \mathcal{P})$ such that

(iii) every set $\bigcap_{t \in A \cup \{x\}} C_t$, $x \in X$, is connected for $\mathcal{Q}$, and
(iv) every cosection $D_y^*, y \in Y$, is convex.

Then $\min_{y \in Y} \sup_{x \in X} a(x, y) = \sup_{x \in X} \inf_{y \in Y} a(x, y)$.

Proof. For $A \in \mathcal{E}(a) \cup \{\emptyset\}, \mu \in B$ and $H := \bigcap_{t \in A}\{a(t, \cdot) < \mu\} (\in \mathcal{K}(a))$ the correspondence $x \mapsto \{a(x, \cdot) < \mu\} \cap H$, $x \in X$, is $\mathcal{P} - \mathcal{Q}-$ quartercontinuous according to Lemma 7. Hence, the assumptions of Theorem 2 (b) are satisfied.□

A topological version of the following example is due to Kindler and Trost [9]:

Example 6. Let $(X, \mathcal{P}, \langle \cdot, \cdot \rangle)$ and $(Y, \mathcal{Q}, [\cdot, \cdot])$ be interval spaces, and let $\mathcal{C}$ and $\mathcal{D}$ be the systems of convex subsets $P \in \mathcal{P}$ and $Q \in \mathcal{Q}$, respectively. Suppose that $\mathcal{P}$ is $\cap_a-$ closed, $\mathcal{Q}$ is $\cap_f-$ closed, $\mathcal{D}$ is compact, and for all $\lambda > \sup_{x \in X} \inf_{y \in Y} a(x, y)$

$$\{a(\cdot, y) \ge \lambda\} \in \mathcal{C}\ \forall y \in Y \text{ and } \{a(x, \cdot) \le \lambda\} \in \mathcal{D}\ \forall x \in X.$$

Then $\min_{y\in Y}\sup_{x\in X} a(x,y) = \sup_{x\in X}\inf_{y\in Y} a(x,y)$.

Proof. By Remarks 5 and 9 d), Corollary 6 applies with $\mathcal{P}$ and $\mathcal{Q}$ replaced by $\mathcal{C}$ and $\mathcal{D}$ and with $C = D = \{a \leq \lambda\}$. □

A paved space $(S, \mathcal{C})$ is a *K–convexity* iff

$$A, B \in \mathcal{C},\ A\cap B \neq \emptyset \ \Rightarrow A \cap B \in \mathcal{C} \text{ and}$$
$$A, B, C \in \mathcal{C},\ A \cup B = C,\ A \cap B = \emptyset \Rightarrow A = C \text{ or } B = C.$$

K-convexities were introduced under the name "convexities" by Komiya [13].

Remark 10. A paved space $(S, \mathcal{C})$ is a K–convexity iff the paving $\mathcal{P} = \mathcal{C} \cup \{\emptyset\}$ is connected and $\cap_f$–closed.

Remark 11. Let $(X, \mathcal{P})$ be a paved space and D a subset of $X \times Y$ such that every set $\langle x_1, x_2\rangle := \bigcap\{D_y^* : y \in Y - D_{x_1} \cup D_{x_2}\}$, $(x_1, x_2) \in X \times X$, is connected for $\mathcal{P}$. Then $\langle\cdot,\cdot\rangle$ is an interval function for $(X, \mathcal{P})$, and every cosection D_y^*, $y \in Y$, is convex.

Example 7 (cf. Komiya [13]). Let $(X, \mathcal{C})$ and $(Y, \mathcal{D})$ be two K–convexities such that $\mathcal{C}$ is $\cap_a$–closed and $\mathcal{D}$ is compact, and let $a : X \times Y \to I\!R$ with border set B. Suppose that for all $\lambda \in B$ we have

$$\{a(\cdot, y) \geq \lambda\} \in \mathcal{C},\ y \in Y, \text{ and } \{a(x,\cdot) \leq \lambda\} \in \mathcal{D},\ x \in X.$$

Then $\min_{y\in Y}\sup_{x\in X} a(x,y) = \sup_{x\in X}\inf_{y\in Y} a(x,y)$.

Proof. By Remarks 10 and 11, Corollary 6 can be applied with $\mathcal{P} = \mathcal{C}\cup\{\emptyset\}$, $\mathcal{Q} = \mathcal{D} \cup \{\emptyset\}$, $\mu \in B$, $\mu < \lambda$, $C = \{a \leq \mu\}$, $D = \{a < \lambda\}$ and $\langle x_1, x_2\rangle = \bigcap\{D_y^* : y \in Y - D_{x_1} \cup D_{x_2}\}(\in \mathcal{C})$. □

5.2. TOPOLOGICAL MINIMAX THEOREMS

The following example is a slight generalization of a topological minimax theorem due to Komiya [12]:

Example 8. Let X and Y be topological spaces, and let $a : X \times Y \to I\!R$ such that

(i) every set $\{x \in X : a(x,\cdot) \geq a(x_1,\cdot) \wedge a(x_2,\cdot)\}, (x_1, x_2) \in X \times X$, is connected.

Suppose that for some border set B for a

(ii) every nonvoid set $\bigcap_{t\in A}\{a(t,\cdot) \leq \lambda\}, A \in \mathcal{E}(X),\ \lambda \in B$, is connected, compact and closed, and

(iii) for any $\lambda \in B$ and any nonvoid $F \in \mathcal{F}(Y)$ with $\sup_{x\in X} \inf_{y\in F} a(x,y) = \sup_{x\in X} \inf_{y\in Y} a(x,y)$ the correspondence Λ with $\Lambda(x) = \{a(x,\cdot) \leq \lambda\} \cap F$ is $\mathcal{F}(X) - \mathcal{F}(Y) \cap \mathcal{K}(Y)$–quartercontinuous.

Then $\min_{y\in Y} \sup_{x\in X} a(x,y) = \sup_{x\in X} \inf_{y\in Y} a(x,y)$.

Proof. Apply Example 5 with $\mathcal{P} = \mathcal{F}(X), \langle x_1, x_2\rangle = \{x \in X : a(x,\cdot) \geq a(x_1,\cdot) \wedge a(x_2,\cdot)\}$ and $\mathcal{Q} = \mathcal{F}(Y) \cap \mathcal{K}(Y)$. □

Remark 12. Instead of condition (i) above Komiya used the equivalent condition

$\forall x_1, x_2 \in X\ \exists\ C \supset \{x_1, x_2\} : C$ is connected for $\mathcal{P}$ and $a(x,y) \geq a(x_1,y) \wedge a(x_2,y)\ \forall x \in C,\ y \in Y$.

Example 9. Let X and Y be topological spaces, and let $a : X \times Y \to \mathbb{R}$ such that either

(i) every function $a(\cdot,y), y \in Y$, is upper semicontinuous and every function $a(x,\cdot), x \in X$, is lower semicontinuous, or

(i)* a is lower semicontinuous on $X \times Y$.

Suppose moreover that for some border set B for a

(ii) every nonvoid set $\bigcap_{x\in A}\{a(x,\cdot) \leq \lambda\},\ A \in \mathcal{E}(X),\ \lambda \in B$, is connected and compact, and

(iii) for every $\lambda \in B$ there exists an interval function $\langle\cdot,\cdot\rangle$ for $(X, \mathcal{F}(X))$ such that every set $\{a(\cdot,y) > \lambda\},\ y \in Y$, is convex.

Then $\min_{y\in Y} \sup_{x\in X} a(x,y) = \sup_{x\in X} \inf_{y\in Y} a(x,y)$.

Proof. Take $\mathcal{P} = \mathcal{F}(X),\ \mathcal{Q} = \mathcal{K}(Y) \cap \mathcal{F}(Y),\ \mu = \lambda$ and $C = D = \{a \leq \lambda\}$. Then obviously assumptions (i) – (iv) of Theorem 2 are satisfied. But assumption (v) is satisfied as well. This follows from Lemma 7 in case (i) and from Lemma 5 together with Remark 9 b) in case (i)*. □

Example 9 contains minimax theorems of Ha [4], König [11], Komiya [12], Komornik [14] and Sion [17] as special cases. For example, we obtain:

Example 10 (Komiya [12]). Example 9 remains true, when condition (iii) is replaced by "Komiya's condition"

Every set $\{x \in X : a(x,\cdot) \geq a(x_1,\cdot) \wedge a(x_2,\cdot)\}, x_1, x_2 \in X$, is connected.

Proof. Apply Example 9 with the interval function $\langle x_1, x_2\rangle = \{x \in X : a(x,\cdot) \geq a(x_1,\cdot) \wedge a(x_2,\cdot)\}$. □

The following example generalizes König's "topological minimax theorems". In contrast to [11] we neither need any Hausdorff condition nor convexity of a border set.

Example 11 (König [11]). Let X and Y be topological spaces, and let $a : X \times Y \to I\!R$ such that either

(i) every function $a(\cdot, y), y \in Y$, is upper semicontinuous and every function $a(x, \cdot), x \in X$, is lower semicontinuous, or

(i)* a is lower semicontinuous on $X \times Y$.

Suppose moreover that for some border set B for a either

(ii) every nonvoid set $\bigcap_{x\in A}\{a(x,\cdot) \le \lambda\}$, $A \in \mathcal{E}(X)$, $\lambda \in B$, is connected and compact, or

(ii)* every nonvoid set $\bigcap_{x\in A}\{a(x,\cdot) < \lambda\}$, $A \in \mathcal{E}(X)$, $\lambda \in B$, is connected and relatively compact,

and either

(iii) every nonvoid set $\bigcap_{y\in H}\{a(\cdot, y) \ge \lambda\}$ with $H = \{a(x_1,\cdot) \ge \lambda\} \cap \{a(x_2,\cdot) \ge \lambda\}$, $\lambda \in B$, $(x_1, x_2) \in X \times X$, is connected, or

(iii)* every nonvoid set $\bigcap_{y\in K}\{a(\cdot, y) > \lambda\}$ with $K = \{a(x_1,\cdot) > \lambda\} \cap \{a(x_2,\cdot) > \lambda\}$, $\lambda \in B$, $(x_1, x_2) \in X \times X$, is connected.

Then $\min_{y\in Y} \sup_{x\in X} a(x, y) = \sup_{x\in X} \inf_{y\in Y} a(x, y)$.

Proof. Let $\mathcal{P} = \mathcal{F}(X)$, $\mathcal{Q} = \mathcal{K}(Y) \cap \mathcal{F}(Y)$ and B as above. For $\lambda \in B$ take $\mu \in B$ with $\mu < \lambda$ and choose C and D as follows:
Case (ii) and (iii): Take $C = \{a \le \mu\}$ and $D = \{a < \lambda\}$.
Case (ii) and (iii)*: Take $C = D = \{a \le \lambda\}$.
Case (ii)* and (iii): Take $C = D = \{a < \lambda\}$.
Case (ii)* and (iii)*: Take $C = \{a < \lambda\}$ and $D = \{a \le \lambda\}$.
Finally, let $\langle x_1, x_2\rangle = \bigcap\{D_y^* : y \in Y - D_{x_1} \cup D_{x_2}\}$, $(x_1, x_2) \in X \times X$. Then obviously conditions (i), (ii) and (iii) of Theorem 2 are satisfied. By Remark 11, $\langle\cdot,\cdot\rangle$ is an interval function for $(X, \mathcal{F}(X))$ and condition (iv) of Theorem 2 holds. As in the proof of Example 9 it follows that also condition (v) of Theorem 2 is satisfied. □

We close with a slightly generalized version of Tuy's minimax theorem:

Example 12 (Tuy [19, 20]). Let X and Y be topological spaces and $a : X \times Y \to I\!R$. Suppose that there exists a border set B for a such that for every $\lambda \in B$

(i) every set $\{a(\cdot, y) \le \lambda\}$, $y \in Y$, is closed,

(ii) every nonvoid set $\bigcap_{x\in A}\{a(x,\cdot) \le \lambda\}$, $A \in \mathcal{E}(X)$, is closed, compact and connected, and

(iii) for all $x_1, x_2 \in X$ with $x_1 \neq x_2$ there exists a continuous map $h : [0,1] \to X$ with $h(0) = x_1, h(1) = x_2$ and $\{a(h(t),\cdot) \leq \lambda\} \subset \{a(h(t_1),\cdot) \leq \lambda\} \cup \{a(h(t_2),\cdot) \leq \lambda\}$ for all $0 \leq t_1 \leq t \leq t_2 \leq 1$.

Then $\min_{y\in Y} \sup_{x\in X} a(x,y) = \sup_{x\in X} \inf_{y\in Y} a(x,y)$.

Proof. This follows from Corollary 5 together with Remark 8. □

Acknowledgement

I would like to thank Paul Deguire, Gabriele Greco, Charles Horvath and Marc Lassonde for stimulating discussions and valuable hints.

References

1. K. Border, *Fixed point theorems with applications to economics and game theory*, Cambridge University Press, 1985.
2. R. Correa, J.B. Hiriart-Urruty and J.-P. Penot, *A note on connected set–valued mappings*, Boll. Un. Mat. Ital. C (6) **5** (1986), 357-366.
3. I. Del Prete and M. Di Iorio, *Semicontinuous set valued mappings and connected sets*, Le Matematiche **46** (1991), 609-624.
4. Ch.-W. Ha, *Minimax and fixed point theorems*, Math. Ann. **248** (1980), 73-77.
5. I. Joó, *A simple proof for von Neumann's minimax theorem*, Acta Sci. Math. **42** (1980) , 91-94.
6. J. Kindler, *On a minimax theorem of Terkelsen's*, Arch. Math. **66** (1990), 573-583.
7. J. Kindler, *Intersection theorems and minimax theorems based on connectedness*, J. Math. Anal. Appl. **178** (1993), 529-546.
8. J. Kindler, *Intersecting sets in midset spaces I, II*, Arch. Math. **62** (1994), 49-57 and 168-176.
9. J. Kindler and R. Trost, *Minimax theorems for interval spaces*, Acta Math. Hungarica **54** (1989), 39-49.
10. E. Klein and A.C. Thomson, *Theory of correspondences*, Wiley, New York, 1984.
11. H. König, *A general minimax theorem based on connectedness*, Arch. Math. **59** (1992), 55-64. Addendum **64** (1995), 139-143.
12. H. Komiya, *On minimax theorems without linear structure*, Hiyoshi Review of Natural Science **8** (1990), 74-78.
13. H. Komiya, *A set theoretical approach to minimax theory*, Proc. Second Internat. Conf. on fixed point theory and applications, Kok–Keong Tan, ed., World Scientific, 1992.
14. V. Komornik, *Minimax theorems for upper semicontinuous functions*, Acta Math. Acad. Sci. Hungar. **40** (1982), 159-163.
15. S. Simons, *A flexible minimax theorem*, Acta Math. Hungar. **63** (1994) , 119-132. Addendum (1995) **69**, 359-360.
16. S. Simons, *Minimax theorems and their proofs*, In D.–Z. Du and P. M. Pardalos (eds.), Minimax and applications, 1-23, Kluwer, Dordrecht, 1995.
17. M. Sion, *On general minimax theorems*, Pac. J. Math. **8** (1958), 171-176.
18. L.L. Stachó, *Minimax theorems beyond topological vector spaces*, Acta Sci. Math. **42** (1980) , 157-164.
19. H. Tuy, *On a general minimax theorem*, Soviet Math. Dokl. **15** (1974), 1689-1693.
20. H. Tuy, *On the general minimax theorem*, Colloq. Math. **33** (1975), 145-158.
21. Wen–Tsün Wu, *A remark on the fundamental theorem in the theory of games*, Sci. Rec. (N. S.) **3** (1959), 229-233.

K-K-M-S TYPE THEOREMS IN INFINITE DIMENSIONAL SPACES

HIDETOSHI KOMIYA
Faculty of Business and Commerce
Keio University
Kouhoku-ku, Yokohama 223, Japan
e-mail: hkomiya@fbc.keio.ac.jp

Abstract. We study some extensions of K-K-M-S type covering theorems to infinite dimensional spaces. We mainly use a Boolean space Q in place of the finite set $\{1,...,n\}$ and establish the extensions in the space $M(Q)$ of the regular signed Borel measures on Q. We also discuss the nonemptiness of the core of a nontransferable utility game defined in the space $M(Q)$.

1. Introduction

Since Scarf[6] established the nonemptiness of the core of a balanced non-transferable utility game in 1967, others have attempted to prove it in simple ways. Most of them reduced the problem to that of a covering of a simplex in the Euclidean space, and we have obtained some covering theorems of a simplex .

We must define the terminologies and notations appearing in these theorems before showing them. We denote by P the set $\{1,\ldots,p\}$ and by $\mathcal{P}$ the family of the nonempty subsets of P. The set Δ^P denotes the simplex in R^p spanned by $\{e^i : i \in P\}$, where e^i is the p-vector whose i-th coordinate is 1 and 0 otherwise. Similarly Δ^S denotes the face of Δ^P spanned by $\{e^i : i \in S\}$.

A subfamily $\{S_i\}_{i=1}^q$ of $\mathcal{P}$ is said to be *balanced* if there is a corresponding family $\{\lambda_i\}_{i=1}^q$ of nonnegative numbers such that $\sum_i \lambda_i \chi_{S_i} = \chi_P$, where χ_A denotes the characteristic vector of the set A, i.e. $\chi_A = \sum_{i\in A} e^i$. Note that a subfamily $\{S_i\}_{i=1}^q$ of $\mathcal{P}$ is balanced if and only if the vector χ_P/n is a convex combination of the vectors $\chi_{S_i}/|S_i|$. Geometrically this means the

B. Ricceri and S. Simons (eds.), Minimax Theory and Applications, 121–134.

barycenter of the simplex Δ^P is contained in the polytope spanned by the barycenters of the faces Δ^{S_i}.

The following are well-known covering theorems of a simplex.

Theorem A (Ichiishi[3]) *Let $\{C^S\}_{S\in\mathcal{P}}$ be a closed covering of Δ^P such that $\Delta^T \subset \bigcup_{S\supset P\setminus T} C^S$ holds for each $T \in \mathcal{P}$. Then there is a balanced subfamily $\mathcal{B}$ of $\{C^S\}_{S\in\mathcal{P}}$ for which $\bigcap_{S\in\mathcal{B}} C^S \neq \emptyset$.*

Theorem B (Scarf[6]) *Let $\{C^S\}_{S\in\mathcal{P}}$ be a closed covering of Δ^P such that $\Delta^{P\setminus\{j\}} \subset C^{\{j\}}$ holds for each $j \in P$. Then there is a balanced subfamily $\mathcal{B}$ of $\{C^S\}_{S\in\mathcal{P}}$ for which $\bigcap_{S\in\mathcal{B}} C^S \neq \emptyset$.*

Theorem C (Shapley[7]) *Let $\{C^S\}_{S\in\mathcal{P}}$ be a family of closed subsets of Δ^P such that $\Delta^T \subset \bigcup_{S\subset T} C^S$ holds for each $T \in \mathcal{P}$. Then there is a balanced subfamily $\mathcal{B}$ of $\{C^S\}_{S\in\mathcal{P}}$ for which $\bigcap_{S\in\mathcal{B}} C^S \neq \emptyset$.*

We shall extend these theorems to the case of infinite dimensional case and apply these results to game theory with infinite players in this paper.

2. Selection of Base Spaces and Preliminaries

We would like to find spaces which are suitable to describe sets of infinite players and their coalitions, and have good properties shared with finite sets. We list such properties:

- the spaces are compact topologically;
- the sets representing coalitions have common topological properties;
- the families representing coalitions are algebras set-theoretically;

We chose Boolean spaces, that is, compact totally disconnected spaces as the base space of our discussion instead of the finite set P because they satisfy the conditions mentioned above and has good structures which can represent coalitions properly.

When we regard the finite set P as a modulus group Z_p with the discrete topology, the vector χ_P/n which appeared in the necessary and sufficient conditions for balancedness in section 1 can be seen as the Haar measure on P. Any coalition, that is, any subset of P has positive measure with respect to the Haar measure. One of the fundamental properties of Haar measures is the fact that any open set has positive measure and we would like to treat open sets as meaningful coalitions. Moreover, the complement of a coalition should be a coalition, and hence closed sets also must be coalitions. The smallest family satisfying these conditions is the family of all clopen sets, but most of topological spaces do not have sufficient clopen sets. Boolean spaces satisfy these requirements and we shall also require topological group structures to ensure the existence of Haar Measure.

We proceed to prepare mathematical background necessary for the arguments hereafter and define balanced families in totally disconnected compact Hausdorff topological group Q in the next section. Let $C(Q)$ be the Banach space of all continuous real valued functions on Q with the supremum norm $\|\xi\| = \max_{q\in Q} |\xi(q)|$. Let $M(Q)$ be the Banach space of all regular signed Borel measures on Q with the norm $\|x\| = |x|(Q)$, where $|x|$ denotes the total variation of the regular signed Borel measure x on Q. Then we can regard $M(Q)$ as the dual Banach space $C(Q)'$ of $C(Q)$ by the bijection $x \mapsto \tilde{x}$ from $M(Q)$ onto $C(Q)'$ defined by

$$\tilde{x}(\xi) = \int \xi dx, \qquad \xi \in C(Q).$$

The space $M(Q)$ is equipped with the weak-star topology throughout this paper. We shall write $x(\xi)$ in place of $\int \xi dx$ when no confusion is likely to arise. We denote by $\mathbf{\Sigma}$ the σ-field of the Borel sets in Q. We denote the *support* of an element x of $M(Q)$ by $\text{supp}(x)$, that is,

$$\text{supp}(x) = Q \setminus \bigcup \{G : x(G) = 0,\ G \text{ is open}\}.$$

We introduce two binary relations $\geq$ and $\gg$ in $M(Q)$ by

$$x \geq y \text{ if } x(A) \geq y(A) \text{ for all } A \in \mathbf{\Sigma},$$

$$x \gg y \text{ if } x \geq y \text{ and } \text{supp}(x - y) = Q,$$

respectively. We shall use the symbol Δ to denote the convex subset

$$\{x \in M(Q) : \|x\| = x(1) = 1\}$$

of $M_+(Q) = \{x \in M(Q) : x \geq 0\}$.

Recall that Δ is compact and $M_+(Q)$ is closed. Moreover, if we correspond a point q in Q to the mass measure $\widehat{q}$ at q on Q, then the correspondence is into-homeomorphism. For any nonempty subsets A of Q, let Δ^A be the closed convex hull of $\{\widehat{q} : q \in A\}$. We shall use the same symbols as the finite dimensional case, but no confusion may occur.

3. Balanced Families

Let Q be a totally disconnected compact Hausdorff topological group, $\bar{x}$ the Haar measure on Q, and $\mathcal{C}$ the family of all clopen sets of Q. (cf. [2] and [5]) For any $S \in \mathcal{C}$, define a measure $\bar{x}^S$in Δ by

$$\bar{x}^S(A) = \bar{x}(A \cap S)/\bar{x}(S), \quad A \in \mathbf{\Sigma}.$$

Note that $\bar{x}^S$ belongs to Δ^S and $\bar{x}^S(\xi) = \dfrac{\int_S \xi d\bar{x}}{\bar{x}(S)}$ for any $\xi \in C(Q)$.

As mentioned in Section 1, a subfamily $\{S_i\}_{i=1}^q$ of $\mathcal{P}$ is balanced if and only if the vector χ_P/n is a convex combination of the vectors $\chi_{S_i}/|S_i|$. We adopt this property of balanced families as the definition of balancedness in infinite dimensional case as follows:

Definition 1 Let Q be a totally disconnected compact Hausdorff topological group and $\bar{x}$ be the Haar measure on Q. A subfamily $\mathcal{B}$ of $\mathcal{C}$ is said to be *balanced* if $\bar{x}$ belongs to the closed convex hull of the set $\{\bar{x}^S : S \in \mathcal{B}\}$.

We probe the balancedness just defined hereafter. Let N be the set of all positive integers as usual. At first, we give fundamental properties of clopen sets in 2^N, which is a typical example of totally disconnected compact Hausdorff topological group if we regard $2 = \{0, 1\}$ as the topological group with modulus group structure and discrete topology, and 2^N as the ordinary product space of countably infinitely many 2's.

Proposition 1 *(1) For any positive integer n, there is a partition of 2^N whose elements are just n different clopen sets.*
(2) There is a disjoint family $\{S_i\}$ consisting of countably infinite clopen sets in 2^N such that $\bar{x}(\cup S_i) = 1$.

Proof. For any two disjoint finite subsets A and B of N, define the subset $H^{A,B}$ of 2^N by

$$H^{A,B} = \{q \in 2^N : q(n) = 0 \text{ for } n \in A, q(n) = 1 \text{ for } n \in B\}.$$

Then it is easily seen that $\bar{x}(H^{A,B}) = 1/2^{|A|+|B|}$, where $\bar{x}$ is the Haar measure on 2^N.

For each positive integer n, define a clopen subsets of 2^N by

$$\begin{aligned} S_1 &= H^{\{1\},\emptyset}; \\ S_n &= H^{\{n\},\{1,\dots,n-1\}} \quad \text{if } n \geq 2. \end{aligned}$$

Then S_i's are mutually disjoint and $\bar{x}(S_n) = 1/2^n$. Therefore we have $\bar{x}(\cup S_i) = 1$. The family $\{S_i\}$ is the desired family described in (2).

As for the statement (1), set $T_n = (S_1 \cup \cdots \cup S_{n-1})^c$. Then the family $\{S_1, \dots, S_{n-1}, T_n\}$ is the partition of Q whose elements are disjoint n different clopen sets. □

Now we give some balance families in 2^N.

Example 1 *Take a clopen partition $\{S_1, S_2, S_3\}$ of 2^N by means of (1) of Proposition 1 and define $R_1 = S_1 \cup S_2$, $R_2 = S_1 \cup S_3$, $R_3 = S_2 \cup S_3$. Then the family $\{R_1, R_2, R_3\}$ is balanced.*

Example 2 *Take a clopen sets $\{S_1, S_2, \cdots\}$ with the property (2) of Proposition 1 and define $R_1 = S_1$, and $R_n = S_{n-1} \cup S_n$ for $n \geq 2$. Then the infinite family $\{R_1, R_2, \cdots\}$ is balanced.*

The following is the infinite dimensional version of the proposition obtained in Ichiishi[3].

Proposition 2 *$\mathcal{B} = \{S_1, \ldots, S_p\}$ be a finite subfamily of the family $\mathcal{C}$ of clopen sets of totally disconnected compact Hausdorff topological group Q. Then $\mathcal{B}$ is balanced if and only if the family $\mathcal{B}' = \{Q \setminus S_1, \ldots, Q \setminus S_p\}$ is balanced.*

Proof. We need to prove only the "only if" part because of the symmetry of the statement. There are nonnegative numbers $\lambda_1, \ldots, \lambda_p$ such that

$$\bar{x} = \sum_{i=1}^{p} \lambda_i \bar{x}^{S_i} \quad \text{and} \quad \sum_{i=1}^{p} \lambda_i = 1$$

by the hypothesis. Then we have $\sum_{i=1}^{p} \lambda_i(\bar{x} - \bar{x}^{S_i}) = 0$. On the other hand, we have $\bar{x} = \bar{x}(S_i)\bar{x}^{S_i} + \bar{x}(Q \setminus S_i)\bar{x}^{Q\setminus S_i}$; hence we have

$$\bar{x} - \bar{x}^{S_i} = -\frac{\bar{x}(Q \setminus S_i)}{\bar{x}(S_i)}(\bar{x} - \bar{x}^{Q\setminus S_i}).$$

Therefore we have

$$\sum_{i=1}^{p} \frac{\lambda_i \bar{x}(Q \setminus S_i)}{\bar{x}(S_i)}(\bar{x} - \bar{x}^{Q\setminus S_i}) = 0.$$

If we put $\mu = \sum_{i=1}^{p} \frac{\lambda_i \bar{x}(Q \setminus S_i)}{\bar{x}(S_i)}$ and $\mu_i = \sum_{i=1}^{p} \frac{\lambda_i \bar{x}(Q \setminus S_i)}{\mu \bar{x}(S_i)}$, then we have the desired result $\bar{x} = \sum_{i=1}^{p} \mu_i \bar{x}^{Q\setminus S_i}$. □

We cannot expect the corresponding result for infinite families as shown in Example 3 below. We need some lemmas to show the example. Let Q be a Boolean space. A real-valued function ξ on Q is said to be *very simple* if it can be expressed as $\xi = \sum_{i=1}^{p} c_i \chi_{S_i}$, where $S_1, \ldots, S_p$ are all clopen and form a partition of Q.

Lemma 1 *Let Q be a Boolean space. Then, for any $\xi \in C(Q)$ and any $\varepsilon > 0$, there is a very simple function η such that*

$$\max_{x \in \Delta} |x(\xi) - x(\eta)| < \varepsilon.$$

Proof. It is easily seen that the set A of all the very simple functions on Q forms a subalgebra of $C(Q)$, contains constants, and separates points of Q, and hence the norm closure of A is equal to $C(Q)$ by the Stone-Weierstrass theorem. Therefore, for any $\xi \in C(Q)$ and any $\varepsilon > 0$, there is a very simple function η such that

$$\max_{q \in Q} |\widehat{q}(\xi) - \widehat{q}(\eta)| < \varepsilon/2.$$

Since Δ is the weak-star closed convex hull of Q, we have

$$\max_{x \in \Delta} |x(\xi) - x(\eta)| < \varepsilon.$$

□

Lemma 2 *Let $\{x_\alpha\}$ be a net in Δ and x an element of Δ. Then $x_\alpha(A) \to x(A)$ for every $A \in \mathcal{C}$ implies $x_\alpha \to x$.*

Proof. Let ξ be an element of $C(Q)$. For any $\varepsilon > 0$, there is a very simple function η on Q such that $|y(\xi) - y(\eta)| < \varepsilon$ for all $y \in \Delta$ by Lemma 1. Since $x_\alpha(\eta) \to x(\eta)$ by the hypothesis, there is α_0 such that $|x_\alpha(\eta) - x(\eta)| < \varepsilon/3$ for $\alpha \geq \alpha_0$. Therefore, for any $\alpha \geq \alpha_0$, we have

$$\begin{aligned} |x_\alpha(\xi) - x(\xi)| &= |x_\alpha(\xi) - x_\alpha(\eta)| + |x_\alpha(\eta) - x(\eta)| + |x(\eta) - x(\xi)| \\ &< \varepsilon. \end{aligned}$$

□

Example 3 Let $S_1, S_2, \cdots$ be the subsets defined in the proof of Proposition 1:

$$\begin{aligned} S_1 &= H^{\{1\},\emptyset}; \\ S_n &= H^{\{n\},\{1,\dots,n-1\}} \quad \text{if } n \geq 2. \end{aligned}$$

Define a sequence T_n by $T_n = S_1 \cup \cdots \cup S_n$, $n = 1, 2, \dots$. Then we have $\bar{x}(T_n) = 1 - 1/2^n$ and $T_n \nearrow 2^N \setminus \{(1, 1, \dots)\}$. Therefore, we have

$$\bar{x}^{T_n}(A) = \frac{\bar{x}(A \cap T_n)}{\bar{x}(T_n)} \to \bar{x}(A) \quad \text{for all } A \in \mathcal{C};$$

and hence, $\bar{x}^{T_n}$ converges to $\bar{x}$ by Lemma 2. Therefore the family $\{T_n\}$ is balanced. On the other hand, since $Q \setminus T_n = H^{\emptyset,\{1,\dots,n\}} \subset 2^N \setminus S_1 \subset H^{\emptyset,\{1\}}$, $\bar{x}^{Q \setminus T_n}$ belongs to $\Delta^{H^{\emptyset,\{1\}}}$, i.e. $\text{supp}(\bar{x}^{Q \setminus T_n}) \subset H^{\emptyset,\{1\}}$ for all $n = 1, 2, \dots$. Therefore, every point of $\overline{\text{co}}\{\bar{x}^{Q \setminus T_n} : n = 1, 2, \dots\}$ has the support in $H^{\emptyset,\{1\}}$. However, since $\text{supp}(\bar{x}) = 2^N$, we have $\bar{x} \notin \overline{\text{co}}\{\bar{x}^{Q \setminus T_n} : n = 1, 2, \dots\}$ and $\mathcal{B}' = \{2^N \setminus T_n : n = 1, 2, \dots\}$ is not $\bar{x}$-balanced.

Let Q be a totally disconnected compact Hausdorff topological group again. Note that there is no partition of Q consisting of infinitely many clopen sets because Q is compact, but there is a family of infinite clopen sets which is almost a partition of Q. The following proposition assures us our definition of balancedness is appropriate.

Proposition 3 *Let $\{S_i\}$ be a countable disjoint family of clopen sets in Q with $\bar{x}(\cup S_i) = 1$. Then the family $\{S_i\}$ is balanced.*

Proof. Note that

$$\begin{aligned} \bar{x}(A) &= \sum \bar{x}(A \cap S_i) \\ &= \sum \bar{x}(S_i)\bar{x}^{S_i}(A). \end{aligned}$$

Since $\{S_i\}$ is disjoint, we have $\sum \bar{x}(S_i) = 1$. If the sum is essentially finite, then the proof is complete. Suppose the sum has infinitely many nonzero terms. We can assume $\bar{x}(S_1) \neq 0$ without loss of generality. For any $n = 1, 2, \ldots$, define an element x_n of $\text{co}\{\bar{x}^{S_i} : i = 1, 2, \ldots\}$ by $x_n = \sum_{i=1}^{n}(\bar{x}(S_i)/\lambda_n)\bar{x}^{S_i}$, where $\lambda_n = \sum_{i=1}^{n} \bar{x}(S_i)$. Then we have the equations

$$\begin{aligned} \bar{x}(A) &= (\lambda_n x_n)(A) + \sum_{i>n} \bar{x}(S_i)\bar{x}^{S_i}(A) \\ &= x_n(A) + (\lambda_n - 1)x_n(A) + \sum_{i>n} \bar{x}(S_i)\bar{x}^{S_i}(A). \end{aligned}$$

for any $A \in \mathcal{C}$. Hence we have

$$\begin{aligned} |\bar{x}(A) - x_n(A)| &\leq (1 - \lambda_n)x_n(A) + \sum_{i>n} \bar{x}(S_i) \\ &\leq 2(1 - \lambda_n). \end{aligned}$$

We can conclude $x_n \to \bar{x}$ from Lemma 2 since $\lambda_n \to 1$. Therefore we have $\bar{x} \in \overline{\text{co}}\{\bar{x}^{S_i} : i = 1, 2, \ldots\}$. □

4. K-K-M-S Type Theorems in Infinite Dimensional Spaces

We study K-K-M-S type covering theorems in a totally disconnected compact Hausdorff topological group in this section. The following is an infinite dimensional covering theorem extending Theorem A and Theorem B simultaneously.

Theorem 1 *Let Q be a totally disconnected compact Hausdorff topological group. Let $\mathcal{C}$ be the family of all nonempty and proper clopen subsets in Q. Suppose that $\{C^S : S \in \mathcal{C}\}$ is a closed covering of Δ satisfying*

1. *for each* $T \in \mathcal{C}$ *and each* $x \in \Delta^T$, *there is a countable disjoint family* $\{S_i\}$ *in* $\mathcal{C}$ *such that* $Q \setminus T \subset \bigcup_i S_i$ *and* $x \in \bigcap_i C^{S_i}$;
2. *for each* $x \in \Delta$, $x \notin \overline{\bigcup\{C^S : x \notin C^S\}}$.

Then there is a balanced subfamily $\mathcal{B}$ *of* $\mathcal{C}$ *such that*

$$\bigcap_{S \in \mathcal{B}} C^S \neq \emptyset.$$

The method of the proof of Theorem 1 is borrowed from Komiya[4]. We need to establish the infinite dimensional versions of the tools used at the proof of K-K-M-S theorem in [4]. We start with some lemmas.

Lemma 3 *Suppose* Q *is a compact Hausdorff space. Let* $\xi \in C(Q)$ *and* $T = \{q \in Q : \xi(q) = \max_{r \in Q} \xi(r)\}$. *Then we have*

$$\Delta^T = \{y \in \Delta : y(\xi) = \max_{x \in \Delta} x(\xi)\}.$$

Proof. Denote by M the right hand side of the equation in the lemma. If $q \in T$, then $\xi(q) \geq \xi(r)$ for all $r \in Q$, that is, $\widehat{q}(\xi) \geq \widehat{r}(\xi)$ for all $r \in Q$. Hence we have $\widehat{q}(\xi) \geq x(\xi)$ for all $x \in \Delta$. This means $\widehat{q} \in M$ for all $q \in T$. Since M is closed and convex, we have $\Delta^T \subset M$. Suppose there would be a point y such that $y \in M \setminus \Delta^T$. It is easily seen that $y(T) = 1$ because $y \in M$. By the separation theorem for convex sets, there exists a point $\eta \in C(Q)$ and a real number c such that $y(\eta) > c > \widehat{q}(\eta)$ for $q \in T$. But this leads the contradiction

$$y(\eta) = \int_Q \eta dy = \int_T \eta dy \leq c < y(\eta).$$

□

Lemma 4 *Let* Q *be a compact Hausdorff space. Suppose that a function* $f : \Delta \to C(Q)$ *is continuous. Then the set-valued function* $G : \Delta \to 2^{\Delta}$ *defined by*

$$G(x) = \{y \in \Delta : y(f(x)) = \max\{z(f(x)) : z \in \Delta\}\}$$

is upper semicontinuous, closed and convex-valued.

Proof. It is easily seen that G is convex-valued. To prove that G is upper semicontinuous and closed-valued, it is sufficient to show the graph of G is closed. Let $\{(x_\alpha, y_\alpha)\}$ be a net in the graph of G converging to $(x_0, y_0) \in$

$\Delta \times \Delta$. For any $z \in \Delta$, we have $y_\alpha(f(x_\alpha)) \geq z(f(x_\alpha))$. Moreover we have the inequalities

$$\begin{aligned} & |y_\alpha(f(x_\alpha)) - y_0(f(x_0))| \\ \leq\ & |y_\alpha(f(x_\alpha)) - y_\alpha(f(x_0))| + |y_\alpha(f(x_0)) - y_0(f(x_0))| \\ \leq\ & \|y_\alpha\| \|f(x_\alpha) - f(x_0)\| + |y_\alpha(f(x_0)) - y_0(f(x_0))| \\ =\ & \|f(x_\alpha) - f(x_0)\| + |y_\alpha(f(x_0)) - y_0(f(x_0))| \end{aligned}$$

Hence $y_0(f(x_0)) \geq z(f(x_0))$ for all $z \in \Delta$, and we have $y_0 \in G(x_0)$. This proves the closedness of the graph of G. □

Proof of Theorem 1. For each $x \in \Delta$, let $I(x) = \{T \in \mathcal{C} : C^T \ni x\}$. Since $\{C^T : T \in \mathcal{C}\}$ is a covering of Δ, $I(x)$ is nonempty for each $x \in \Delta$. Suppose the conclusion is false. For each $x \in \Delta$, define an open neighborhood $O(x)$ of x in Δ by

$$O(x) = Q \setminus \overline{\bigcup\{C^S : x \notin C^S\}}.$$

Then $\{O(x)\}_{x \in \Delta}$ is an open covering of Δ. Hence there is a finite open subcovering $\{O(x_k)\}_{k=1}^p$ of $\{O(x)\}_{x\in\Delta}$. Let $\{\beta_k\}_{k=1}^p$ be a partition of unity subordinate to $\{O(x_k)\}_{k=1}^p$. Since $\bar{x} \notin \overline{\text{co}}\{\bar{x}^S : S \in I(x_k)\}$ for $k = 1, \ldots, p$, for any k there is $\xi_k \in C(Q)$ such that $\bar{x}(\xi_k) < \bar{x}^S(\xi_k)$ for any $S \in I(x_k)$ by the separation theorem for convex sets. We can assume that ξ_k is very simple for $k = 1, \ldots, p$ by Lemma 1. Define the continuous function $f : \Delta \to C(Q)$ by

$$f(x) = \sum_{k=1}^{p} \beta_k(x)\xi_k.$$

Then, it is easily seen that, for each $x \in \Delta$, we have $\bar{x}(f(x)) < \bar{x}^S(f(x))$ for $S \in I(x)$. Moreover, define the set-valued function $G : \Delta \to 2^\Delta$ by

$$G(x) = \{y \in \Delta : y(f(x)) = \max\{z(f(x)) : z \in \Delta\}\}.$$

Then G is upper semicontinuous, closed and convex-valued by Lemma 4.

There is $x_0 \in \Delta$ such that $x_0 \in G(x_0)$ by [1, Theorem 1]. If $T_0 = \{q \in Q : f(x_0)(q) = \max_{r \in Q} f(x_0)(r)\}$, then we have $\Delta^{T_0} = G(x_0)$ by Lemma 3.

It is obvious that T_0 is not empty and $T_0 \neq Q$. If $T_0 = Q$, then $f(x_0)$ is a constant function on Q. But $f(x_0)$ separates $\bar{x}$ and $\overline{\text{co}}\{\bar{x}^S : S \in I(x_0)\}$ strictly. This is a contradiction. The set T_0 is clopen because $f(x_0)$ is very simple. Therefore we have a countable subfamily $\{S_i\}$ of $\mathcal{C}$ corresponding to x_0 with the properties described in the theorem. Let $U_0 = Q \setminus \bigcup_i S_i$.

From the definition of T_0, we have $\bar{x}(f(x_0)) \leq f(x_0)(q)$ for all $q \in U_0$. On the other hand, since $x_0 \in C^{S_i}$, we have $S_i \in I(x_0)$ for all i. Hence we have $\bar{x}(f(x_0)) < \bar{x}^{S_i}(f(x_0))$ for all i by the definition of f. Therefore we have

$$\begin{aligned} \bar{x}(f(x_0)) &= \int_{U_0} f(x_0)d\bar{x} + \sum_i \int_{S_i} f(x_0)d\bar{x} \\ &> \int_{U_0} \bar{x}(f(x_0))d\bar{x} + \sum_i \bar{x}(S_i)\bar{x}(f(x_0)) \\ &= \bar{x}(f(x_0)), \end{aligned}$$

which is a contradiction. □

The following corollary of Theorem 1 will be used in the next section.

Corollary 1 *Let Q be a totally disconnected compact Hausdorff topological group. Let $\mathcal{C}$ be the family of all nonempty and proper clopen subsets in Q. Suppose that $\{C^S : S \in \mathcal{C}\}$ is a closed covering of Δ satisfying*

1. *for each $T \in \mathcal{C}$, $\Delta^T \subset C^{Q \setminus T}$;*
2. *for each $x \in \Delta$, $x \notin \overline{\bigcup\{C^S : x \notin C^S\}}$.*

Then there is a balanced subfamily $\mathcal{B}$ of $\mathcal{C}$ such that

$$\bigcap_{S \in \mathcal{B}} C^S \neq \emptyset.$$

5. Application to Game Theory

We continue to assume that Q is a totally disconnected compact Hausdorff topological group and $\mathcal{C}$ is the family of all nonempty clopen subsets of Q in this section. For $u \in M(Q)$ and $S \in \mathcal{C}$, define the *restriction* $u|_S$ of u to S by $u|_S(A) = u(A \cap S)$, $A \in \Sigma$.

We define a nontransferable utility game in our setting.

Definition 2 A *nontransferable utility game* is a set-valued mapping $V : \mathcal{C} \to M(Q)$ satisfying

1. if $u \in V(S)$ and $v|_S = u|_S$ and $v \in M(Q)$ then $v \in V(S)$;
2. $V(S) - M_+(Q) = V(S)$ for every $S \in \mathcal{C}$.

We introduce a weaker notion, say ε-core, than the usual core of games, but note that it coincides with the usual core when the game is finite dimensional and ε is sufficiently small.

Definition 3 For any $\varepsilon > 0$, define the *ε-core* $C_\varepsilon(V)$ of V by

$$C_\varepsilon(V) = \{u \in V(Q) : \forall S \in \mathcal{C},\ \forall v \in V(S),\ \neg[v \gg (1+\varepsilon)u]\}.$$

Let A be a subset of $M(Q)$ and u be a point of $M(Q)$. It is said that A *absorbs* u if there is $\lambda > 0$ such that $u \in \lambda A$ and that A *emits* u if there is $\lambda > 0$ such that $u \notin \lambda A$.

Definition 4 A nontransferable utility game V is called *balanced* if it follows that $\bigcap_{S \in \mathcal{B}} V(S) \subset V(Q)$ for any balanced subfamily $\mathcal{B}$ of $\mathcal{C}$.

Definition 5 Let A be a subset of $M(Q)$ which both absorbs and emits each u with $u > 0$. Then we can define a set $\partial^r A$ called the *positive radial boundary* of A by

$$\partial^r A = \{ru : u > 0 \text{ and } r = \sup\{\lambda : \lambda u \in A\}\}$$

The following is an example of the sets which both absorb and emit any u with $u > 0$.

Proposition 4 Let ξ be an element of $C(Q)$ such that $\xi(q) > 0$ for all $q \in Q$. Let $V = \{u \in M(Q) : \exists v \geq 0; v(\xi) = 1,\ v \geq u\}$. Then V is closed, there is $\varepsilon > 0$ such that the closed ball of radius ε about 0 is contained in V, and $V \cap M_+(Q)$ is norm bounded. Moreover, the positive radial boundary $\partial^r V$ is compact.

Proof. At first, we prove $V \cap M_+(Q)$ is norm bounded. Let $F = \{v \in M_+(Q) : v(\xi) = 1\}$. Choose any u in $V \cap M_+(Q)$. Then there is $v \in F$ such that $0 \leq u \leq v$. On the other hand, there is a number ε such that $0 < \varepsilon \leq \xi(q)$ for all $q \in Q$. Therefore, we have $\varepsilon\|u\| = u(\varepsilon 1) \leq u(\xi) \leq v(\xi) = 1$; hence, $\|u\| \leq 1/\varepsilon$. This means that $V \cap M_+(Q)$ is norm bounded.

In particular, since F is a closed subset of $V \cap M_+(Q)$, F is compact. We can write $V = F - M_+(Q)$, and $M_+(Q)$ is closed; hence, we have V is closed. Therefore, we can define the positive radial boundary $\partial^r V$ of V. We claim that $\partial^r V = F$. Let $v \in \partial^r V$; then, there is $u > 0$ and $r > 0$ such that $r = \sup\{\lambda : \lambda u \in V\}$ and $v = ru$. It is obvious that $v(\xi) > 0$. If $v(\xi) < 1$, then there $\mu > 1$ such that $\mu v(\xi) = 1$. Then, $\mu v = (\mu r)u \in F \subset V$, but this contradicts the definition of r. If $v(\xi) > 1$, then v does not belong to V. Therefore, we have $v(\xi) = 1$ and v belongs to F. Conversely, if v is an element of F, then $v = 1v \in F \subset V$. If $\mu > 1$, then μv does not belong to V. Therefore, v belongs to $\partial^r V$.

The $\|\cdot\|$ is continuous on $M_+(Q)$ because $\|u\| = u(1)$ for all $u \in M_+(Q)$. Hence, there is a number δ such that $0 < \delta \leq \|u\|$ for all $u \in F$. We claim that the closed ball of radius δ about 0 is contained in V. Take an element u with $\|u\| \leq \delta$. Then we have $u \leq |u|$ and $\||u|\| \leq \delta$. If $|u|(\xi) > 1$, then there is λ such that $0 < \lambda < 1$ and $\lambda|u|(\xi) = 1$; hence, $\lambda|u|$ belongs to F, but $\|\lambda|u|\| < \||u|\| \leq \delta$. This is a contradiction; therefore, we have $|u|(\xi) \leq 1$. Then there is a number μ such that $\mu \geq 1$ and $\mu|u| = 1$; hence, $u \leq |u| \leq \mu|u|$ and $\mu|u| \in F$, and hence we have $u \in V$. □

Theorem 2 *Let $V : \mathcal{C} \to M(Q)$ be a nontransferable utility game. Let ε be a fixed positive number. Suppose that*

1. *$V(S)$ is closed and contains an element u with $u \gg 0$ for every $S \in \mathcal{C}$;*
2. *$V(Q)$ both absorbs and emits each u with $u > 0$, and the radial boundary $\partial^r V(Q)$ is compact;*
3. *for any $u \in \partial^r V(Q)$, the family $\{S \in \mathcal{C} : \forall v \in V(S),\ \neg[v \gg (1+\varepsilon)u]\}$ is finite;*
4. *V is balanced.*

Then the ε-core $C_\varepsilon(V)$ is nonempty.

Proof. Since the norm of $M(Q)$ restricted to $M_+(Q)$ is continuous, the function $h : M_+(Q) \setminus \{0\} \to \Delta$ defined by $h(u) = u/\|u\|$ is continuous. Let $F = (1+\varepsilon)\partial^r V(Q)$. Then F is compact and the restriction of h to F is homeomorphism onto Δ.

Suppose the ε-core $C_\varepsilon(Q)$ is empty. Define a subset G^S of Δ by

$$G^S = \{x \in \Delta : \exists v \in V(S);\ v \gg h^{-1}(x)\}.$$

We can derive $\Delta \subset \cup_{S \in \mathcal{C}} G^S$ from $C_\varepsilon(Q) = \emptyset$, and also $G^Q = \emptyset$. Moreover we have $\Delta^S \subset G^{Q \setminus S}$ for all $S \in \mathcal{C} \setminus \{Q\}$. Put $C^S = \overline{G^S}$ for $S \in \mathcal{C}$. Then the family $\{C^S\}_{S \in \mathcal{C}}$ satisfies the condition of Corollary 1 and there is a point x_0 and a balanced subfamily $\mathcal{B}$ of $\mathcal{C}$ such that $x_0 \in C^S$ for all $S \in \mathcal{B}$. Since $V(S)$ is closed and h is homeomorphism, we have $u_0 = h^{-1}(x_0) \in V(S)$ for $S \in \mathcal{B}$. Therefore we have $u_0 \in V(Q)$ by the balancedness of V. This is a contradiction because u_0 belongs to F. □

6. Extensions of K-K-M-S Theorem

We have the same type of extension of K-K-M-S theorem(Theorem C) as Theorem 1. The proof of the extension is similar to that of Theorem 1, but we provide the whole proof for clarity.

Theorem 3 *Let Q be a totally disconnected compact Hausdorff topological group. Let $\mathcal{C}$ be the family of all nonempty and proper clopen subsets in Q. Suppose that $\{C^S : S \in \mathcal{C}\}$ is a closed covering of Δ satisfying*

1. *for each $T \in \mathcal{C}$, $\Delta^T \subset \bigcup_{S \subset T} C^S$;*
2. *for each $x \in \Delta$, $x \notin \overline{\bigcup\{C^S : x \notin C^S\}}$.*

Then there is a balanced subfamily $\mathcal{B}$ of $\mathcal{C}$ such that

$$\bigcap_{S \in \mathcal{B}} C^S \neq \emptyset.$$

Proof. For each $x \in \Delta$, let $I(x) = \{T \in \mathcal{C} : C^T \ni x\}$. Since $\{C^T : T \in \mathcal{C}\}$ is a covering of Δ, $I(x)$ is nonempty for each $x \in \Delta$. Suppose the conclusion is false. For each $x \in \Delta$, define an open neighborhood $O(x)$ of x in Δ by

$$O(x) = Q \setminus \overline{\bigcup\{C^S : x \notin C^S\}}.$$

Then $\{O(x)\}_{x\in\Delta}$ is an open covering of Δ. Hence there is a finite open subcovering $\{O(x_k)\}_{k=1}^p$ of $\{O(x)\}_{x\in\Delta}$. Let $\{\beta_k\}_{k=1}^p$ be a partition of unity subordinate to $\{O(x_k)\}_{k=1}^p$. Since $\bar{x} \notin \overline{\mathrm{co}}\{\bar{x}^S : S \in I(x_k)\}$ for $k = 1, \ldots, p$, for any k there is $\xi_k \in C(Q)$ such that $\bar{x}(\xi_k) > \bar{x}^S(\xi_k)$ for any $S \in I(x_k)$ by the separation theorem for convex sets. We can assume that ξ_k is very simple for $k = 1, \ldots, p$ by Lemma 1. Define the continuous function $f : \Delta \to C(Q)$ by

$$f(x) = \sum_{k=1}^{p} \beta_k(x)\xi_k.$$

Moreover, define the set-valued function $G : \Delta \to 2^\Delta$ by

$$G(x) = \{y \in \Delta : y(f(x)) = \max\{z(f(x)) : z \in \Delta\}\}.$$

Then G is upper semicontinuous, closed and convex-valued by Lemma 4. Therefore, there is $x_0 \in \Delta$ such that $x_0 \in G(x_0)$ by [1, Theorem 1]. If $T_0 = \{q \in Q : f(x_0)(q) = \max_{r\in Q} f(x_0)(r)\}$, then we have $\Delta^{T_0} = G(x_0)$ by Lemma 3.

It is obvious that T_0 is not empty and $T_0 \neq Q$. If $T_0 = Q$, then $f(x_0)$ is a constant function on Q. But $f(x_0)$ separates $\bar{x}$ and $\overline{\mathrm{co}}\{\bar{x}^S : S \in I(x_0)\}$ strictly. This is a contradiction. The set T_0 is clopen because $f(x_0)$ is very simple.

Therefore we have

$$x_0 \in \Delta^{T_0} \subset \bigcup_{S\subset T_0} C^S$$

by the hypothesis. Then there is a subset $S_0 \in \mathcal{C}$ of T_0 with $x_0 \in C^{S_0}$, that is, $S_0 \in I(x_0)$. Noting $\{\beta_k\}_{k=1}^p$ are subordinate to the open covering $\{O(x_k)\}$, we have

$$\bar{x}(f(x_0)) > \bar{x}^{S_0}(f(x_0))$$

by the definition of f. On the other hand, $\bar{x}^{S_0} \in \Delta^{S_0} \subset \Delta^{T_0} = G(x_0)$, that is,

$$\bar{x}^{S_0}(f(x_0)) = \max\{z(f(x_0)) : z \in \Delta\},$$

which contradicts the inequality above. □

References

1. K. Fan, *Fixed-point and Minimax Theorems in Locally Convex Topological Linear Spaces,* Proc. Nat. Acad. Sci. USA **38** (1952), 121-126.
2. P.R. Halmos, *Boolean Algebra,* Stevens & Co., New York, 1959.
3. T. Ichiishi, *Alternative Version of Shapley's Theorem on Closed Coverings of a Simplex,* Proc. Amer. Math. Soc. **104** (1988), 759-763.
4. H. Komiya, *A Simple Proof of K-K-M-S Theorem,* Economic Theory **4** (1994), 463-466.
5. W. Rudin, *Functional Analysis,* McGraw-Hill, New York, 1973.
6. H. Scarf, *The Core of an n Person Game,* Econometrica **35** (1967) , 50-69.
7. L.S. Shapley, *On Balanced Games without Side Payments,* in T.C Hu and S.M. Robinson (eds.), Mathematical Programming, Academic Press, New York, 1973, 261-290.

HAHN-BANACH THEOREMS FOR CONVEX FUNCTIONS

MARC LASSONDE
Université des Antilles et de la Guyane
Mathématiques
97159 Pointe-à-Pitre, Guadeloupe, France
e-mail: lassonde@univ-ag.fr

We start from a basic version of the Hahn-Banach theorem, of which we provide a proof based on Tychonoff's theorem on the product of compact intervals. Then, in the first section, we establish conditions ensuring the existence of affine functions lying between a convex function and a concave one in the setting of vector spaces — this directly leads to the theorems of Hahn-Banach, Mazur-Orlicz and Fenchel. In the second section, we caracterize those topological vector spaces for which certain convex functions are continuous — this is connected to the uniform boundedness theorem of Banach-Steinhaus and to the closed graph and open mapping theorems of Banach. Combining both types of results readily yields topological versions of the theorems of the first section.

In all the text, X stands for a real vector space. For $A \subset X$, we denote by $\operatorname{cor}(A)$ the core (algebraic interior) of A : $a \in \operatorname{cor}(A)$ if and only if $A - a$ is absorbing. Given a function $f : X \to \mathbb{R} \cup \{+\infty\}$, we call *domain of* f the set $\operatorname{dom} f = \{ x \in X \mid f(x) < +\infty \}$ and we declare f *convex* if $f(tx + (1-t)y) \leq tf(x) + (1-t)f(y)$ for all x, y in $\operatorname{dom} f$ and $0 \leq t \leq 1$. A real-valued function $p : X \to \mathbb{R}$ is said to be *sublinear* if it is convex and positively homogeneous.

We use standard abbreviations and notations: *TVS* for topological vector space, *LCTVS* for locally convex topological vector space, *lcs* for lower semicontinuous, X^* for the algebraic dual of X, X' for its topological dual, $\sigma(X^*, X)$ for the topology of pointwise convergence on X^*, etc.

Prologue

The following basic theorem is the starting point, and crucial part, of the theory. It retains the essence of both the Hahn-Banach theorem — non-emptiness assertion — and the Banach-Alaoglu theorem — $\sigma(X^*, X)$-

B. Ricceri and S. Simons (eds.), Minimax Theory and Applications, 135–145.

compactness assertion. Its proof combines the key arguments of the proofs of these theorems.

Basic Theorem *For any sublinear function* $p : X \to \mathbb{R}$, *the set* $\{ x^* \in X^* \mid x^* \leq p \}$ *is non-empty and* $\sigma(X^*, X)$*-compact.*

Proof. In the space $E = \mathbb{R}^X$ supplied with the product topology, the set X^ν of all sublinear forms is closed and the set

$$K := \{ q \in X^\nu \mid q \leq p \} = \prod_{x \in X} [-p(-x), p(x)] \cap X^\nu$$

is compact by Tychonoff's theorem. For $x \in X$, put

$$F(x) := \{ q \in K \mid q(x) + q(-x) = 0 \}.$$

Clearly $\bigcap_{x \in X} F(x) = \{ x^* \in X^* \mid x^* \leq p \}$. Since each $F(x)$ is closed in the compact set K, to obtain statement (2) it only remains to show that for any finite family $\{x_0, x_1, \ldots, x_n\}$ in X, the intersection of the $F(x_i)$'s is not empty. We first observe that $F(x_0)$ is not empty; that is, we observe that there exists $q_0 \in X^\nu$ verifying

$$q_0 \leq p \quad \text{and} \quad q_0(x_0) + q_0(-x_0) = 0.$$

Indeed, it suffices to take for q_0 the sublinear hull of p and of the function equal to $-p(x_0)$ at $-x_0$ and to $+\infty$ elsewhere, namely:

$$q_0(x) := \inf_{\lambda \geq 0} (p(x + \lambda x_0) - \lambda p(x_0)).$$

We then apply the argument again, with q_0 and x_1 in lieu of p and x_0, to obtain q_1 in $F(x_0) \cap F(x_1)$, and so forth until obtaining q_n in $F(x_0) \cap F(x_1) \cap \ldots \cap F(x_n)$. □

The non-emptiness assertion is Theorem 1 in Banach [2]. Its original proof, as well as the proof given in most textbooks, relies on the axiom of choice. The fact that it can also be derived from Tychonoff's theorem on the product of compact intervals was observed for the first time by Loś and Ryll-Nardzewski [9]. It is now well-known, after the works of Luxemburg [10] and Pincus [12], that Banach's theorem (and hence the Hahn-Banach theorem) is logically weaker than Tychonoff's theorem on the product of compact intervals (and hence the Banach-Alaoglu theorem and the above theorem), which itself is weaker than the axiom of choice. From this point of view, the above statement is therefore optimal with respect to its proof.

1. Separation of Convex Functions

We first extend the basic theorem to the case of convex functions. For $f : X \to \mathbb{R} \cup \{+\infty\}$, we denote by $p_f : x \mapsto \inf_{t>0} \frac{1}{t} f(tx)$ the *homogeneous hull* of f, and we set

$$\begin{aligned} S(f) &:= \{ x^* \in X^* \mid x^* \le f \} \\ &= \{ x^* \in X^* \mid 0 \le \inf(f - x^*) \}. \end{aligned}$$

Because of the following elementary facts, the extension is a straightforward consequence of the basic theorem.

Lemma 1 *If $f : X \to \mathbb{R} \cup \{+\infty\}$ is convex such that $f(0) \ge 0$ and $0 \in \operatorname{cor}(\operatorname{dom} f)$, then the function p_f is real-valued and sublinear.*

Lemma 2 *For any $f : X \to \mathbb{R} \cup \{+\infty\}$, $S(f) = S(p_f)$.*

Theorem 1 (Minoration of convex functions) *Let $f : X \to \mathbb{R} \cup \{+\infty\}$ be convex such that $f(0) \ge 0$. If $0 \in \operatorname{cor}(\operatorname{dom} f)$, then $S(f)$ is non-empty and $\sigma(X^*, X)$-compact.*

Proof. Apply the basic theorem to p_f. □

Theorem 1' (ε-subdifferential) *Let $f : X \to \mathbb{R} \cup \{+\infty\}$ be convex. If $x_0 \in \operatorname{cor}(\operatorname{dom} f)$, then for every $\varepsilon \ge 0$ the set*

$$\{ x^* \in X^* \mid x^*(x - x_0) \le f(x) - f(x_0) + \varepsilon \quad \textit{for all } x \in X \}$$

is non-empty and $\sigma(X^*, X)$-compact.

Proof. Apply Theorem 1 to the function $\tilde{f}(x) := f(x + x_0) - f(x_0) + \varepsilon$. □

Corollary *If in Theorem 1 we suppose further that X is a TVS and that f is continuous at some point of its domain, then $S(f)$ is non-empty, equicontinuous and $\sigma(X', X)$-compact.*

Proof. The set $S(f)$ is clearly equicontinuous, hence contained in X', and it is also clearly $\sigma(X', X)$-closed. □

More generally, we now search for separating a convex function from a concave one by an affine form. For $f, g : X \to \mathbb{R} \cup \{+\infty\}$, we denote by $f +_e g : x \mapsto \inf_{y \in X} (f(y) + g(x - y))$ the *epi-sum* (or *inf-convolution*) of f and g, and we set

$$\begin{aligned} S(f, g) &:= \{ x^* \in X^* \mid -g \le x^* + r \le f \quad \text{for some } r \in \mathbb{R} \} \\ &= \{ x^* \in X^* \mid 0 \le \inf(f - x^*) + \inf(g + x^*) \}. \end{aligned}$$

As above, two elementary facts similarly reduce the argument to a simple invocation of the previous theorem.

Lemma 3 *If* $f, g : X \to \mathbb{R} \cup \{+\infty\}$ *are convex such that* $(f +_e g)(0)$ *is finite and* $0 \in \operatorname{cor}(\operatorname{dom} f + \operatorname{dom} g)$*, then* $f +_e g$ *takes its values in* $\mathbb{R} \cup \{+\infty\}$ *and is convex.*

Lemma 4 *For any* $f, g : X \to \mathbb{R} \cup \{+\infty\}$, $S(f, g) = S(f +_e g^-)$*, where* $g^- : x \mapsto g(-x)$.

Theorem 2 (Separation of convex functions) *Let* $f, g : X \to \mathbb{R} \cup \{+\infty\}$ *be convex such that* $-g \leq f$*. If* $0 \in \operatorname{cor}(\operatorname{dom} f - \operatorname{dom} g)$*, then* $S(f, g)$ *is non-empty and* $\sigma(X^*, X)$*-compact.*

Proof. Apply Theorem 1 to $f +_e g^-$. □

Theorem 2' (Decomposition of the infimum of a sum) *Let* $f, g : X \to \mathbb{R} \cup \{+\infty\}$ *be convex such that* $\inf(f + g)$ *is finite. If* $0 \in \operatorname{cor}(\operatorname{dom} f - \operatorname{dom} g)$*, then for every* $\varepsilon \geq 0$ *the set*

$$\{ x^* \in X^* \mid \inf(f + g) \leq \inf(f - x^*) + \inf(g + x^*) + \varepsilon \}$$

is non-empty and $\sigma(X^*, X)$*-compact.*

Proof. Apply Theorem 2 to the functions $\tilde{f} := f - \inf(f + g) + \varepsilon$ and g. □

Corollary *If in Theorem 2 we suppose further that* X *is a TVS and that* $f +_e g$ *is continuous at some point of its domain (this is the case if* f *is continuous at some point of its domain), then* $S(f, g)$ *is non-empty, equicontinuous and* $\sigma(X', X)$*-compact.*

Proof. If f is continuous at some point of its domain, it is actually continuous on the non-empty set $\operatorname{int}(\operatorname{dom} f)$, and since 0 belongs to $\operatorname{cor}(\operatorname{dom} f - \operatorname{dom} g) = \operatorname{int}(\operatorname{dom} f) - \operatorname{dom} g$, we infer that f is continuous at some point of $\operatorname{dom} f \cap \operatorname{dom} g$, which implies at once that $f +_e g^-$ is continuous at 0. The result now follows from the corollary of Theorem 1 because $S(f, g) = S(f +_e g^-)$. □

The literature on the Hahn-Banach theorem is too broad to give any fair account in this short article. We refer to Buskes [5] for a comprehensive survey and an extensive bibliography, and to König [8] for a deep discussion on the theorem and its various applications. For variants of the above results, see, e.g., Holmes [6, p. 23 and p. 42], Vangeldère [21], Théra [19].

Before proceeding, we mention several simple consequences.

Banach-Alaoglu Theorem – algebraic version *If $C \subset X$ is convex and absorbing, then the set $\{ x^* \in X^* \mid x^*(x) \leq 1 \text{ for all } x \in C \}$ is $\sigma(X^*, X)$-compact.*

Proof. Apply Theorem 1 to the function equal to 1 on C and to $+\infty$ elsewhere. □

When X is a TVS and C a convex neighborhood of 0, we recover the classical Banach-Alaoglu Theorem.

Hahn-Banach Theorem – sandwich version *Let $p : X \to \mathbb{R}$ be sublinear, $C \subset X$ be convex and $\tau : C \to \mathbb{R}$ be concave. If $\tau \leq p|C$, then there exists $x^* \in X^*$ such that $\tau \leq x^*|C$ and $x^* \leq p$.*

Proof. Apply Theorem 2, with $f = p$ and g equal to $-\tau$ on C and to $+\infty$ elsewhere, to obtain $x^* \in X^*$ et $r \in \mathbb{R}$ such that

$$\tau \leq x^*|C + r \quad \text{and} \quad x^* + r \leq p.$$

The second inequality implies $r \leq 0$, so that $\tau \leq x^*|C$. Since on the other hand p is positively homogeneous, we also have $x^* \leq p$. □

The result above is Theorem 1.7 in König [8]. Of course, if C is a vector subspace and if τ is linear, we get the classical Hahn-Banach Theorem.

Mazur-Orlicz Theorem – convex version *Let $f : X \to \mathbb{R} \cup \{+\infty\}$ be convex, $A \subset X$ and $\beta : A \to \mathbb{R}$. If $0 \in \operatorname{cor}(\operatorname{dom} f - \operatorname{conv} A)$, then the following two statements are equivalent :*

(1) *There exist $x^* \in X^*$ and $r \in \mathbb{R}$ such that*

$$\beta(a) \leq x^*(a) + r, \text{ for all } a \in A, \quad \text{and} \quad x^* + r \leq f\,;$$

(2) *There exists $\gamma : \operatorname{conv} A \to \mathbb{R}$ such that*

$$\sum_{k=1}^{n} \lambda_k \beta(a_k) \leq \gamma(\sum_{k=1}^{n} \lambda_k a_k) \leq f(\sum_{k=1}^{n} \lambda_k a_k),$$

whenever $\{a_1, \ldots, a_n\} \subset A$, $\lambda_1 \geq 0, \ldots, \lambda_n \geq 0$, $\sum_{k=1}^{n} \lambda_k = 1$.

Proof. Clearly, (1) implies (2). Conversely, let g be the convex hull of $-\beta$, namely : if $x \notin \operatorname{conv} A$, $g(x) := +\infty$, while if $x \in \operatorname{conv} A$,

$$g(x) := \inf\{ -\sum_{k=1}^{n} \lambda_k \beta(a_k) \mid x = \sum_{k=1}^{n} \lambda_k a_k,\ a_k \in A, \lambda_k \geq 0, \sum_{k=1}^{n} \lambda_k = 1 \}.$$

We derive from (2) that $g(x) \geq -\gamma(x)$ for every $x \in \operatorname{conv} A$, hence g takes its values in $\mathbb{R} \cup \{+\infty\}$ and $\operatorname{dom} g = \operatorname{conv} A$; moreover, $-g \leq f$. We then deduce from Theorem 2 that $S(f, g)$ is non-empty, which is equivalent to statement (1). □

In the classical Mazur-Orlicz Theorem, f is sublinear (with finite values), so we can take $r = 0$ in statement (1) and $\gamma(x) = f(x)$ in statement (2). See also Sikorski [18] and Pták [13] for other simple proofs of Mazur-Orlicz's theorem starting from Banach's theorem. Note that the above theorem immediately yields Theorem 2: if g is as in Theorem 2, put $A = \operatorname{dom} g$ and $\beta = -g|A$.

Fenchel Theorem – algebraic version *Let* $f, g : X \to \mathbb{R} \cup \{+\infty\}$ *be convex. If* $0 \in \operatorname{cor}(\operatorname{dom} f - \operatorname{dom} g)$, *then there exists* $x^* \in X^*$ *such that*

$$\inf(f + g) = \inf(f - x^*) + \inf(g + x^*).$$

Proof. We always have $\inf(f+g) \geq \inf(f - x^*) + \inf(g + x^*)$. If $\inf(f+g) = -\infty$, the result is obvious; otherwise, it suffices to invoke Theorem 2' with $\varepsilon = 0$. □

The Fenchel Duality Theorem corresponds to the case $X = \mathbb{R}^n$. On the other hand, when X is a TVS and f is continuous at some point of its domain, we obtain the theorem of Moreau [11] and Rockafellar [16].

2. Continuity of Convex Functions

From now on, X denotes a TVS. It follows from the corollary of Theorem 1 (Theorem 2, resp.) that the continuity of f ($f +_e g$, resp.) on the core of its domain is a sufficient condition for the set $S(f)$ ($S(f, g)$, resp.) to be non-empty and equicontinuous. In general this is also a necessary condition: if $f : X \to \mathbb{R} \cup \{+\infty\}$ is lsc convex and if $S(f)$ is non-empty and equicontinuous, then f is continuous at 0 (see Moreau [11, Proposition 8.d]). It is therefore interesting to determine those TVS for which the continuity of certain classes of convex functions is automatic.

For example, for the finest locally convex topology, any convex function is continuous on the core of its domain: this topology is such that the core of any convex set is equal to its interior. Let us consider the following less drastic properties:

(T) For any closed convex set $C \subset X$, $\operatorname{cor}(C) = \operatorname{int}(C)$.

(T+) For any closed convex sets $C, D \subset X$, $\operatorname{cor}(C + D) = \operatorname{int}(C + D)$.

Property (T) is well known: a Hausdorff locally convex TVS verifying (T) is said to be *tonnelé* (barrelled).

Proposition 1 *Let X be a TVS. Then:*

(1) *X verifies* (T) *if and only if every lsc convex function on X is continuous on the core of its domain.*

(2) *X verifies* (T+) *if and only if every epi-sum of lsc convex functions on X is continuous on the core of its domain.*

Proof. Let us show (2) for example. Suppose first that X verifies (T+). If the core of the domain of $f +_e g$ is empty, there is nothing to prove. If $f +_e g$ takes the value $-\infty$ at some point of this core, it is equal to $-\infty$ everywhere on it and so continuous. Otherwise we may assume that 0 lies in $\mathrm{cor}(\mathrm{dom}\, f + \mathrm{dom}\, g)$ and that $f +_e g$ is finite at 0. We have to show that $f +_e g$ is continuous at 0. We first easily see that 0 lies in $\mathrm{cor}(f^{\leq r} + g^{\leq r})$ for some $r \in \mathbb{R}$, where $f^{\leq r} := \{x \in X \mid f(x) \leq r\}$ and $g^{\leq r} := \{x \in X \mid g(x) \leq r\}$ are closed convex sets. We therefore derive from (T+) that 0 belongs to $\mathrm{int}(f^{\leq r} + g^{\leq r})$. Now, the convex function $f +_e g$ being bounded above on this neighborhood of 0, we conclude that it is continuous at that point.

The converse is evident for property (T+) precisely expresses that the epi-sum of indicator functions of closed convex sets is continuous on the core of its domain. □

The next proposition provides examples of spaces enjoying these properties:

Proposition 2 (1) *Every Baire TVS verifies* (T).

(2) *Every metrizable complete TVS verifies* (T+).

Proof. (1) is a classical result. We briefly show (2) in the particular case of Fréchet spaces, the adaptations for the non locally convex case being left to the reader. As usual, we may assume $0 \in \mathrm{cor}(C + D)$ and $0 \in C \cap D$. We must show that 0 belongs to $\mathrm{int}(C + D)$. Denote by (U_n) a countable basis of closed convex neighborhoods of 0 such that $U_{n+1} \subset U_n$. From Baire's theorem, we rapidly obtain that 0 belongs to $\mathrm{int}(\overline{C \cap U_n + D \cap U_n})$ for every n. Let then (U_{k_n}) be a subsequence of (U_n) such that

$$2U_{k_n} \subset C \cap U_n + D \cap U_n + U_{k_{n+1}},$$

from which follows

$$U_{k_1} \subset \sum_{i=1}^{n} \frac{1}{2^i}(C \cap U_i + D \cap U_i) + \frac{1}{2^n} U_{k_{n+1}}.$$

Any point x_0 in U_{k_1} can therefore be written as

$$x_0 = \sum_{i=1}^{\infty} \frac{1}{2^i}(c_i + d_i)$$

where $c_i \in C \cap U_i$ and $d_i \in D \cap U_i$. Since the U_i's are convex, we have

$$\sum_{i=p}^{q} \frac{c_i}{2^i} \in U_p,$$

so, from the completeness of X we derive that the point

$$c := \sum_{i=1}^{\infty} \frac{c_i}{2^i}$$

exists and belongs to U_1. In the same manner we have

$$d := \sum_{i=1}^{\infty} \frac{d_i}{2^i} \in U_1.$$

Now, C and D being closed convex, we also have $c \in C$ and $d \in D$. Finally, $x_0 = c + d$ belongs to $C \cap U_1 + D \cap U_1$. We thus have shown that U_{k_1} is contained in $C \cap U_1 + D \cap U_1$, which proves that 0 belongs to $\operatorname{int}(C + D)$. □

The above proof is an adaptation of the original proof of Banach's open mapping theorem.

Before concluding, let us show how the theorem of Banach-Steinhaus and the closed graph and open mapping theorems of Banach can be derived from Propositions 1 and 2.

Banach-Steinhaus Theorem *Let X be a TVS verifying* (T), *Y be a Hausdorff LCTVS, and $\mathcal{H}$ be a set of continuous linear mappings from X into Y. If for every $x \in X$ and every continuous seminorm p on Y we have*

$$\sup_{h \in \mathcal{H}} p\,(h(x)) < \infty,$$

then $\mathcal{H}$ is equicontinuous.

Proof. For every continuous seminorm p on Y, the function $f : X \to \mathbb{R} \cup \{+\infty\}$ given by

$$f(x) := \sup_{h \in \mathcal{H}} p\,(h(x))$$

is everywhere finite, lsc, and convex, hence, by Proposition 1 (1), it is everywhere continuous, which exactly means that $\mathcal{H}$ is equicontinuous. □

We recall that a relation $A \subset X \times Y$ between two topological spaces X and Y is said to be *lower semicontinuous* (LSC) at $x_0 \in \operatorname{dom} A := \{ x \in X \mid Ax \neq \emptyset \}$ if for every open set $U \subset Y$ which meets Ax_0 the set $A^{-1}(U) := \{ x \in X \mid Ax \cap U \neq \emptyset \}$ is a neighborhood of x_0. It is clear that A is LSC at every point of $\operatorname{dom} A$ if and only if for every open set $U \subset Y$, the set $A^{-1}(U)$ is open in X.

Example : Let $T : X \to Y$ be a mapping. Then T is continuous at $x_0 \in X$ if and only if T, considered as a relation $T \subset X \times Y$, is LSC at x_0; T is open if and only if T^{-1}, considered as a relation $T^{-1} \subset Y \times X$, is LSC at every point of $\operatorname{dom} T^{-1} = \operatorname{Im} T$.

Theorem 3 *Let X, Y be Hausdorff LCTVS such that $X \times Y$ verifies* (T+).

(1) *If $h : X \times Y \to \mathbb{R} \cup \{+\infty\}$ is lsc convex, then the marginal function $\varphi : x \in X \mapsto \inf_{y \in Y} h(x, y)$ is continuous on the core of its domain.*

(2) *If $A \subset X \times Y$ is a relation with closed convex graph, then A is LSC on the core of its domain.*

Proof. (1) By Proposition 1 (2), the epi-sum of h and of the indicator function ψ of the closed convex set $\{0\} \times Y$ is continuous on the core of its domain. But $\varphi(x) = (h +_e \psi)(x, 0)$, so φ is continuous at every point x such that $(x, 0) \in \operatorname{cor}(\operatorname{dom}(h +_e \psi))$, that is, at every point of $\operatorname{cor}(\operatorname{dom} \varphi)$.

(2) Let $x_0 \in \operatorname{cor}(\operatorname{dom} A)$ and let $y_0 \in U \cap Ax_0$ where U is open in Y. We must show that x_0 belongs to $\operatorname{int}(A^{-1}(U))$. We may assume that $x_0 = 0$ and $y_0 = 0$. Let $V \subset U$ be a closed convex neighborhood of 0 in Y and let h be the indicator function of the closed convex set $A \cap (X \times V)$. By (1), the marginal function φ is continuous on $\operatorname{cor}(\operatorname{dom} \varphi)$. But $\operatorname{dom} \varphi = A^{-1}(V)$ and it is immediat that 0 belongs to $\operatorname{cor}(A^{-1}(V))$. Whence 0 belongs to $\operatorname{int}(A^{-1}(V)) \subset \operatorname{int}(A^{-1}(U))$. □

When X and Y are Banach spaces, the above theorem is due to Robinson [15]; see also Jameson [7], Ursescu [20], Borwein [3]. For a converse of (2), see Ricceri [14].

Banach Theorems *Let X and Y be Fréchet spaces, $T : X \to Y$ be a linear mapping with closed graph. Then T is continuous and, if it is onto, it is open.*

Proof. By the preceding theorem applied to $T \subset X \times Y$, the *relation* T is LSC on $\operatorname{cor}(\operatorname{dom} T) = X$, which amounts to saying that the *mapping* T

is continuous. By the same theorem applied to $T^{-1} \subset Y \times X$, the relation T^{-1} is LSC on $\operatorname{cor}(\operatorname{dom} T^{-1}) = Y$, hence T is open. $\square$

Epilogue

By combining Proposition 1 with the algebraic theorems of Section 1 we immediatly obtain topological versions of these theorems. For instance:

Theorem 1 – topological version *If X verifies* (T), *then for any lsc convex function $f : X \to \mathbb{R} \cup \{+\infty\}$ such that $f(0) \geq 0$ and $0 \in \operatorname{cor}(\operatorname{dom} f)$ the set*

$$S(f) = \{ x^* \in X^* \mid x^* \leq f \}$$

is non-empty and equicontinuous.

Theorem 2 – topological version *If X verifie* (T+), *then for any lsc convex functions $f, g : X \to \mathbb{R} \cup \{+\infty\}$ such that $-g \leq f$ and $0 \in$* $\operatorname{cor}(\operatorname{dom} f - \operatorname{dom} g)$ *the set*

$$S(f, g) = \{ x^* \in X^* \mid -g \leq x^* + r \leq f \quad \textit{for some } r \in \mathbb{R} \}$$

is non-empty and equicontinuous.

Fenchel Theorem – topological version *If X verifies* (T+), *then for any lsc convex functions $f, g : X \to \mathbb{R} \cup \{+\infty\}$ such that* $0 \in \operatorname{cor}(\operatorname{dom} f - \operatorname{dom} g)$ *there exists $x' \in X'$ such that*

$$\inf(f + g) = \inf(f - x') + \inf(g + x').$$

For X a Banach space, the above theorem is due to Attouch-Brezis [1] (see also Borwein [4, p. 421]); for X a Fréchet space, it is proved in Rodrigues-Simons [17].

References

1. H. Attouch and H. Brezis, *Duality for the sum of convex functions in general Banach spaces,* in J. Barroso (ed.), Aspects of Mathematics and its Applications, Elsevier Science Publishers, Amsterdam, 1986, 125-133.
2. S. Banach, *Sur les fonctionnelles linéaires II,* Studia Math. **1** (1929), 223-239.
3. J. Borwein, *Convex relations in analysis and optimization,* in S. Schaible and T. Ziemba (eds.), Generalized Concavity in Optimization and Economics, Academic Press, New York, 1981, 335-376.
4. J. Borwein, *Adjoint process duality,* Math. Oper. Res. **8** (1983), 403-434.
5. G. Buskes, *The Hahn-Banach theorem surveyed,* Dissertationes Math. **327** (1993), 49

6. R.B. Holmes, *Geometric functional analysis and its applications,* Graduate texts in mathematics, v. 24, Springer-Verlag, New-York, 1975.
7. G.J.O. Jameson, *Convex series,* Proc. Camb. Phil. Soc. **72** (1972) , 37-47.
8. H König, *On some basic theorems in convex analysis,* in B. Korte (ed.), Modern Applied Mathematics – Optimization and Operations Research, North-Holland, Amsterdam, 1982, 107-144.
9. J. Łoś and and C. Ryll-Nardzewski, *On the application of Tychonoff's theorem in mathematical proofs,* Fund. Math. **38** (1951), 233-237.
10. W.A. Luxemburg, *Reduced powers of the real number system and equivalents of the Hahn-Banach extension theorem,* in W.A.J. Luxemburg (ed.), Applications of model theory to algebra, analysis and probability, Holt, Rinehart and Winston, New York, 1969, 123-137.
11. J.-J. Moreau, *Fonctionnelles convexes,* Séminaire "Équations aux dérivées partielles", Collège de France, Paris, 1966.
12. D. Pincus, *Independence of the prime ideal theorem from the Hahn-Banach theorem,* Bull. Amer. Math. Soc. **78** (1972), 766-769.
13. V. Pták, *On a theorem of Mazur and Orlicz,* Studia Math. **15** (1956), 365-366.
14. B. Ricceri, *Remarks on multifunctions with convex graph,* Arch. Math. **52** (1989), 519-520.
15. S.M. Robinson, *Regularity and stability for convex multivalued functions,* Math. Oper. Res. **1** (1976), 130-143.
16. R.T. Rockafellar, *Extension of Fenchel's duality theorem for convex functions,* Duke Math. J. **33** (1966), 81-89.
17. B. Rodrigues and S. Simons, *Conjugate functions and subdifferentials in nonnormed situations for operators with complete graphs,* Nonlinear Analysis **12** (1988), 1069-1078.
18. R. Sikorski, *On a theorem of Mazur and Orlicz,* Studia Math. **13** (1953), 180-182.
19. M. Théra, *Subdifferential calculus for convex operators,* J. Math. Anal. Appl. **80** (1981), 78-91.
20. C. Ursescu, *Multifunctions with convex closed graph,* Czechoslovak Math. J. **25** (1975), 438-441.
21. J. Vangeldère, *Frank separation of two convex sets and the Hahn-Banach theorem,* J. Math. Anal. Appl. **60** (1977), 36-46.

TWO FUNCTIONS GENERALIZATION OF HORVATH'S MINIMAX THEOREM

BOR-LUH LIN AND FENG-SHUO YU
Department of Mathematics
University of Iowa
Iowa City, IA 52242, U.S.A.
e-mail: bllin@math.uiowa.edu

Recently, by following the method of Joó [4] , Horvath [3] gave a general result in the theory of minimax theorem, which require neither Hahn-Banach theorem nor Brouwer's fixed point theorem. Horvath's minimax theorem has been generalized to non-compact case by Kim in [5]. The main purpose of this note is to give a two functions version of Horvath's minimax theorem and consequently, we obtain a genelization of Sion-von Neumann's minimax theorem [7]. Also, by using the main result and the idea of Ha in [2], Lin-Quan in [6] and Kim in [5], we give a two functions, non-compact version of our main result.

Through out this note, we assume all topological spaces are Hausdorff topological spaces. Let X be a convex set of a vector space and f be a real-valued function on X. Then f is said to be quasi-convex on X if the set $\{x \in X : f(x) \leq \lambda\}$ is convex for each $\lambda \in R$. Similarly, f is quasi-concave if $\{x \in X : f(x) \geq \lambda\}$ is convex for each $\lambda \in R$. Let X be a non-empty set in a topological space, Y a non-empty convex set in a vector space and $f : X \times Y \to R$ be a real-valued function. Then f is said to be lower semicontinuous on the line segment of Y if for each $x \in X$ and y_1, $y_2 \in Y$, the function

$$t \to f(x, ty_1 + (1-t)y_2)$$

is a lower semicontinuous function of t on $[0, 1]$. If f is a lower semicontinuus function on Y, then f is clearly lower semicontinuous on the line segment of Y.

Theorem 1. *Let X be a non-empty compact topological space and let Y be a non-empty convex set. Let f, g be real-valued functions defined on $X \times Y$*

B. Ricceri and S. Simons (eds.), Minimax Theory and Applications, 147–156.

with the following properties:

(0) $f(x,y) \leq g(x,y)$ *for all* $(x,y) \in X \times Y$;

(i) $f(\cdot,y)$ *and* $g(\cdot,y)$ *are upper semicontinuous (u.s.c.) on* X *for each* $y \in Y$;

(ii) For any y_0, $y_1 \in Y$,

$$g(x,(1-t)y_0 + ty_1) \leq \max\{g(x,y_0), f(x,y_1)\}$$

for all $x \in X$, *for all* $t \in [0,1]$;

(iii) f *is lower semicontinuous on the line segment of* Y;

(iv) For any $y_1, \ldots, y_n \in Y$, *any* $\lambda \in R$, *the set* $\cap_{i=1}^n \{x \in X : g(x,y_i) \geq \lambda\}$ *is either connected or empty;*

then, for any $\lambda \in R$, *we have the the following alternative:*

Either there exist $x_0 \in X$ *such that* $g(x_0,y) \geq \lambda$ *for all* $y \in Y$, *or there exists* $y_0 \in Y$ *such that* $f(x,y_0) < \lambda$ *for all* $x \in X$.

Proof. For each $\lambda \in R$, $y \in Y$, let

$$U_g(y) = \{x \in X : g(x,y) \geq \lambda\},$$

$$U_f(y) = \{x \in X : f(x,y) \geq \lambda\}.$$

Fixed $\lambda \in R$. If for some $y_0 \in Y$, $U_f(y_0) = \emptyset$, then $f(x,y_0) < \lambda$ for all $x \in X$. Thus, we may assume that

$$U_f(y) \neq \emptyset. \tag{1}$$

for all $y \in Y$.

We are going to show that

$$\bigcap_{y \in Y} U_g(y) \neq \emptyset.$$

Since X is a compact topological space, so we need only to show that the family $\{U_g(y)\}$ has the finite intersection property, that is,

$$\bigcap_{y \in F} U_g(y) \neq \emptyset \quad \text{for all finite subsets } F \subset Y.$$

If card(F)=1. By (1), $U_f(y) \neq \emptyset$ for all $y \in Y$. From condition (0), $U_g(y) \supset U_f(y)$, so $U_g(y) \neq \emptyset$ for all $y \in Y$.

Let $F = \{y_0, y_1\} \subset Y$, $y_0 \neq y_1$. We want to show that

$$U_g(y_0) \bigcap U_g(y_1) \neq \emptyset.$$

By condition (0) and

$$U_g(y_0) \bigcap U_f(y_1) = \bigcap_{\epsilon>0} \left(U_g(y_0, \lambda - \epsilon) \bigcap U_f(y_1, \lambda - \epsilon)\right),$$

it is sufficent to show that for any $\epsilon > 0$,

$$U_g(y_0, \lambda - \epsilon) \bigcap U_f(y_1, \lambda - \epsilon) \neq \emptyset. \tag{2}$$

Fixed $\epsilon > 0$, for each $t \in [0, 1]$, consider the following sets

$$\begin{aligned} S_g(t) :&= U_g((1-t)y_0 + ty_1, \lambda - \epsilon) \\ &= \{x \in X : g(x, (1-t)y_0 + ty_1) \geq \lambda - \epsilon\}, \end{aligned}$$

and

$$\begin{aligned} S_f(t) :&= U_f((1-t)y_0 + ty_1, \lambda - \epsilon) \\ &= \{x \in X : f(x, (1-t)y_0 + ty_1) \geq \lambda - \epsilon\}. \end{aligned}$$

Notice that, for all $t \in [0, 1]$, $S_g(t)$ is a nonempty, compact connected subset of X by (1) and condition (i), (iv) and $S_f(t)$ is nonempty closed subset of X by (1) and condition (i).

Let

$$A_0 = \{t \in [0, 1] : S_f(t) \subset S_g(0)\},$$

and

$$A_1 = \{t \in [0, 1] : S_f(t) \subset S_f(1)\}.$$

If $S_g(0) \cap S_f(1)$ were empty set, then

(α) $A_0 \cap A_1 = \emptyset$.

This is clear by the construction of these two sets.

(β) $A_0 \cup A_1 = [0, 1]$.

Indeed, by (ii), we have, for all $x \in X$, for all $t \in [0, 1]$,

$$g(x, (1-t)y_0 + ty_1) \leq \max\{g(x, y_0), f(x, y_1)\}.$$

For each $t \in [0, 1]$, let $x \in S_f(t)$, so $x \in S_g(t)$ by condition (0), then $g(x, (1-t)y_0 + ty_1) \geq \lambda - \epsilon$, thus $g(x, y_0) \geq \lambda - \epsilon$ or $f(x, y_1) \geq \lambda - \epsilon$ by condition (ii). This implies that $x \in S_g(0)$ or $x \in S_f(1)$, hence $S_g(t) \subset S_g(0) \cup S_f(1)$.

Since $S_g(t)$ is nonempty and connected, $S_g(0)$ and $S_f(1)$ are nonempty closed sets and $S_g(0) \cap S_f(1) = \emptyset$, so either $S_g(t) \subset S_g(0)$ or $S_g(t) \subset S_f(1)$

and hence $S_f(t) \subset S_g(0)$ or $S_f(t) \subset S_f(1)$. It follows that $t \in A_0$ or $t \in A_1$ and hence $A_0 \cup A_1 = [0,1]$.

(γ) A_0, A_1 are closed subsets of [0,1].

Let (t_n) be a sequence in A_0 with $t_n \to t_0 \in [0,1]$. Let $x \in S_f(t_0)$. Then

$$f(x,(1-t_0)y_0 + t_0 y_1) \geq \lambda - \epsilon. \tag{3}$$

From (iii), since f is lower semicontinuous on the line segment of Y, so the function F defined by $F(t) = f(x,(1-t)y_0+ty_1)$ is lower semicontinuous on [0,1], thus $\liminf_n F(t_n) \geq F(t_0)$. It follows that there exists $m \in N$ such that $F(t_m) \geq F(t_0)$. That is, $f(x,(1-t_m)y_0 + t_m y_1) \geq f(x,(1-t_0)y_0 + t_0 y_1)$. By (3), this implies that $x \in S_f(t_m)$. Since $t_m \in A_0$, thus $x \in S_g(0)$, therefore, $S_f(t_0) \subset S_g(0)$, it follows that $t_0 \in A_0$, so A_0 is closed. Similarly, we can show that A_1 is closed.

From (α), (β), (γ), we get a contradiction about the connectedness of [0,1], since A_0 and A_1 are nonempty sets ($0 \in A_0$ and $1 \in A_1$). Therefore, $S_g(0) \cap S_f(1) \neq \emptyset$. Hence, (2) holds.

Now, suppose that $\cap_{y\in F} U_g(y) \neq \emptyset$ for all $F \subset Y$, $2 \leq card(F) \leq k$.

Choose $\{y_0,\ldots,y_k\} \subset Y$, $y_i \neq y_j$ if $i \neq j$. From the above argument, in order to show that $\cap_{i=0}^k U_g(y_i) \neq \emptyset$, it is sufficent to show, for each $\epsilon > 0$, that

$$U_g(y_0,\lambda-\epsilon)\bigcap U_f(y_1,\lambda-\epsilon)\bigcap (\cap_{i=2}^k U_g(y_i,\lambda-\epsilon)) \neq \emptyset. \tag{4}$$

Let $\epsilon > 0$ and $H = \cap_{i=2}^k U_g(y_i,\lambda-\epsilon)$. Notice that, by induction hypothesis, since $\cap_{\epsilon>0}[H \cap U_g(y,\lambda-\epsilon)] = (\cap_{i=2}^k U_g(y_i)) \cap U_g(y) \neq \emptyset$ for all $y \in Y$, $H \cap U_g(y,\lambda-\epsilon) \neq \emptyset$ for all $y \in Y$

In particular,

$$H \cap S_g(t) \neq \emptyset \text{ for all } t \in [0,1].$$

Also, by (iv), $H \cap S_g(t)$ is connected for all $t \in [0,1]$.

For $t \in [0,1]$, Let

$$B_0 = \{t \in [0,1] : H \cap S_f(t) \subset H \cap S_g(0)\},$$

and

$$B_1 = \{t \in [0,1] : H \cap S_f(t) \subset H \cap S_f(1)\}.$$

If $(H\cap S_g(0))\cap(H\cap S_f(1))$ were empty set, then by the same argument as above, we would have

(α') $B_0 \cap B_1 = \emptyset$.

(β') $B_0 \cup B_1 = [0,1]$.
(γ') B_0 and B_1 are colsed subsets of [0,1].

This again contradicts the connectedness of $[0,1]$ since B_0 and B_1 are not empty sets. Therefore, (4) holds and hence $\cap_{i=0}^{k} U_g(y_i) \neq \emptyset$.

□

Remark 1. The condition (iv) of theorem 1 is a weakening introducted by Geraghty and Lin in [1] of the condition first introduced by Terkelson in [8].

Theorem 2. *Let X be a non-empty compact topological space and let Y be a non-empty convex set. Let f, g be real-valued functions defined on $X \times Y$ with the following properties:*
(0) $f(x,y) \leq g(x,y)$ for all $(x,y) \in X \times Y$;
(i) $f(\cdot,y)$ and $g(\cdot,y)$ are u.s.c. on X for each $y \in Y$;
(ii) For any y_0, $y_1 \in Y$,

$$g(x,(1-t)y_0 + ty_1) \leq \max\{g(x,y_0), f(x,y_1)\}$$

for all $x \in X$, for all $t \in [0,1]$;
(iii) f is lower semicontinuous on the line segment of Y;
(iv) For any $y_1, \ldots, y_n \in Y$, any $\lambda \in R$, the set $\cap_{i=1}^{n}\{x \in X : g(x,y_i) \geq \lambda\}$ is either connected or empty;
then,

$$\inf_{y \in Y} \sup_{x \in X} f(x,y) \leq \sup_{x \in X} \inf_{y \in Y} g(x,y).$$

Proof. Let $\lambda \in R$ with

$$\sup_{x \in X} \inf_{y \in Y} g(x,y) < \lambda.$$

Thus, there exists $y_0 \in Y$ such that $g(x,y_0) < \lambda$ for all $x \in X$. By Theorem 1, there exists $y' \in Y$ such that $f(x,y') < \lambda$ for all $x \in X$, this implies that $\inf_{y \in Y} \sup_{x \in X} f(x,y) < \lambda$. It follows that

$$\inf_{y \in Y} \sup_{x \in X} f(x,y) \leq \sup_{x \in X} \inf_{y \in Y} g(x,y).$$

□

Corollary 1. *Let X be a non-empty compact convex set and let Y be a non-empty convex set. Let f, g be two real-valued functions defined on $X \times Y$ with the following properties:*
(0) $f(x,y) \leq g(x,y)$ for all $(x,y) \in X \times Y$;

(i) $f(\cdot, y)$ *and* $g(\cdot, y)$ *are u.s.c. on* X *for each* $y \in Y$*;*
(ii) For any y_0, $y_1 \in Y$,

$$g(x, (1-t)y_0 + ty_1) \leq \max\{g(x, y_0), f(x, y_1)\}$$

for all $x \in X$, *for all* $t \in [0, 1]$*;*
(iii) f *is lower semicontinuous on the line segment of* Y*;*
(iv) $g(\cdot, y)$ *is quasi-concave on* X *for each* $y \in Y$*;*
then,

$$\inf_{y \in Y} \sup_{x \in X} f(x, y) \leq \sup_{x \in X} \inf_{y \in Y} g(x, y).$$

Proof. Since $g(\cdot, y)$ is quasi-concave on X for each $y \in Y$, thus for any $y_1, \ldots, y_n \in Y$, any $\lambda \in R$, the set $\cap_{i=1}^n \{x \in X : g(x, y_i) \geq \lambda\}$ is either convex or empty and hence either connected or empty. By Theorem 2, we have the desired conclusion.

□

Corollary 2. *Let* X *be a non-empty compact topological space and let* Y *be a non-empty convex set. Let* f, g *be two real-valued functions defined on* $X \times Y$ *with the following properties:*
(0) $f(x, y) \leq g(x, y)$ *for all* $(x, y) \in X \times Y$*;*
(i) $f(\cdot, y)$ *and* $g(\cdot, y)$ *are u.s.c. on* X *for each* $y \in Y$*;*
(ii) For any y_0, $y_1 \in Y$,

$$g(x, (1-t)y_0 + ty_1) \leq \max\{g(x, y_0), f(x, y_1)\}$$

for all $x \in X$, *for all* $t \in [0, 1]$*;*
(iii) $f(x, \cdot)$ *is lower semicontinuous on* Y *for each* $x \in X$*;*
(iv) For any $y_1, \ldots, y_n \in Y$, *any* $\lambda \in R$, *the set* $\cap_{i=1}^n \{x | g(x, y_i) \geq \lambda\}$ *is either connected or empty;*
then,

$$\inf_{y \in Y} \sup_{x \in X} f(x, y) \leq \sup_{x \in X} \inf_{y \in Y} g(x, y).$$

Proof. By condition (iii), $f(x, \cdot)$ is lower semicontinuous on Y for each $x \in X$, so f is lower semicontinuous on line segment of Y. Apply Theorem 2, we obtain the desired conclusion.

□

Corollary 3. *Let* X *be a non-empty compact convex set and let* Y *be a non-empty convex set. Let* f, g *be two real-valued functions defined on* $X \times Y$ *with the following properties:*
(0) $f(x, y) \leq g(x, y)$ *for all* $(x, y) \in X \times Y$*;*
(i) $f(\cdot, y)$ *and* $g(\cdot, y)$ *are u.s.c. on* X *for each* $y \in Y$*;*

(ii) For any y_0, $y_1 \in Y$,

$$g(x,(1-t)y_0+ty_1) \le \max\{g(x,y_0), f(x,y_1)\}$$

for all $x \in X$, *for all* $t \in [0,1]$;

(iii) $f(x,\cdot)$ *is lower semicontinuous on* Y *for each* $x \in X$;

(iv) $g(\cdot,y)$ *is quasi-concave on* X *for each* $y \in Y$;

then,

$$\inf_{y\in Y}\sup_{x\in X} f(x,y) \le \sup_{x\in X}\inf_{y\in Y} g(x,y).$$

Corollary 4 [3]. *Let* X *be a non-empty compact topological space and let* Y *be a non-empty convex set. Let* f *be real-valued functions defined on* $X \times Y$ *with the following properties:*

(i) $f(\cdot,y)$ *is u.s.c. for each* $y \in Y$;

(ii) $f(x,\cdot)$ *is quasi-convex on* Y *for each* $x \in X$;

(iii) f *is lower semicontinuous on the line segment of* Y;

(iv) For any $y_1,\ldots,y_n \in Y$, *any* $\lambda \in R$, *the set* $\cap_{i=1}^n\{x \in X : f(x,y_i) \ge \lambda\}$ *is either connected or empty;*

then,

$$\inf_{y\in Y}\sup_{x\in X} f(x,y) = \sup_{x\in X}\inf_{y\in Y} f(x,y).$$

Proof. Let $f = g$ in Theorem 2.

□

Corollary 5 [7]. *Let* X *be a non-empty compact convex set in a topological vector space and let* Y *be a non-empty convex set in a vector space. Let* f *be real-valued functions defined on* $X \times Y$ *with the following properties:*

(i) $f(\cdot,y)$ *is u.s.c. on* X *for each* $y \in Y$;

(ii) $f(x,\cdot)$ *is lower semicontinuous on* Y *for each* $x \in X$;

(iii) $f(\cdot,y)$ *is quasi-concave on* X *for each* $y \in Y$;

(iv) $f(x,\cdot)$ *is quasi-convex on* Y *for each* $x \in X$;

then,

$$\inf_{y\in Y}\sup_{x\in X} f(x,y) = \sup_{x\in X}\inf_{y\in Y} f(x,y).$$

As we mention in introduction, by using the main result and the idea of Ha in [2], Lin-Quan in [6] and Kim in [5], we give a two functions, non-compact version of our main result which extend the result of Kim [5].

Theorem 3. *Let X be a non-empty topological space and let Y be a non-empty convex set. Let f, g be two real-valued functions defined on $X \times Y$ with the following properties:*

(0) $f(x,y) \leq g(x,y)$ *for all* $(x,y) \in X \times Y$;

(i) $f(\cdot, y)$ *and* $g(\cdot, y)$ *are u.s.c. on* X *for each* $y \in Y$;

(ii) For any y_0, $y_1 \in Y$,

$$g(x, (1-t)y_0 + ty_1) \leq \max\{g(x, y_0), f(x, y_1)\}$$

for all $x \in X$, *for all* $t \in [0,1]$;

(iii) f *is lower semicontinuous on the line segment of* Y;

(iv) For each non-empty finite set $F \subset Y$, *there exists a non-empty compact set* C_F *in* X *such that for each* $\lambda \in R$ *and each finite set* G *in* $co(F)$,

$$\bigcap_{y \in G} \{x \in C_F : g(x,y) \geq \lambda\}$$

is either connected or empty, and for each $y \in co(F)$,

$$\sup_{x \notin C_F} g(x,y) \leq \min\{\sup_{x \in C_F} f(x,y), \sup_X \inf_Y g(x,y)\}.$$

Then,

$$\inf_{y \in Y} \sup_{x \in X} f(x,y) \leq \sup_{x \in X} \inf_{y \in Y} g(x,y).$$

Proof. Let $\beta = \sup_X \inf_Y g(x,y)$ and $\alpha = \inf_Y \sup_X f(x,y)$. If $\alpha > \beta$. Choose $\epsilon > 0$ such that $\alpha - \epsilon > \beta$. For each $y \in Y$, Let

$$A_g(y) = \{x \in X : g(x,y) \geq \alpha - \epsilon\},$$

and

$$A_f(y) = \{x \in X : f(x,y) \geq \alpha - \epsilon\}.$$

By condition (0), $A_f(y) \subset A_g(y)$. Also, by the definition of α, $A_f(y) \neq \emptyset$ for each $y \in Y$.

Notice that $A_g(y)$ is a non-empty compact set in X for each $y \in Y$. To see this, by condition (iv), for each $y \in Y$, there exists a non-empty compact set C_y of X such that

$$\alpha - \epsilon > \beta = \sup_X \inf_Y g(x,y) \geq \sup_{x \notin C_y} g(x,y).$$

Hence $\emptyset \neq A_f(y) \subset A_g(y) \subset C_y$.

Let $F = \{y_1, \ldots, y_n\}$ be any finite set in Y and C_F be the non-empty compact subset of X satisfying condition (iv). Now, consider the restriction $f|_{C_F \times co(F)}$ of f and $g|_{C_F \times co(F)}$ of g on $C_F \times co(F) \subset X \times Y$, respectively. Then the assumptions of Theorem 2 are clearly satisfied, so we have

$$\inf_{y \in co(F)} \sup_{x \in C_F} f(x,y) \leq \sup_{x \in C_F} \inf_{y \in co(F)} g(x,y).$$

Notice that by condition (iv), $\sup_{x \in C_F} f(x,y) = \sup_{x \in X} f(x,y)$. Therefore,

$$\begin{aligned}
\sup_{x \in C_F} \inf_{y \in co(F)} g(x,y) &\geq \inf_{y \in co(F)} \sup_{x \in C_F} f(x,y) \\
&= \inf_{y \in co(F)} \sup_{x \in X} f(x,y) \\
&\geq \inf_{y \in Y} \sup_{x \in X} f(x,y) \\
&\geq \alpha - \epsilon.
\end{aligned}$$

This implies that there exists $x_0 \in C_F$ such that $g(x_0, y) \geq \alpha - \epsilon$ for all $y \in co(F)$. In particular, $x_0 \in \cap_{i=1}^n A_g(y_i)$, so $\cap_{i=1}^n A_g(y_i) \neq \emptyset$. Since $\{A_g(y)\}_{y \in Y}$ is a family of non-empty compact subsets having the finite intersection property, hence $\cap_{y \in Y} A_g(y) \neq \emptyset$. Thus, there exists $x' \in X$ such that $g(x', y) \geq \alpha - \epsilon$ for all $y \in Y$ and this implies that $\sup_{x \in X} \inf_{y \in Y} g(x,y) \geq \alpha - \epsilon > \beta$. This contradicts the definition of β and hence $\alpha \leq \beta$. That is,

$$\inf_{y \in Y} \sup_{x \in X} f(x,y) \leq \sup_{x \in X} \inf_{y \in Y} g(x,y).$$

□

Corollary 6 [5]. *Let X be a non-empty topological space and let Y be a non-empty convex set. Let f be a real-valued functions defined on $X \times Y$ with the following properties:*

(i) $f(\cdot, y)$ is u.s.c. on X for each $y \in Y$;

(ii) $f(x, \cdot)$ is quasi-convex on Y for each $x \in X$;

(iii) f is lower semicontinuous on the line segment of Y;

(iv) For each non-empty finite set $F \subset Y$, there exists a non-empty compact set C_F in X such that for each $\lambda \in R$ and each finite set G in $co(F)$,

$$\bigcap_{y \in G} \{x \in C_F : f(x,y) \geq \lambda\}$$

is either connected or empty, and for each $y \in co(F)$,

$$\sup_{x \notin C_F} f(x,y) \leq \min\{\sup_{x \in C_F} f(x,y), \sup_X \inf_Y f(x,y)\}.$$

Then,

$$\inf_{y \in Y} \sup_{x \in X} f(x,y) = \sup_{x \in X} \inf_{y \in Y} f(x,y).$$

Proof. Let $f = g$ in Theorem 3. □

References

1. M. A. Geraghty and Bor-Luh Lin, *Topological minimax theorems,* Proc. Amer. Math. Soc. **91** (1984), 377-380.
2. Chung-Wei Ha, *A non-compact minimax theorem,* Pacific J. Math. **97** (1981), 115-117.
3. C. Horvath, *Qulques théorèms en théorie des Mini-Max,* C. R. Acad. Sci. Paris, Série I, **310** (1990), 269-272.
4. I. Joó, *A simple proof for von Neumann's minimax theorem,* Acta Sci. Math. **42** (1980), 91-94.
5. Won Kyu Kim, *A non-compact generalization of Horvath's intersection theorem,* Bull. Korean Math. Soc. **32** (1995), 153-162.
6. Bor-Luh Lin and Xiu-Chi Quan, *A noncompact topological minimax theorem,* J. Math. Anal. Appl. **161** (1991), 587-590.
7. M. Sion, *On general minimax theorem,* Pacific J. Math. **8** (1958), 171-176.
8. F. Terkelson, *Some minimax theorems,* Math. Scand. **31** (1972), 405-413.

SOME REMARKS ON A MINIMAX FORMULATION OF A VARIATIONAL INEQUALITY

GIANDOMENICO MASTROENI
Department of Mathematics
University of Pisa
Via Buonarroti 2
56127 Pisa, Italy
e-mail: mastroen@mail.dm.unipi.it

Abstract. Saddle point conditions of suitable functions are equivalent to particular classes of variational inequalities. Some existence theorems for variational inequalities, based on monotonicity assumptions on the operator F, allow to prove these saddle point conditions. Applications to the generalized complementarity problem and extensions to vector variational inequalities are considered.

1. Saddle Point Conditions and Variational Inequalities

Consider the variational inequality:

$$\text{find } y^* \in K \text{ s.t. } \langle F(y^*), x - y^* \rangle \geq f(y^*) - f(x), \quad \forall x \in K, \qquad VI(f, K)$$

where $F : K \longrightarrow H$,$f : K \longrightarrow R$, H is a real Hilbert space, K is a nonempty convex subset of H and $\langle \cdot, \cdot \rangle$ is the inner product in H.

Following the approach considered in [8], let us introduce the function $\phi : K \times K \longrightarrow R$,

$$\phi(x, y) := \langle F(y), y - x \rangle + f(y) - f(x).$$

Suppose that ϕ has a saddle point (x^*, y^*) on $K \times K$ that is

$$\phi(x, y^*) \leq \phi(x^*, y^*) \leq \phi(x^*, y), \quad \forall (x, y) \in K \times K.$$

If we compute the previous inequalities for $(x, y) := (y^*, x^*)$ we obtain $\phi(x^*, y^*) = 0$. Therefore the existence of a saddle point of ϕ guarantees

B. Ricceri and S. Simons (eds.). Minimax Theory and Applications. 157–166.

that

$$\phi(x, y^*) \leq 0, \quad \forall x \in K, \tag{1}$$

$$\phi(x^*, y) \geq 0, \quad \forall y \in K. \tag{2}$$

The inequality (1) states that y^* is a solution of $VI(f, K)$.

We are interested in studying the relationships between $VI(f, K)$ and the problem which arises from inequality (2) that is

$$\text{find } x^* \in K \ \text{ s.t. } \langle F(y), y - x^* \rangle \geq f(x^*) - f(y), \quad \forall y \in K. \quad VI^*(f, K)$$

More precisely we want to investigate under which conditions the existence of a solution of $VI(f, K)$ guarantees the existence of a solution of $VI^*(f, K)$ and viceversa. We observe that if both problems have a solution, that is (1) and (2) hold, then the point (x^*, y^*) is a saddle point for $\phi(x, y)$ on $K \times K$. The previous cosiderations lead to state the following result:

Proposition 1.1 *Let* $\phi(x, y) := \langle F(y), y - x \rangle + f(y) - f(x)$.

1. $VI(f, K)$ *admits a solution* y^* *if and only if*

$$\inf_{y \in K} \sup_{x \in K} \phi(x, y) = 0$$

and the infimum is attained at y^**;*

2. $VI^*(f, K)$ *admits a solution* x^* *if and only if*

$$\sup_{x \in K} \inf_{y \in K} \phi(x, y) = 0$$

and the supremum is attained at x^**;*

3. y^* *and* x^* *are the respective solutions of the problems* $VI(f, K)$ *and* $VI^*(f, K)$ *if and only if* (x^*, y^*) *is a saddle point for* ϕ *on* $K \times K$.

Proof. *1.* Let $h(y) := \sup_{x \in K} \phi(x, y)$. Since $\phi(x, x) = 0, \ \forall x \in K$ then $h(y) \geq 0, \ \forall y \in K$. Moreover $y^* \in K$ is a solution of $VI(f, K)$ if and only if $h(y^*) = 0$. Therefore $0 = h(y^*) = \min_{y \in K} \sup_{x \in K} \phi(x, y)$.

2. Let $q(x) := \inf_{y \in K} \phi(x, y)$. It is immediate that $q(x) \leq 0, \forall x \in K$ and that $x^* \in K$ is a solution for $VI^*(f, K)$ if and only if $q(x^*) = 0$. Therefore $0 = q(x^*) = \max_{x \in K} \inf_{y \in K} \phi(x, y)$.

3. Let y^* and x^* the respective solutions of the problems $VI(f, K)$ and $VI^*(f, K)$; from *1* and *2* it follows that the point (x^*, y^*) is a saddle point for ϕ on $K \times K$.

Viceversa, let $(x^*, y^*) \in K \times K$ be a saddle point for ϕ on $K \times K$ that is:

$$\langle F(y^*), y^* - x \rangle + f(y^*) - f(x) \leq \langle F(y^*), y^* - x^* \rangle + f(y^*) - f(x^*) \leq$$
$$\langle F(y), y - x^* \rangle + f(y) - f(x^*) \quad \forall (x, y) \in K \times K.$$

Evaluating the previous inequalities at the point $(x, y) := (y^*, x^*)$ we obtain $\langle F(y^*), y^* - x^* \rangle + f(y^*) - f(x^*) = 0$ from which the thesis follows since the first inequality states that y^* is a solution of $VI(f, K)$,while the second that x^* is a solution of $VI^*(f, K)$.

2. Applications to the Classical Variational Inequality

In this section we will consider the particular case in which f is a constant function on the set K so that $VI(f, K)$ becomes the classical variational inequality

$$\text{find } y^* \in K \text{ s.t. } \langle F(y^*), x - y^* \rangle \geq 0, \quad \forall x \in K. \tag{VI}$$

We will denote by VI^* the problem associated to VI by means of the saddle point condition for the function $\phi(x, y) := \langle F(y), y - x \rangle$,

$$\text{find } x^* \in K \text{ s.t. } \langle F(y), y - x^* \rangle \geq 0, \quad \forall y \in K. \tag{VI^*}$$

Proposition 1.1 states that y^* and x^* are the respective solutions of the problems VI and VI^* if and only if (x^*, y^*) is a saddle point for the function ϕ on $K \times K$.

Let us recall some notations that will be used in what follows: int (M), $\partial(M)$ will denote the interior and the boundary of the set $M \subseteq H$, respectively. If $K \subseteq H$, int $_K(M)$, $\partial_K(M)$ denote the relative interior and relative boundary of M in K, $M^* := \{y \in H : \langle y, x \rangle \geq 0, \ \forall x \in M\}$ will denote the positive polar of the set M. Moreover we will say that the mapping F is pseudomonotone on K if:

$$\langle F(y), x - y \rangle \geq 0 \text{ implies } \langle F(x), x - y \rangle \geq 0 \quad \forall x, y \in K.$$

We will see how the main results concerning the existence of a solution of the variational inequality VI, based on monotonicity assumptions on the operator F, also allow to prove the existence of a saddle point for the function ϕ on $K \times K$.

Proposition 2.1 *[1] Let K be a closed convex subset in the real Hilbert space H. Let F be a pseudomonotone mapping from K into H which is continuous on finite-dimensional subspaces. Then $y^* \in K$ solves VI if and only if $x^* := y^*$ solves VI^*.*

Remark 2.1 If we assume that K is a bounded set in H, then the hypotheses of Proposition 2.1 are sufficient to guarantee the existence of a solution of both problems VI and VI^* (see [1]).

The next result states necessary and sufficient conditions for the existence of a solution of the problem VI under the hypothesis of pseudomonotonicity of the mapping F:

Theorem 2.1 *[1] Let K be a closed convex subset in the real Hilbert space H. Let F be a pseudomonotone mapping from K into H which is continuous on finite-dimensional subspaces. Then the following statements are equivalent:*

1. *There exists a solution y^* for VI;*
2. *There exist $u \in K$ and a constant $r > \|u\|$ such that $\langle y - u, F(y)\rangle \geq 0$, $\forall y \in K$ with $\|y\| = r$;*
3. *There exists $r > 0$ such that the set $\{y \in K : \|y\| \leq r\}$ is nonempty and such that, for each $y \in K$ with $\|y\| = r$, there exists $u \in K$ with $\|u\| < r$ and $\langle F(y), y - u\rangle \geq 0$.*
4. *There exists a nonempty closed, bounded, convex subset B of K with $\text{int}_K(B)$ that satisfies the following condition:*

$$\forall y \in \partial_K(B), \textit{there exists } u \in \text{ int }_K(B) \textit{ such that } \langle F(y), y - u\rangle \geq 0.$$

5. *There exists a convex set $E \subset H$ with $\text{int } E \neq \emptyset$ such that $\emptyset \neq K \cap E$ is bounded and, for each $y \in K \cap \partial(E)$, there exists $u \in K \cap \text{ int } (E)$ such that $\langle F(y), y - u\rangle \geq 0$.*

Remark 2.2 In the previous theorem it is possible to add the further statements, recalling Propositions 2.1 and 1.1:

6. There exists a solution y^* for VI^*;
7. The point (y^*, y^*) is a saddle point for $\phi(x, y) := \langle F(y), y - x\rangle$ on $K \times K$.

Therefore any of the conditions *2,3,4* and *5* of Theorem 2.1 not only guarantees the existence of a solution of VI but also turns out to be necessary and sufficient for the existence of a saddle point of the function ϕ on $K \times K$.

Another interesting result about the existence of solutions of VI, based on monotonicity assumptions on the operator F, is the following:

Proposition 2.2 *[1] Let K be a closed convex subset in the real Hilbert space H. Let F be a pseudomonotone mapping from K into H which is continuous on finite-dimensional subspaces. Suppose that there exists $u \in K$ such that $F(u) \in \text{ int } (K^*)$. Then there exists a solution for VI.*

Remark 2.3 Once again we observe that the hypotheses of the previous proposition are sufficient to guarantee the existence of a saddle point of the function ϕ on $K \times K$.

In the next theorem we summarize the results so far obtained about the existence of a saddle point of the function ϕ in the hypotheses of pseudo-monotonicity and continuity on finite-dimensional subspaces of the operator F.

Theorem 2.2 *Let K be a closed convex subset in the real Hilbert space H. Let F be a pseudomonotone mapping from K into H which is continuous on finite-dimensional subspaces. Then*

1. *If K is a bounded set in H then $\phi(x, y)$ admits a saddle point on $K \times K$.*
2. *Any of the conditions 1,2,3,4,5 of Theorem 2.1 is necessary and sufficient for $\phi(x, y)$ to admit a saddle point on $K \times K$.*
3. *If there exists $u \in K$ such that $F(u) \in \text{int } (K^*)$ then $\phi(x, y)$ admits a saddle point on $K \times K$.*

Proof. We preliminarily observe that the hypotheses guarantee that the solutions of both problems VI and VI^* coincide, as stated in Proposition 2.1.

1. It follows from Remark 2.1 and 3 of Proposition 1.1;
2. It follows from Theorem 2.1 recalling Remark 2.2 and 3 of Proposition 1.1;
3. It follows from Proposition 2.2 recalling 3 of Proposition 1.1.

3. Connections with Complementarity Problems

In this section we will consider the particular case in which the feasible set K is a closed convex cone in H with apex at the origin.

Remark 3.1 In [1, 7] it is proved that if K is a convex cone then y^* solves VI if and only if y^* is a solution of the generalized complementarity problem:

$$\text{find } y \in K \text{ s.t. } F(y) \in K^* \text{ and } \langle F(y), y \rangle = 0. \qquad (GCP)$$

It is possible to express GCP as a constrained extremum problem:

Proposition 3.1 *[2] $y^* \in K$ solves GCP if and only if y^* is a solution of the problem:*

$$\min \langle F(y), y \rangle \quad s.t. \ F(y) \in K^*, \ y \in K \qquad (3)$$

and $\langle F(y^*), y^* \rangle = 0$.

We will show that the problem VI^* is closely related to the Lagrangean Dual of the problem (3) defined by

$$\sup_{\lambda \in K} \inf_{y \in K} \langle F(y), y \rangle - \langle \lambda, F(y) \rangle. \qquad (4)$$

The following result holds:

Proposition 3.2 *Let K be a closed convex cone in the real Hilbert space H. Let F be a pseudomonotone mapping from K into H which is continuous on finite-dimensional subspaces and suppose that*

$$\text{there exists } y' \in K \text{ such that } F(y') \in \text{ int } (K^*). \tag{5}$$

Then y^* solves (3) if and only if y^* solves (4).

Proof. Let $y^* \in K$ be a solution of (3). For Proposition 2.2 VI has a solution. Since K is a cone, then GCP has a solution and therefore $\langle F(y^*), y^* \rangle = 0$.

The following equivalences allow to achieve the thesis:

y^* solves (3) and $\langle F(y^*), y^* \rangle = 0 \iff y^*$ solves $GCP \iff y^*$ solves $VI \iff y^*$ solves $VI^* \iff \sup_{\lambda \in K} \inf_{y \in K} \langle F(y), y \rangle - \langle \lambda, F(y) \rangle = 0$ and the maximum is attained at $\lambda := y^*$.

These equivalences follow from Proposition 3.1, Remark 3.1, Proposition 2.1 and *2* of Proposition 1.1, respectively.

Viceversa, let λ^* a solution of (4). Propositions 2.2 and 2.1 imply that the problem VI^* has a solution. By *2* of proposition 1.1 it follows that

$$\sup_{\lambda \in K} \inf_{y \in K} \langle F(y), y \rangle - \langle \lambda, F(y) \rangle = 0.$$

From the previous equivalences the thesis follows.

Remark 3.2 In the Proposition 3.2 it is possible to replace condition (5) with any of the conditions *2,3,4,5,* stated in Theorem 2.1.

4. Vector Variational Inequalities

The statements obtained in the previous part of the paper for scalar variational inequalities can be generalized, adding suitable assumptions, to vector variational inequalities introduced in [5, 4].

We recall the main definitions of vector optimization that we will use in the sequel.

From now on, K will denote a subset of R^n, C a pointed convex cone in R^p with apex at the origin $\{0\}$ of R^p and nonempty interior. We recall that a cone C is said pointed iff $C \cap (-C) = \emptyset$.

Definition 4.1 *Let $f : R^n \longrightarrow R^p$.*

1. *$y \in K$ is said a C-minimum point of f on the set K iff the following system is impossible:*

$$f(y) - f(x) \in C \setminus \{0\}, \quad x \in K.$$

2. *$y \in K$ is said a C-maximum point of f on the set K iff y is a C-minimum point for $-f$.*

Definition 4.2 *[10] Let $f : X \times Y \longrightarrow R^p$. A point $(x^*, y^*) \in X \times Y$ is said a C-saddle point of f with respect to $X \times Y$ iff*

1. *x^* is a C-minimum point of $f(\cdot, y^*)$ on X;*
2. *y^* is a C-maximum point of $f(x^*, \cdot)$ on Y.*

Consider the following vector variational inequalities introduced in [4].

$$\text{Find } y \in K \text{ s. t. } \langle F(y), y - x\rangle_p \notin C \setminus \{0\}, \quad \forall x \in K. \tag{6}$$

$$\text{Find } x \in K \text{ s. t. } \langle F(y), x - y\rangle_p \notin C \setminus \{0\}, \quad \forall y \in K. \tag{7}$$

where $F : R^n \longrightarrow R^{p \times n}$ and $\langle \cdot, \cdot \rangle_p$ denotes a vector of p scalar products in R^n.

Remark 4.1 If $p = 1$ and $C = R_+$, then (6) and (7) collapse to VI and VI^* respectively.

Following the line adopted in the scalar case it is possible to prove, under a suitable additional assumption, the equivalence between the existence of solutions of the problems (6) and (7) and the existence of a C-saddle point of the function $f(x, y) := \langle F(y), x - y\rangle_p$.

Proposition 4.1 *Let $f : K \times K \longrightarrow R^p, f(x, y) := \langle F(y), x - y\rangle_p$.*

y^* and x^* are the respective solutions of the problems (6) and (7) and $f(x^*, y^*) = 0$ iff (x^*, y^*) is a C-saddle point of $f(x, y)$ on $K \times K$ with $f(x^*, y^*) = 0$.

Proof. y^* and x^* are the respective solutions of the problems (6) and (7) iff

$$-f(x, y^*) \notin C \setminus \{0\}, \quad \forall x \in K$$

$$f(x^*, y) \notin C \setminus \{0\}, \quad \forall y \in K.$$

Since $f(x^*, y^*) = 0$, the previous relations are equivalent to

$$f(x^*, y^*) - f(x, y^*) \notin C \setminus \{0\}, \quad \forall x \in K \tag{8}$$

$$f(x^*, y) - f(x^*, y^*) \notin C \setminus \{0\}, \quad \forall y \in K. \tag{9}$$

Condition (8) states that x^* is a C-minimum point of $f(x,y)$ on K, while condition (9) that y^* is a C-maximum point of $f(x^*,\cdot)$ on K, that is (x^*,y^*) is a C-saddle point of $f(x,y)$ on $K \times K$.

Let us consider now an application of Proposition 4.1 in the particular case in which $C = \text{ int } (R^p_+)$ and the operator F is continuous and pseudomonotone. The following result has been proved in [4]:

Proposition 4.2 *Let $C := \text{ int } (R^p_+)$, K a convex set in R^n and F continuous and pseudomonotone on K. Then y is a solution of (6) iff it is a solution of (7).*

The next result is a direct consequence of the previous propositions:

Corollary 4.1 *Let $f : K \times K \longrightarrow R^p$, K a convex set in R^n, $f(x,y) := \langle F(y), x-y\rangle_p, F$ continuous and pseudomonotone on K, $C := \text{ int } R^p_+$. Then the following statements are equivalent:*

1. *y^* is a solution of (6);*
2. *y^* is a solution of (7);*
3. *(y^*,y^*) is a C-saddle point of $f(x,y)$ on $K \times K$.*

The existence of C-saddle points for a vector valued function has been analysed in [10] where a generalization of the classical results of K. Fan [3] for scalar functions, has been proven.

5. Further Developments

As we have observed, the minimax setting provides a natural field in which embedding variational inequalities in their more general formulations.

Recently F. Giannessi has proposed [6] the following generalized variational inequality:

$$\text{find } y^* \in A(K),\ F \in \mathcal{F}(y) \text{ s.t. } \langle F, x - y^*\rangle \geq 0, \quad \forall x \in K,$$

where $A : 2^{R^n} \longrightarrow 2^Y$, $\mathcal{F} : Y \longrightarrow 2^{R^n}$, $K \subseteq R^n$, $Y \subseteq R^n$.

A first existence theorem for the previous problem, in the particular case in which $\mathcal{F}$ is a single valued operator, has been proposed by B. Ricceri.

Theorem 5.1 *Let X and D two subsets in R^n. Put*

$$I := \{x \in X \ : \ D \subseteq \bigcup_{\lambda > 0} \lambda(x - X)\}.$$

Suppose that X is convex and compact and $F : R^n \longrightarrow R^n$ is a continuous function such that $\langle F(x), x\rangle = 0, \ \ \forall x \in X$ and

$$\sup_{y \in X} \langle F(x), y\rangle > 0, \quad \forall x \in X \setminus I.$$

Then there exists $x^* \in X$ such that

$$\langle F(x^*), y - x^*\rangle \geq 0 \quad \forall y \in D.$$

Remark 5.1 The previous theorem descends from a general result (Theorem 2.1 of [9]) stated for D-regular operators in a real vector space.

Taking into account the results of the previous sections it is possible to prove the following proposition in which X is supposed to be a closed convex cone.

Proposition 5.1 *Let X be a closed convex cone and D a subset in R^n. Put*

$$I := \{x \in X \ : \ D \subseteq \bigcup_{\lambda > 0} \lambda(X - x)\}$$

Suppose that

1. *$F : R^n \longrightarrow R^n$ is a continuous pseudomonotone function and*

$$\text{there exists } y' \in X \text{ such that } F(y') \in \text{ int } (X^*); \tag{10}$$

2. *$\sup_{y \in X} \langle F(x), x - y\rangle > 0, \quad \forall x \in X \setminus I$;*

then there exists $x^* \in X$ such that

$$\langle F(x^*), y - x^*\rangle \geq 0, \quad \forall y \in D.$$

Proof. From *1.* (see Proposition 2.2) it follows that there exists $x^* \in X$ such that

$$\langle F(x^*), y - x^*\rangle \geq 0, \quad \forall y \in X \tag{11}$$

(11) is equivalent to $\sup_{y \in X} \langle F(x^*), x^* - y\rangle \leq 0$ and, therefore the hypothesis *2.* implies that $x^* \in I$. Then $D \subseteq \bigcup_{\lambda > 0} \lambda(X - x^*)$ that is

$$D \subseteq \{z \in R^n \ : \ z = \lambda(x - x^*) \text{ for suitable } \lambda > 0, \ x \in X\}.$$

We recall that, since X is a closed convex cone, $\langle F(x^*), x^*\rangle = 0$ (see Remark 3.1). We obtain:

$$\begin{aligned} \inf_{y \in D} \langle F(x^*), y - x^*\rangle &\geq \inf_{\lambda > 0, \ x \in X} \langle F(x^*), \lambda(x - x^*) - x^*\rangle = \\ &= \inf_{\lambda > 0, \ x \in X} \lambda \langle F(x^*), x - x^*\rangle \geq 0. \end{aligned}$$

Remark 5.2 We observe that, in the previous proposition, it is possible to replace the condition (10) with any of the conditions 2,3,4,5 of Theorem 2.1.

References

1. R.W. Cottle and J.C. Yao, *Pseudomonotone Complementarity Problems in Hilbert Space,* Journal of Optimization Theory and Applications **75** (1992), 281-295.
2. R.W. Cottle, *Nonlinear Programs with positively bounded Jacobians,* SIAM Journal on Applied Mathematics **14** (1966), 147-158.
3. K. Fan, *A Minimax Inequality and Applications,* in "Inequalities III" O.Shisha,(eds.), Academic Press, New York-London, 1972.
4. F. Giannessi, *On Minty Variational Principle,* to appear in "New Trends in Mathematical Programming", Giannessi, Komlosi, Rapcsak (eds.);
5. F. Giannessi, *Theorems of the Alternative, Quadratic Programs and Complementarity Problems,* in "Variational Inequalities and Complementarity Problems",R.W. Cottle, F. Giannessi and J.L. Lions (eds.), Wiley, New York, 1980, 151-186.
6. F. Giannessi and A.A. Khan, *On a Separation Approach to Regularization for Generalized Variational and Quasi-Variational Inequalities,* To appear.
7. S. Karamardian, *Generalized Complementary Problem,* Journal of Optimization Theory and Applications **8** (1971), 161-167.
8. G. Mastroeni, *Some relations between Duality theory for extremum problems and Variational Inequalities,* Le Matematiche **49** (1994), 295-304.
9. B. Ricceri, *Existence Theorems for Nonlinear Problems,* Rendiconti Accademia Nazionale delle Scienze detta dei XL **11** (1987), 77-99.
10. T. Tanaka,*Generalized Quasiconvexities, Cone Saddle Points, and Minimax Theorem for Vector-Valued Functions,* Journal of Optimization Theory and Applications **81** (1994), 355-377.

NETWORK ANALYSIS

MICHAEL M. NEUMANN
Department of Mathematics and Statistics
Mississippi State University
Mississippi State, MS 39762, U.S.A.
e-mail: neumann@deanas.msstate.edu

AND

MARIA VICTORIA VELASCO
Departamento de Análisis Matemático
Universidad de Granada
18071 Granada, Spain

1. Introduction

In this paper, a uniform functional-analytic framework is developed to establish a variety of feasibility theorems on the existence of flows and potentials in an abstract setting. The approach is based on certain Hahn-Banach type theorems and leads to a general theory of measure-theoretic flavor. The existence of flows and potentials in this context is characterized in terms of simple conditions, which arise naturally from the theory of finite networks. These results not only subsume some of the basic theorems on classical networks, but also admit further applications, for instance to marginal problems in measure theory.

2. From Finite to Infinite Networks

In the classical theory of finite networks, a natural and rather general starting point is to consider a finite set of nodes, together with all their connecting directed arcs, and to assume that both the nodes and the arcs are endowed with certain lower and upper capacities. Thus, given a finite set S of nodes and the corresponding set $S \times S$ of all directed arcs on S, the basic setting is to consider two systems of real numbers $\lambda = (\lambda_j)_{j \in S}$ and $\mu = (\mu_j)_{j \in S}$ with $\lambda_j \leq \mu_j$ for all $j \in S$, to be interpreted as the

B. Ricceri and S. Simons (eds.), Minimax Theory and Applications, 167–189.

node capacities, as well as two systems of real numbers $\sigma = (\sigma_{jk})_{j,k\in S}$ and $\tau = (\tau_{jk})_{j,k\in S}$ with $\sigma_{jk} \le \tau_{jk}$ for all $j, k \in S$, to be interpreted as the arc capacities.

In the network corresponding to these data, a *flow* is then a system of real numbers $\nu = (\nu_{jk})_{j,k\in S}$ which satisfies

$$\sigma_{jk} \le \nu_{jk} \le \tau_{jk} \quad \text{and} \quad \lambda_j \le \sum_{k\in S} \nu_{jk} - \sum_{k\in S} \nu_{kj} \le \mu_j \qquad \text{for all } j, k \in S\,.$$

A related notion is that of a *circulation* which is a system of real numbers $\omega = (\omega_{jk})_{j,k\in S}$ with the property that

$$\sum_{k\in S} \omega_{jk} = \sum_{k\in S} \omega_{kj} \qquad \text{for all } j \in S\,.$$

Finally recall that a *potential* is defined as a system of real numbers $\pi = (\pi_j)_{j\in S}$ which satisfies $\sigma_{jk} \le \pi_k - \pi_j \le \tau_{jk}$ for all $j, k \in S$.

Large portions of the basic theory of networks center around the question of the existence of flows and potentials in this or closely related situations. For instance, a celebrated theorem, dating back to Gale and Hoffman, states that a flow exists precisely when

$$\sum_{j\in A} \lambda_j\,, - \sum_{k\in A^c} \mu_k \le \sum_{j\in A}\sum_{k\in A^c} \tau_{jk} - \sum_{k\in A^c}\sum_{j\in A} \sigma_{kj} \qquad \text{for all sets } A \subseteq S\,,$$

where $A^c := S\backslash A$ denotes the complement of A in S. Note that this result is an equivalent version of the fundamental theorem on maximal flows and minimal cuts due to Ford and Fulkerson. Moreover, it is well known that the existence of a potential is equivalent to the condition that

$$\sum_{j\in S}\sum_{k\in S} \omega_{jk}^{+}\sigma_{jk} \le \sum_{j\in S}\sum_{k\in S} \omega_{jk}^{-}\tau_{jk} \qquad \text{for every circulation } \omega = (\omega_{jk})_{j,k\in S}.$$

We refer to [2] and [20] for a comprehensive account of the classical theory.

To see how these results may be extended to the case of not necessarily finite networks, we observe that the classical theory is dominated by certain canonical set functions. More specifically, given two finite systems of real numbers $\pi = (\pi_j)_{j\in S}$ and $\nu = (\nu_{jk})_{j,k\in S}$, let $\widetilde{\pi} : \mathfrak{P}(S) \to \mathbb{R}$ and $\widetilde{\nu} : \mathfrak{P}(S) \times \mathfrak{P}(S) \to \mathbb{R}$ be given by

$$\widetilde{\pi}(A) := \sum_{j\in A} \pi_j \quad \text{and} \quad \widetilde{\nu}(A, B) := \sum_{j\in A}\sum_{k\in B} \nu_{jk} \qquad \text{for all } A, B \in \mathfrak{P}(S)\,,$$

where $\mathfrak{P}(S)$ denotes the power set of the given finite set S. Set functions of this type not only govern many definitions and results of the classical theory of finite networks, but also enjoy important structural properties. In fact, the set function $\widetilde{\pi}$ is *additive* in the sense that the identity $\widetilde{\pi}(A \cup B) = \widetilde{\pi}(A) + \widetilde{\pi}(B)$ holds for all disjoint $A, B \subseteq S$, while $\widetilde{\nu}$ is *biadditive* in the sense that, for every fixed $A \subseteq S$, both set functions $\widetilde{\nu}(A, \cdot)$ and $\widetilde{\nu}(\cdot, A)$ are additive on $\mathfrak{P}(S)$.

It is clear that a number of basic results on flows and potentials in finite networks may be completely expressed in the language of additive and biadditive set functions. The purpose of the present paper is to show that, in this reformulation, such results remain valid in a considerably more general setting, which is not restricted to the case of finite sets, and which turns out to be very suitable for applications to measure theory and mathematical economics.

Thus, in our abstract context, flows and potentials are identified with certain biadditive and additive set functions and hence correspond to linear functionals on appropriate spaces of functions.

Our principal tool is the following version of the Hahn - Banach theorem: if a sublinear functional ϑ on a real vector space X dominates a concave functional ρ on a convex subset K of X, then there exists a linear functional ξ on X which satisfies both $\xi \leq \vartheta$ on X and $\rho \leq \xi$ on K. This sandwich theorem is an equivalent version of the interpolation theorem due to Mazur - Orlicz [12]. We refer to [19] for an elegant short proof of this basic result of convex analysis, to [5], [10], and [21] for a thorough discussion of results of this kind, and to [16] for a recent simple proof of a generalized version of the sandwich theorem.

There are intimate connections to minimax theory. In fact, as demonstrated in [9], [10], and [13], certain convex-concave type generalizations of the von Neumann minimax theorem are immediate consequences of the sandwich theorem. Moreover, in the simple, but important special case of a constant function ρ, the sandwich theorem may be reformulated in terms of the minimax identity

$$\inf_{u \in K} \max_{\xi \in L} \xi(x) = \max_{\xi \in L} \inf_{x \in K} \xi(x),$$

where L denotes the convex set of all linear functionals ξ on X with the property that $\xi \leq \vartheta$ on X.

In the following section, the sandwich theorem is used to establish necessary and sufficient conditions for the existence of linear functionals on a vector lattice which satisfy certain linear constraints. The results of this section are fundamental for the rest of the paper and lead to a uniform and natural approach to the existence of various kinds of flows and potentials. One of the applications is Theorem 5, which extends the classical feasibility theorem due to Gale and Hoffman to the case of infinite networks. As recorded in [5] and [10], certain versions of this flow theorem have been obtained earlier by Fuchssteiner, König, and the first-named author, with methods close in spirit to the Strassen disintegration theorem [3], [4]. The present approach seems to be considerably more direct, and entails a variety of new applications. In particular, we are able to handle, in Section 6, the cases of symmetric and antisymmetric flows for different types of capacity constraints. We conclude with a brief discussion of some connections and applications to measure theory in Sections 7 and 8.

3. Tools from Functional Analysis

This section is to derive some general results on the existence of certain linear functionals on real vector spaces, both in an algebraic and a topological setting.

Given a real vector space X, let X^* stand for the algebraic dual of X, *i.e.* the space of all real-valued linear functionals on X. For a linear mapping $T : X \to Y$ between vector spaces X and Y, let $T^* : Y^* \to X^*$ denote the algebraic adjoint of T, given by $T^*\psi := \psi T$ for all $\psi \in Y^*$. Similarly, if X and Y are normed spaces and $T : X \to Y$ is a continuous linear mapping, then the topological adjoint $T' : Y' \to X'$ is given by $T'\psi := \psi T$ for all $\psi \in Y'$, where X' and Y' consist of all continuous linear functionals on X and Y, respectively.

If X is a vector lattice with positive cone X_+, we consider the canonical order on X^*, *i.e.* for $\varphi_1, \varphi_2 \in X^*$ we write $\varphi_1 \leq \varphi_2$ precisely when $\varphi_1(x) \leq \varphi_2(x)$ for all $x \in X_+$. Also, recall that every $x \in X$ has a decomposition $x = x^+ - x^-$, where $x^+ = x \vee 0$ and $x^- = (-x) \vee 0$ are the positive and negative part of x, respectively.

Theorem 1 *Suppose that $T : X \to Y$ is a linear mapping from a vector lattice X to a vector space Y, and consider $\varphi_1, \varphi_2 \in X^*$ such that $\varphi_1 \leq \varphi_2$. Then there exists some $\xi \in Y^*$ such that $\varphi_1 \leq T^*\xi \leq \varphi_2$ if and only if $\varphi_1(x^+) \leq \varphi_2(x^-)$ for all $x \in \ker T$. Moreover, if X is a Banach lattice, Y is a Banach space, T is a continuous linear operator with closed range, and φ_1, φ_2 belong to X', then the same equivalence holds with some $\xi \in Y'$.*

Proof. If there exists a $\xi \in Y^*$ with $\varphi_1 \leq T^*\xi \leq \varphi_2$, then we obtain that $\varphi_1(x^+) - \varphi_2(x^-) \leq (T^*\xi)(x^+) - (T^*\xi)(x^-) = (T^*\xi)(x) = \xi(Tx) = 0$ for all $x \in \ker T$. This proves the necessity of the indicated condition in both the algebraic and the topological setting.

Conversely, suppose that $\varphi_1(x^+) \leq \varphi_2(x^-)$ for all $x \in \ker T$, and observe that this condition is equivalent to $\varphi_1(x^-) \leq \varphi_2(x^+)$ for all $x \in \ker T$. Now consider the functional $\vartheta : X \to \mathbb{R}$ given by $\vartheta(x) := \varphi_2(x^+) - \varphi_1(x^-)$ for all $x \in X$. Since $\varphi_1 \leq \varphi_2$ and $(u+v)^+ \leq u^+ + v^+$ for all $u, v \in X$, we obtain that

$$\varphi_1(u^- + v^- - (u+v)^-) = \varphi_1(u^+ + v^+ - (u+v)^+) \leq \varphi_2(u^+ + v^+ - (u+v)^+)$$

and therefore $\vartheta(u+v) \leq \vartheta(u) + \vartheta(v)$ for all $u, v \in X$. We conclude that ϑ is sublinear and satisfies $\vartheta \geq 0$ on $\ker T$. Hence, by the Hahn-Banach theorem, there exists some $\zeta \in X^*$ such that $\zeta \leq \vartheta$ on X and $\zeta = 0$ on $\ker T$.

The first estimate ensures that $\varphi_1 \leq \zeta \leq \varphi_2$, since $\zeta(x) \leq \vartheta(x) = \varphi_2(x)$ and $-\zeta(x) = \zeta(-x) \leq \vartheta(-x) = -\varphi_1(x)$ for all $x \in X_+$. Now let $\xi(y) := \zeta(x)$ for all $y \in Y$ and $x \in X$ with $Tx = y$. Since $\zeta = 0$ on $\ker T$, we obtain a well-defined linear functional on $T(X)$ which, by Zorn's lemma, may be extended to a linear functional on the entire space Y, also denoted by ξ. Because $T^*\xi = \zeta$, the functional $\xi \in Y^*$ has the desired properties, which completes the proof in the algebraic case.

In the topological setting, we first observe that the continuity of the lattice operations in the Banach lattice X and the continuity of φ_1 and φ_2 imply that the sublinear functional ϑ is continuous. Since the linear functional ζ obtained in the preceding paragraph satisfies $-\vartheta(-x) \leq \zeta(x) \leq \vartheta(x)$ for all $x \in X$, we conclude that ζ is continuous as well. Moreover, since T has closed range, the open mapping theorem provides a constant $c > 0$ such that, for every y in the range of T, there exists some $x \in X$ with $Tx = y$ and $\|x\| \leq c\,\|y\|$. Consequently the linear functional ξ on $T(X)$, given by $\xi(y) := \zeta(x)$ for all $y \in Y$ and $x \in X$ with $Tx = y$, is continuous of norm less or equal to $c\,\|\zeta\|$. Hence, by the Hahn-Banach theorem, this functional may be extended to a continuous linear functional ξ on Y. Since $\xi \in Y'$ and $T'\xi = \zeta$, we conclude that the assertion of the theorem remains valid in the topological setting.

Note that the last step of the proof also follows from the closed range theorem, since the identity $T'(Y') = (\ker T)^{\perp}$ holds for every continuous linear operator T with closed range; see Theorem IV.1.2 of [6]. □

In connection with the continuity assertion of Theorem 1, we note that, for any pair of linear functionals φ_1 and φ_2 on a Banach lattice, the estimate $\varphi_1 \leq \varphi_2$ implies that the continuity of φ_1 is equivalent to that of φ_2. In fact, this observation is immediate from the automatic continuity of all positive linear functionals on Banach lattices; see Proposition 1.3.5 of [11]. It follows that, in the last statement of Theorem 1, the continuity of φ_1 and φ_2 is not only sufficient, but also necessary for the existence of some $\xi \in Y'$ with the property that $\varphi_1 \leq T'\xi \leq \varphi_2$.

The following variant of Theorem 1 will be our main source for the existence of flows and potentials in infinite networks.

Theorem 2 *Let $T : X \to Y$ be a linear mapping between the vector lattices X and Y, and consider functionals $\varphi_1, \varphi_2 \in X^*$ and $\psi_1, \psi_2 \in Y^*$ such that $\varphi_1 \leq \varphi_2$ and $\psi_1 \leq \psi_2$. Then the following assertions are equivalent:*

(a) *there exists some $\xi \in Y^*$ for which $\psi_1 \leq \xi \leq \psi_2$ and $\varphi_1 \leq T^*\xi \leq \varphi_2$;*

(b) *$\varphi_1(x^+) - \varphi_2(x^-) \leq \psi_2((Tx)^+) - \psi_1((Tx)^-)$ for all $x \in X$.*

Moreover, if Y is a Banach lattice and the linear functionals ψ_1 and ψ_2 are continuous, then the same equivalence holds with some $\xi \in Y'$.

Proof. Assertion (a) immediately implies that

$$\varphi_1(x^+) - \varphi_2(x^-) \leq (T^*\xi)(x^+) - (T^*\xi)(x^-) = (T^*\xi)(x) = \xi(Tx)$$

$$= \xi\left((Tx)^+\right) - \xi\left((Tx)^-\right) \leq \psi_2\left((Tx)^+\right) - \psi_1\left((Tx)^-\right)$$

for all $x \in X$.

Conversely, if assertion (b) holds, let us introduce the sublinear functional $\vartheta : Y \to \mathbb{R}$ by $\vartheta(y) := \psi_2(y^+) - \psi_1(y^-)$, and observe that a linear functional $\xi \in Y^*$ satisfies the estimate $\xi \leq \vartheta$ on Y precisely when $\psi_1 \leq \xi \leq \psi_2$. Moreover, for all $x \in X$ and $y \in Y$ with $Tx = y$, we obtain from assertion (b) that $\varphi_1(x^+) - \varphi_2(x^-) \leq \vartheta(y)$. It follows that the definition

$$\rho(y) := \sup\left\{\varphi_1(x^+) - \varphi_2(x^-) : x \in X \text{ with } Tx = y\right\} \quad \text{for all } y \in T(X)$$

yields a real-valued superlinear functional ρ on $T(X)$ for which $\rho \leq \vartheta$ on $T(X)$. Hence the sandwich theorem [16], [19], [21] provides a linear functional ξ on Y for which $\xi \leq \vartheta$ on Y and $\rho \leq \xi$ on $T(X)$.

The first inequality means exactly that $\psi_1 \leq \xi \leq \psi_2$. And from $\rho \leq \xi$ on $T(X)$ we conclude that every $x \in X_+$ satisfies $\varphi_1(x) \leq \rho(Tx) \leq \xi(Tx)$ and $-\varphi_2(x) \leq \rho(-Tx) \leq \xi(-Tx)$. This shows that $\varphi_1 \leq T^*\xi \leq \varphi_2$, and hence settles the assertion in the algebraic case.

Finally, to obtain the result in the topological setting, it suffices to recall from above that every linear functional ξ on the Banach lattice Y with the property that $\psi_1 \leq \xi \leq \psi_2$ is automatically continuous if and only if one of the functionals ψ_1 or ψ_2 is continuous. □

Remark 3 (i) It is perhaps interesting to note that, alternatively, the algebraic version of Theorem 2 may also be obtained as a consequence of Theorem 1. Indeed, in the setting of Theorem 2, we consider the linear mapping $\widetilde{T}$ from the vector lattice $X \times Y$ into the vector space Y, given by $\widetilde{T}(x, y) := Tx - y$ for all $(x, y) \in X \times Y$. We also introduce the functionals $\widetilde{\varphi}_1, \widetilde{\varphi}_2 \in (X \times Y)^*$ by $\widetilde{\varphi}_1(x, y) := \varphi_1(x) - \psi_2(y)$ and $\widetilde{\varphi}_2(x, y) := \varphi_2(x) - \psi_1(y)$ for all $(x, y) \in X \times Y$. Since the kernel of $\widetilde{T}$ coincides with the graph of T, it is easily seen that Theorem 1 now implies the equivalence of the assertions (a) and (b) in Theorem 2.

(ii) We mention in passing that a similar reasoning may be used to handle the case of constraints given by an arbitrary number of adjoint operators. In addition to the functionals ψ_1 and ψ_2 on the vector lattice Y with the property that $\psi_1 \leq \psi_2$, let us consider an arbitrary index set A, and, for each $\alpha \in A$, let T_α be a linear operator from a vector lattice X_α into Y, and let $\varphi_{1,\alpha}$ and $\varphi_{2,\alpha}$ be linear functionals on X_α such that $\varphi_{1,\alpha} \leq \varphi_{2,\alpha}$. Then an obvious adaptation of the preceding arguments leads to the following characterization: there exists a functional $\xi \in Y^*$ such that $\psi_1 \leq \xi \leq \psi_2$ and $\varphi_{1,\alpha} \leq T_\alpha^* \xi \leq \varphi_{2,\alpha}$ for all $\alpha \in A$ precisely when the condition

$$\sum_{j=1}^{n} \varphi_{1,\alpha_j}(x_j^+) - \sum_{j=1}^{n} \varphi_{2,\alpha_j}(x_j^-) \leq \psi_2\left((\sum_{j=1}^{n} T_{\alpha_j} x_j)^+ \right) - \psi_1\left((\sum_{j=1}^{n} T_{\alpha_j} x_j)^- \right)$$

holds for every finite collection of indices $\alpha_1, \ldots, \alpha_n \in A$ and corresponding vectors $x_j \in X_{\alpha_j}$ for $j = 1, \ldots, n$.

(iii) In certain cases, the feasibility condition of Theorem 2 can be considerably simplified. The most obvious case is probably that of a lattice homomorphism. Recall from Proposition 1.3.11 of [11] that the linear operator $T : X \to Y$ is a lattice homomorphism precisely when the identities $(Tx)^+ = T(x^+)$ and $(Tx)^- = T(x^-)$ hold for all $x \in X$. It is clear that, in this case, condition (b) of Theorem 2 reduces to the pair of estimates $\varphi_1 \leq T^*\psi_2$ and $T^*\psi_1 \leq \varphi_2$. This observation will lead to a simple solution of one of the marginal problems in Section 7, but does not apply to the theory of flows in networks in Section 4.

(iv) We finally note that our approach is flexible enough to work also in the case of one-sided constraints. For example, in the setting of Theorem 2, we obtain that there exists a $\xi \in Y^*$ such that $\psi_1 \leq \xi \leq \psi_2$ and $\varphi_1 \leq T^*\xi$ if and only if the estimate $\varphi_1(x) \leq \psi_2((Tx)^+) - \psi_1((Tx)^-)$ holds for all $x \in X_+$. For the proof, one has to apply the sandwich theorem [16], [19] to the sublinear functional ϑ on Y, given as before by $\vartheta(y) := \psi_2(y^+) - \psi_1(y^-)$ for all $y \in Y$, and to the superlinear functional ρ on the convex cone $K = T(X_+)$, now given by $\rho(y) := \sup\{\varphi_1(x) : x \in X_+ \text{ with } Tx = y\}$ for all $y \in K$. Of course, the case of the upper constraint $T^*\xi \leq \varphi_2$ may be handled by a similar argument.

4. Existence of Flows

Throughout this section, we fix an arbitrary non-empty set S, endowed with an algebra Σ of subsets, and consider a pair of additive set functions

$\lambda, \mu : \Sigma \to \mathbb{R}$ such that $\lambda \le \mu$ on Σ and a pair of biadditive set functions $\sigma, \tau : \Sigma \times \Sigma \to \mathbb{R}$ such that $\sigma \le \tau$ on $\Sigma \times \Sigma$. In this setting, our first goal is to obtain the following natural generalizations of the corresponding feasibility theorems from the theory of finite networks. As explained in the introduction, the biadditive set functions ν provided by the next two theorems may be interpreted as *flows* without or with arc capacity constraints for the abstract network given by the set functions λ, μ, σ, and τ.

For an additive set function $\pi : \Sigma \to \mathbb{R}$, boundedness means, as usual, that for some suitable constant c we have that $|\pi(A)| \le c$ for all $A \in \Sigma$. On the other hand, a biadditive set function $\nu : \Sigma \times \Sigma \to \mathbb{R}$ is said to be *bounded* if there exists a constant c such that

$$\left| \sum_{k=1}^{n} \nu(A_k, B_k) \right| \le c$$

for every finite collection of pairwise disjoint rectangles $A_k \times B_k$ for which $A_k, B_k \in \Sigma$. The significance of this condition will soon become apparent.

Theorem 4 *There exists a biadditive set function $\nu : \Sigma \times \Sigma \to \mathbb{R}$ such that*

$$\lambda(A) \le \nu(A, A^c) - \nu(A^c, A) \le \mu(A) \qquad \text{for all } A \in \Sigma$$

if and only if the condition $\lambda(S) \le 0 \le \mu(S)$ is fulfilled. Moreover, in this case, there exists a bounded flow ν if and only if λ and μ are bounded.

Theorem 5 *There exists a biadditive set function $\nu : \Sigma \times \Sigma \to \mathbb{R}$ such that*

$$\sigma(A, B) \le \nu(A, B) \le \tau(A, B) \quad \text{and} \quad \lambda(A) \le \nu(A, A^c) - \nu(A^c, A) \le \mu(A)$$

for all $A, B \in \Sigma$ if and only if the following conditions are fulfilled:

$$\lambda(A)\,,\ -\mu(A^c) \le \tau(A, A^c) - \sigma(A^c, A) \qquad \text{for all } A \in \Sigma\,. \tag{1}$$

Moreover, in this case, all these flows ν are bounded precisely when σ and τ are bounded.

To derive these theorems from the general results of the preceding section, we have to introduce a suitable functional-analytic setting.

First, let E denote the space of all real-valued Σ-measurable simple functions on S. Clearly, E is a normed vector lattice with respect to pointwise operations and the supremum norm. The algebraic dual E^* may be canonically identified with the vector space $\mathfrak{A}(\Sigma)$ of all real-valued additive set functions on Σ. More precisely, every $\pi \in \mathfrak{A}(\Sigma)$ induces a functional $\widehat{\pi} \in E^*$ via the elementary integral $\widehat{\pi}(f) := \int_S f\, d\pi$ for all $f \in E$, and conversely, for every $\varphi \in E^*$, the corresponding $\pi \in \mathfrak{A}(\Sigma)$ is given by $\pi(A) := \varphi(\chi_A)$ for all $A \in \Sigma$, where χ_A denotes the characteristic function of A. Note that the canonical identification of E^* and $\mathfrak{A}(\Sigma)$ preserves the natural order structure of these spaces.

Obviously, every positive $\pi \in \mathfrak{A}(\Sigma)$ induces a continuous linear functional $\widehat{\pi} \in E'$. Moreover, for an arbitrary $\pi \in \mathfrak{A}(\Sigma)$, the following four statements are equivalent:

(i) the functional $\widehat{\pi}$ is continuous on E;

(ii) π is bounded;

(iii) π is bounded above (or below);

(iv) π is the difference of two positive additive set functions on Σ.

In fact, the implications (iv) $\Rightarrow$ (i) $\Rightarrow$ (ii) $\Rightarrow$ (iii) are immediate, while (iii) entails that the definition

$$\pi^+(A) := \sup\{\pi(B) : B \in \Sigma \text{ with } B \subseteq A\} \qquad \text{for every } A \in \Sigma$$

yields a real-valued additive set function π^+ on Σ with the properties $\pi^+ \geq 0$ and $\pi^+ \geq \pi$, so that (iv) holds with the decomposition $\pi = \pi^+ - (\pi^+ - \pi)$.

Next, let $\Sigma \circ \Sigma$ denote the algebra of subsets of the product space $S \times S$ that is generated by the measurable rectangles $A \times B$ for all $A, B \in \Sigma$. Thus $\Sigma \circ \Sigma$ consists precisely of all finite unions of measurable rectangles. We observe that the space $\mathfrak{B}(\Sigma \times \Sigma)$ of all real-valued biadditive set functions on $\Sigma \times \Sigma$ may be canonically identified with the space $\mathfrak{A}(\Sigma \circ \Sigma)$. Indeed, for every additive $\pi : \Sigma \circ \Sigma \to \mathbb{R}$, the set function ν on $\Sigma \times \Sigma$, given by $\nu(A, B) := \pi(A \times B)$ for all $A, B \in \Sigma$, is clearly biadditive. Conversely, given an arbitrary $\nu \in \mathfrak{B}(\Sigma \times \Sigma)$ and any $C \in \Sigma \circ \Sigma$, let

$$\pi(C) := \sum_{k=1}^{n} \nu(A_k, B_k) \qquad \text{for } C = \bigcup_{k=1}^{n} (A_k \times B_k) \text{ with } A_k, B_k \in \Sigma,$$

where C is represented as an arbitrary union of finitely many pairwise disjoint measurable rectangles $A_k \times B_k$. Exactly as in the elementary theory of product measures, it follows that this definition provides an additive set function π on $\Sigma \circ \Sigma$ with the property that $\pi(A \times B) = \nu(A, B)$ for all $A, B \in \Sigma$. From this and the identification of additive set functions mentioned earlier, we conclude that the real-valued biadditive set functions on $\Sigma \times \Sigma$ correspond canonically, via integration, with the linear functionals on F, where F stands for the normed vector lattice of all real-valued $\Sigma \circ \Sigma$-measurable simple functions on $S \times S$. In particular, the characterization of the continuous linear functionals from the preceding paragraph leads to the following equivalence: a biadditive set function on $\Sigma \times \Sigma$ is bounded in the sense defined above precisely when the corresponding linear functional on F is continuous.

Proof of Theorem 4. We first apply Theorem 1 to the normed vector lattices E and F and the continuous linear operator $T : E \to F$, given by $(Tf)(s,t) := f(s) - f(t)$ for all $f \in E$ and $s, t \in S$. Note that the kernel of T consists of the real constant functions, and that the identity $T(\chi_A) = \chi_{A\times A^c} - \chi_{A^c\times A}$ holds for all $A \in \Sigma$. With the canonical identification of the algebraic dual spaces F^* and E^* in terms of biadditive and additive set functions, it follows that the algebraic adjoint T^* from $\mathfrak{B}(\Sigma \times \Sigma)$ to $\mathfrak{A}(\Sigma)$ is given by the formula

$$(T^*\nu)(A) = \nu(T\chi_A) = \int_{S\times S} (\chi_{A\times A^c} - \chi_{A^c\times A})\, d\nu = \nu(A, A^c) - \nu(A^c, A)$$

for all $\nu \in \mathfrak{B}(\Sigma \times \Sigma)$ and $A \in \Sigma$.

Now consider the functionals $\varphi_1, \varphi_2 \in E^*$, given by $\varphi_1(f) := \int_S f\, d\lambda$ and $\varphi_2(f) := \int_S f\, d\mu$ for all $f \in E$. Then it is clear that a set function $\nu \in \mathfrak{B}(\Sigma \times \Sigma)$ satisfies $\varphi_1 \le T^*\nu \le \varphi_2$ precisely when $\lambda(A) \le \nu(A, A^c) - \nu(A^c, A) \le \mu(A)$ holds for all $A \in \Sigma$. Moreover, since $\ker T = \mathbb{R}$, the estimates $\varphi_1(x^+) \le \varphi_2(x^-)$ for all $x \in \ker T$ are equivalent to $\lambda(S) \le 0 \le \mu(S)$. Hence the first assertion of Theorem 4 is immediate from Theorem 1.

Next observe that the set functions λ and μ are obviously bounded if there exists some bounded $\nu \in \mathfrak{B}(\Sigma \times \Sigma)$ such that $\lambda(A) \le \nu(A, A^c) - \nu(A^c, A) \le \mu(A)$ for all $A \in \Sigma$.

To show the converse, we suppose that λ and μ are both bounded and apply the last assertion of Theorem 1 to the Banach lattices $\bar{E}$ and $\bar{F}$, where the closure is taken in the respective Banach space of all real-valued bounded functions with respect to the supremum norm. Evidently, the canonical extension of the continuous linear operator $T : E \to F$ to an operator $\bar{T}$ from $\bar{E}$ to $\bar{F}$ is given by $(\bar{T}f)(s,t) := f(s) - f(t)$ for all $f \in \bar{E}$ and $s, t \in S$. As before, the kernel of $\bar{T}$ consists only of the real constants. To verify that the range of $\bar{T}$ is closed, let $g \in \bar{F}$ and $f_n \in \bar{E}$ such that $\bar{T}f_n \to g$ as $n \to \infty$. We fix some $t \in S$ and consider the functions $h, h_n : S \to \mathbb{R}$, given by $h(s) := g(s,t)$ and $h_n(s) := f_n(s) - f_n(t)$ for all $s \in S$ and $n \in \mathbb{N}$. The functions h_n belong to $\bar{E}$ and satisfy $h_n \to h$ and $\bar{T}h_n = \bar{T}f_n \to g$ as $n \to \infty$. This implies that $h \in \bar{E}$ and therefore $\bar{T}h = g$, by the continuity of $\bar{T}$. Thus $g \in \bar{T}(\bar{E})$, which shows that the operator $\bar{T}$ has closed range. Finally note that, by the boundedness of λ and μ, the functionals $\varphi_1, \varphi_2 \in E^*$ are continuous and hence extend canonically to continuous linear functionals on $\bar{E}$. Since the operators T and $\bar{T}$ have the same kernel and the same topological adjoint, it now follows from the first part of the proof that the conditions of Theorem 4 translate precisely to those considered in Theorem 1. Thus we conclude that a bounded flow exists if and only if $\lambda(S) \le 0 \le \mu(S)$. □

Proof of Theorem 5. We apply Theorem 2 to the operator $T : E \to F$, given as before by $(Tf)(s,t) := f(s) - f(t)$ for all $f \in E$ and $s,t \in S$, and to the linear functionals $\varphi_1, \varphi_2 \in E^*$ and $\psi_1, \psi_2 \in F^*$ which correspond to the set functions $\lambda, \mu \in \mathfrak{A}(\Sigma)$ and $\sigma, \tau \in \mathfrak{B}(\Sigma \times \Sigma)$ via integration.

Since $(T^*\nu)(A) = \nu(A, A^c) - \nu(A^c, A)$ for all $\nu \in \mathfrak{B}(\Sigma \times \Sigma)$ and $A \in \Sigma$, it is clear that, in this setting, assertion (a) of Theorem 2 means precisely the existence of a flow ν in the sense of Theorem 5.

On the other hand, the following argument will show that assertion (b) of Theorem 2 is now equivalent to the last statement of Theorem 5. We first observe that assertion (b) of Theorem 2 translates here to the condition that

$$\int_S f^+ \, d\lambda - \int_S f^- \, d\mu \le \int_{S\times S} (Tf)^+ \, d\tau - \int_{S\times S} (Tf)^- \, d\sigma \qquad \text{for all } f \in E.$$

Since $(T\chi_A)^+ = \chi_{A\times A^c}$ and $(T\chi_A)^- = \chi_{A^c \times A}$ for all $A \in \Sigma$, we immediately obtain from these estimates that

$$\lambda(A)\,,\ -\mu(A^c) \le \tau(A, A^c) - \sigma(A^c, A) \qquad \text{for all } A \in \Sigma\,,$$

which are the feasibility conditions (1) of Theorem 5.

The converse is less obvious and depends on an analysis of the sublinear operator $\vartheta : E \to F$ given by $\vartheta(f) := (Tf)^+$ for all $f \in E$. Clearly, the identity $\vartheta(-f) = (Tf)^-$ holds for all $f \in E$. Moreover, it is straightforward to verify the following properties, which illustrate that ϑ behaves almost like a linear operator. For all $f, g \in E$ with disjoint supports, we have that $\vartheta(f-g) = \vartheta(f) + \vartheta(-g)$, while for every collection of numbers $\alpha_1, \dots, \alpha_n \ge 0$ and sets $A_1, \dots, A_n \in \Sigma$ with $A_1 \subseteq \dots \subseteq A_n$, we obtain that

$$\vartheta\left(\sum_{j=1}^n \alpha_j \chi_{Aj}\right) = \sum_{j=1}^n \alpha_j \vartheta\left(\chi_{A_j}\right) = \sum_{j=1}^n \alpha_j \chi_{Aj\times A_j^c} \quad \text{and}$$

$$\vartheta\left(-\sum_{j=1}^n \alpha_j \chi_{Aj}\right) = \sum_{j=1}^n \alpha_j \vartheta\left(-\chi_{A_j}\right) = \sum_{j=1}^n \alpha_j \chi_{A_j^c\times Aj}\,.$$

Now, to show that the feasibility conditions of Theorem 5 do indeed imply condition (b) of Theorem 2, we consider an arbitrary $f \in E$ and represent the positive and negative part of f in the form

$$f^+ = \sum_{j=1}^n \alpha_j \chi_{Aj} \qquad \text{and} \qquad f^- = \sum_{k=1}^m \beta_k \chi_{B_k}$$

with suitable numbers $\alpha_j, \beta_k \ge 0$ and sets $A_j, B_k \in \Sigma$ such that $A_1 \subseteq \dots \subseteq A_n$ and $B_1 \subseteq \dots \subseteq B_m$. From the feasibility conditions (1) we then conclude that

$$\int_S f^+ \, d\lambda - \int_S f^- \, d\mu = \sum_{j=1}^n \alpha_j \lambda(A_j) - \sum_{k=1}^m \beta_k \mu(B_k)$$

$$\leq \sum_{j=1}^{n} \alpha_j \left(\tau(A_j, A_j^c) - \sigma(A_j^c, A_j)\right) + \sum_{k=1}^{m} \beta_k \left(\tau(B_k^c, B_k) - \sigma(B_k, B_k^c)\right)$$

$$= \int_{S\times S} \vartheta(f^+)\, d\tau - \int_{S\times S} \vartheta(-f^+)\, d\sigma + \int_{S\times S} \vartheta(-f^-)\, d\tau - \int_{S\times S} \vartheta(f^-)\, d\sigma$$

$$= \int_{S\times S} \vartheta(f)\, d\tau - \int_{S\times S} \vartheta(-f)\, d\sigma = \int_{S\times S} (Tf)^+\, d\tau - \int_{S\times S} (Tf)^-\, d\sigma,$$

which proves the desired equivalence.

The final assertion about bounded flows follows immediately from the fact that the boundedness of a biadditive set function on $\Sigma \times \Sigma$ means, by definition, the boundedness of the corresponding additive set function on Σ. □

5. Existence of Potentials

As in the preceding section, we consider a non-empty set S, endowed with an algebra Σ of subsets, as well as set functions $\lambda, \mu \in \mathfrak{A}(\Sigma)$ and $\sigma, \tau \in \mathfrak{B}(\Sigma \times \Sigma)$ such that $\lambda \leq \mu$ on Σ and $\sigma \leq \tau$ on $\Sigma \times \Sigma$. In addition, we fix an arbitrary set function $\gamma \in \mathfrak{A}(\Sigma)$, to be interpreted as a weight function for the set of nodes. A $\Sigma \circ \Sigma$ - measurable simple function $w : S \times S \to \mathbb{R}$ is called a γ - *circulation* if

$$\int_S w(s,t)\, d\gamma(t) = \int_S w(t,s)\, d\gamma(t) \qquad \text{for all } s \in S.$$

If γ is the counting measure on a finite set, this notion obviously coincides with the classical term from the theory of finite networks. Moreover, the additive set functions π provided by the following two theorems may be viewed as *potentials* with respect to γ. Note that these results are, in a sense, dual to the results on flows from the preceding section.

Theorem 6 *There exists an additive set function $\pi : \Sigma \to \mathbb{R}$ such that*

$$\sigma(A,B) \leq \gamma(A)\,\pi(B) - \gamma(B)\,\pi(A) \leq \tau(A,B) \qquad \textit{for all } A, B \in \Sigma$$

if and only if

$$\int_{S\times S} w^+\, d\sigma \leq \int_{S\times S} w^-\, d\tau \qquad \textit{for every } \gamma\textit{-circulation } w : S \times S \to \mathbb{R}.$$

Moreover, if the last condition is fulfilled and if the set function γ is bounded and non-trivial, then all these potentials $\pi \in \mathfrak{A}(\Sigma)$ are bounded precisely when both σ and τ are bounded.

Theorem 7 *There exists an additive set function $\pi : \Sigma \to \mathbb{R}$ such that*

$$\sigma(A,B) \le \gamma(A)\,\pi(B) - \gamma(B)\,\pi(A) \le \tau(A,B) \qquad \text{for all } A,B \in \Sigma \text{ and}$$

$$\lambda(A) \le \pi(A) \le \mu(A) \qquad \text{for all } A \in \Sigma$$

if and only if the estimate

$$\sum_{j=1}^{n}\sum_{k=1}^{n} \alpha_{jk}^{+}\,\sigma(A_j,A_k) - \sum_{j=1}^{n}\sum_{k=1}^{n} \alpha_{jk}^{-}\,\tau(A_j,A_k)$$

$$\le \sum_{j=1}^{n}\left(\sum_{k=1}^{n}(\alpha_{kj}-\alpha_{jk})\gamma(A_k)\right)^{+}\mu(A_j) - \sum_{j=1}^{n}\left(\sum_{k=1}^{n}(\alpha_{kj}-\alpha_{jk})\gamma(A_k)\right)^{-}\lambda(A_j)$$

holds for every collection of pairwise disjoint sets $A_j \in \Sigma$ and real numbers α_{jk} for $j,k = 1, \ldots, n$. Moreover, if the last condition is fulfilled, then all these potentials $\pi \in \mathfrak{A}(\Sigma)$ are bounded precisely when both λ and μ are bounded.

Proof of Theorem 6. As before, let E and F be the normed vector lattices of all real-valued measurable simple functions on S and $S \times S$, respectively, and consider now the linear operator $T_\gamma : F \to E$, given by

$$(T_\gamma w)(s) := \int_S (w(t,s) - w(s,t))\, d\gamma(t) \qquad \text{for all } w \in F \text{ and } s \in S\,.$$

Note that the kernel of T_γ consists precisely of the γ-circulations, and a simple calculation shows that $T_\gamma \chi_{A\times B} = \gamma(A)\,\chi_B - \gamma(B)\,\chi_A$ for all $A, B \in \Sigma$. From this we infer that the algebraic adjoint T_γ^* from $\mathfrak{A}(\Sigma)$ to $\mathfrak{B}(\Sigma\times\Sigma)$ is given by

$$(T_\gamma^*\pi)(A,B) = \pi\,(T_\gamma\chi_{A\times B}) = \gamma(A)\,\pi(B) - \gamma(B)\,\pi(A)$$

for all $\pi \in \mathfrak{A}(\Sigma)$ and $A, B \in \Sigma$, where the algebraic dual spaces E^* and F^* are again canonically identified with $\mathfrak{A}(\Sigma)$ and $\mathfrak{B}(\Sigma \times \Sigma)$, respectively. Hence, if $\varphi_1, \varphi_2 \in F^*$ denote the linear functionals corresponding to σ and τ via integration, then it is clear that the first assertion of Theorem 6 is an immediate consequence of Theorem 1.

It should be noted that the last claim concerning the boundedness of all potentials is *not* contained in the somewhat weaker boundedness assertion of Theorem 1. Instead, we obtain the final statement of Theorem 6 from the following elementary observation. Suppose that the function $\gamma \in \mathfrak{A}(\Sigma)$ is both non-trivial and bounded, and recall that γ may be represented as the difference of two positive additive set functions. This implies that the operator $T_\gamma : F \to E$ is continuous. Consequently, if there exists a bounded

$\pi \in \mathfrak{A}(\Sigma)$ with $\sigma \le T_\gamma^* \pi \le \tau$, we conclude that $T_\gamma^* \pi$ and hence σ and τ are bounded. To see the converse, we fix a set $B \in \Sigma$ for which $\gamma(B) \neq 0$. Since every potential $\pi \in \mathfrak{A}(\Sigma)$ satisfies

$$\sigma(A, B) \le \gamma(A)\,\pi(B) - \gamma(B)\,\pi(A) \le \tau(A, B) \qquad \text{for all } A \in \Sigma\,,$$

the boundedness of σ, τ, and γ obviously ensures that every potential is bounded. This completes the proof of the last assertion of Theorem 6. □

Proof of Theorem 7. We consider the linear operator $T_\gamma : F \to E$ from the preceding proof as well as the linear functionals $\varphi_1, \varphi_2 \in F^*$ and $\psi_1, \psi_2 \in E^*$ which correspond to the set functions $\sigma, \tau \in \mathfrak{B}(\Sigma \times \Sigma)$ and $\lambda, \mu \in \mathfrak{A}(\Sigma)$ via integration. Now observe that every $w \in F$ has a representation of the form

$$w = \sum_{j=1}^{n} \sum_{k=1}^{n} \alpha_{jk}\, \chi_{A_j \times A_k}$$

with suitable numbers $\alpha_{jk} \in \mathbb{R}$ and pairwise disjoint sets $A_j \in \Sigma$. Since

$$T_\gamma w = \sum_{j=1}^{n} \sum_{k=1}^{n} \alpha_{jk} \left(\gamma(A_j)\, \chi_{A_k} - \gamma(A_k)\, \chi_{A_j}\right)$$

$$= \sum_{j=1}^{n} \Big(\sum_{k=1}^{n} (\alpha_{kj} - \alpha_{jk})\, \gamma(A_k)\Big)\, \chi_{A_j}\,,$$

the assertions of Theorem 7 follow immediately from Theorem 2. □

Note that the feasibility condition of Theorem 7 includes the inequalities for the γ-circulations from Theorem 6 as a special case, but it seems unlikely that this condition can be modified to become as simple and short as the corresponding classical conditions (1) for the existence of flows considered in Theorem 5.

6. Symmetric, Antisymmetric and Net Flows

To continue the theory of flows in infinite networks in the spirit of Theorem 5, we assume that Σ is an algebra of subsets of an arbitrary non-empty set S, and consider set functions $\lambda, \mu \in \mathfrak{A}(\Sigma)$ and $\sigma, \tau \in \mathfrak{B}(\Sigma \times \Sigma)$ such that $\lambda \le \mu$ on Σ and $\sigma \le \tau$ on $\Sigma \times \Sigma$. Our first goal is to obtain a symmetric counterpart of Theorem 5. Note that, here and in the following, the desired flows will be automatically bounded provided that the corresponding constraints are bounded.

Theorem 8 *There exists a biadditive set function $\nu : \Sigma \times \Sigma \to \mathbb{R}$ such that*

$$\sigma(A,B) \le \nu(A,B) \le \tau(A,B) \text{ and } \lambda(A) \le \nu(A,S)+\nu(S,A) \le \mu(A) \quad (2)$$

for all $A, B \in \Sigma$ *if and only if the condition*

$$\lambda(A) - \mu(B) \le \tau(A,B^c) + \tau(B^c,A) - \sigma(B,A^c) - \sigma(A^c,B) \quad (3)$$

holds for all disjoint $A, B \in \Sigma$.

Proof. As in the proof of Theorem 5, let E and F denote the spaces consisting of all real-valued measurable simple functions on S and $S \times S$, respectively, and consider the functionals $\varphi_1, \varphi_2 \in E^*$ and $\psi_1, \psi_2 \in F^*$ which correspond to the set functions $\lambda, \mu \in \mathfrak{A}(\Sigma)$ and $\sigma, \tau \in \mathfrak{B}(\Sigma \times \Sigma)$ via integration. We now apply Theorem 2 to the operator $T : E \to F$, given by $(Tf)(s,t) := f(s) + f(t)$ for all $f \in F$ and $s, t \in S$. Since $(T^*\nu)(A) = \nu(A,S) + \nu(S,A)$ for every $\nu \in \mathfrak{B}(\Sigma \times \Sigma)$ and $A \in \Sigma$, assertion (a) of Theorem 2 means, in this setting, the existence of some $\nu \in \mathfrak{B}(\Sigma \times \Sigma)$ which satisfies the conditions given in (2).

We next observe that, for every pair of disjoint sets $A, B \in \Sigma$, the function $f := \chi_A - \chi_B \in E$ satisfies $(Tf)^+ = \chi_{A\times B^c} + \chi_{B^c \times A}$ and $(Tf)^- = \chi_{B\times A^c} + \chi_{A^c\times B}$. Hence condition (b) of Theorem 2 immediately implies that, in the present setting, condition (3) is fulfilled.

To see the converse, we suppose that (3) holds and consider an arbitrary function $f \in F$. Let $0 < r_1 < \ldots < r_n$ be a finite sequence of real numbers that contains all the non-zero values of both f^+ and f^-, and define $\alpha_1 := r_1$ and $\alpha_j := r_j - r_{j-1}$ for $j = 2, \ldots, n$. We also introduce the measurable sets $A_j := \{s \in S : f^+(s) \ge r_j\}$ and $B_j := \{s \in S : f^-(s) \ge r_j\}$ for $j = 1, \ldots,, n$. Then we obtain that

$$f^+ = \sum_{j=1}^{n} \alpha_j \chi_{Aj} \qquad \text{and} \qquad f^- = \sum_{j=1}^{n} \alpha_j \chi_{B_j},$$

where $\alpha_1, \ldots, \alpha_n > 0$, $A_1 \supseteq \ldots \supseteq A_n$, $B_1 \supseteq \ldots \supseteq B_n$ and $A_1 \cap B_1 = \emptyset$. Using these representations, it is easily verified that

$$(Tf)^+(s,t) = (f(s)+f(t))^+ = \left(f^+(s) - f^-(t)\right)^+ + \left(f^+(t) - f^-(s)\right)^+$$
$$= \sum_{j=1}^{n} \alpha_j \left(\chi_{A_j \times B_j^c} + \chi_{B_j^c \times A_j}\right)(s,t)$$

for all $s, t \in S$, and similarly that

$$(Tf)^-(s,t) = (f(s)+f(t))^- = \left(f^-(s) - f^+(t)\right)^+ + \left(f^-(t) - f^+(s)\right)^+$$

$$= \sum_{j=1}^{n} \alpha_j \left(\chi_{B_j \times A_j^c} + \chi_{A_j^c \times B_j}\right)(s,t)$$

for all $s, t \in S$. Together with the feasibility condition (3), we conclude that

$$\psi_2\left((Tf)^+\right) - \psi_1\left((Tf)^-\right) = \int_{S\times S} (Tf)^+ \, d\tau - \int_{S\times S} (Tf)^- \, d\sigma$$

$$= \sum_{j=1}^{n} \alpha_j \left(\tau(A_j, B_j^c) + \tau(B_j^c, A_j) - \sigma(B_j, A_j^c) - \sigma(A_j^c, B_j)\right)$$

$$\geq \sum_{j=1}^{n} \alpha_j \left(\lambda(A_j) - \mu(B_j)\right) = \varphi_1(f^+) - \varphi_2(f^-),$$

which completes the proof. □

Next, given a biadditive set function $\rho : \Sigma \times \Sigma \to \mathbb{R}$, we are interested in the existence of biadditive set functions $\kappa : \Sigma \times \Sigma \to \mathbb{R}$ with the property that $\kappa(A,B) = -\kappa(B,A)$, $\kappa(A,B) \leq \rho(A,B)$ and $\lambda(A) \leq \kappa(A, A^c) \leq \mu(A)$ for all $A, B \in \Sigma$. Such set functions may be interpreted as *antisymmetric flows* with respect to the capacity constraints λ, μ, and ρ. In the following result, we derive a simple condition for the existence of antisymmetric flows.

Theorem 9 *There exists an antisymmetric $\kappa \in \mathfrak{B}(\Sigma \times \Sigma)$ such that*

$$\kappa(A,B) \leq \rho(A,B) \quad \textit{and} \quad \lambda(A) \leq \kappa(A, A^c) \leq \mu(A)$$

for all $A, B \in \Sigma$ if and only if the estimates

$$\rho(A,B) + \rho(B,A) \geq 0 \quad \textit{and} \quad \lambda(A),\ -\mu(A^c) \leq \rho(A, A^c)$$

hold for all $A, B \in \Sigma$.

Proof. The necessity of the latter conditions is obvious. To prove the sufficiency, let $\sigma(A,B) := -\rho(B,A)/2$ and $\tau(A,B) := \rho(A,B)/2$ for all $A, B \in \Sigma$. Then clearly $\sigma(A,B) \leq \tau(A,B)$ and $\tau(A, A^c) - \sigma(A^c, A) = \rho(A, A^c)$ for all $A, B \in \Sigma$. Hence, by Theorem 5, there exists some $\nu \in \mathfrak{B}(\Sigma \times \Sigma)$ such that $\sigma \leq \nu \leq \tau$ on $\Sigma \times \Sigma$ and $\lambda(A) \leq \nu(A, A^c) - \nu(A^c, A) \leq \mu(A)$ for all $A \in \Sigma$. It follows that the corresponding net flow $\kappa \in \mathfrak{B}(\Sigma \times \Sigma)$, given by $\kappa(A,B) := \nu(A,B) - \nu(B,A)$ for all $A, B \in \Sigma$, has the desired properties. □

The feasibility conditions of the preceding result can be considerably simplified in the case $\lambda = \mu$. In fact, as an immediate consequence of Theorem 9, we obtain that there exists some antisymmetric $\kappa \in \mathfrak{B}(\Sigma \times \Sigma)$ such that $\kappa \leq \rho$ on $\Sigma \times \Sigma$ and $\kappa(A, A^c) = \lambda(A)$ for all $A \in \Sigma$ precisely when

$$\lambda(S) = 0,\ \lambda(A) \leq \rho(A, A^c) \text{ and } \rho(A,B) + \rho(B,A) \geq 0 \quad \text{for all } A, B \in \Sigma.$$

Moreover, it is always possible to represent an antisymmetric flow as a suitable net flow. Indeed, if $\kappa \in \mathfrak{B}(\Sigma \times \Sigma)$ is antisymmetric with $\kappa \leq \rho$ on $\Sigma \times \Sigma$, then $\nu := \kappa/2 \in \mathfrak{B}(\Sigma \times \Sigma)$ satisfies $\kappa(A, B) = \nu(A, B) - \nu(B, A)$ for all $A, B \in \Sigma$ as well as $\sigma \leq \kappa \leq \tau$ on $\Sigma \times \Sigma$, where $\sigma, \tau \in \mathfrak{B}(\Sigma \times \Sigma)$ are, as before, given by $\sigma(A, B) := -\rho(B, A)/2$ and $\tau(A, B) := \rho(A, B)/2$ for all $A, B \in \Sigma$.

Similar results may be obtained for biadditive set functions $\kappa : \Sigma \times \Sigma \to \mathbb{R}$ which are *symmetric* in the sense that $\kappa(A, B) = \kappa(B, A)$ for all $A, B \in \Sigma$. In this case, it is appropriate to assume symmetric capacity constraints $\omega, \rho \in \mathfrak{B}(\Sigma \times \Sigma)$ such that $\omega \leq \rho$ on $\Sigma \times \Sigma$. From Theorem 8 we obtain the following feasibility result for symmetric flows.

Theorem 10 *There exists a symmetric $\kappa \in \mathfrak{B}(\Sigma \times \Sigma)$ such that*

$$\omega(A, B) \leq \kappa(A, B) \leq \rho(A, B) \quad \textit{and} \quad \lambda(A) \leq \kappa(A, S) \leq \mu(A)$$

for all $A, B \in \Sigma$ if and only if the estimate

$$\lambda(A) - \mu(B) \leq \rho(A, B^c) - \omega(B, A^c)$$

holds for all disjoint $A, B \in \Sigma$.

We finally turn to the existence of antisymmetric and symmetric flows for a different kind of capacity constraints. For the following two results, we consider arbitrary set functions $\omega, \rho, \sigma, \tau \in \mathfrak{B}(\Sigma \times \Sigma)$ with the property that $\omega \leq \rho$ and $\sigma \leq \tau$ on $\Sigma \times \Sigma$.

Theorem 11 *There is a biadditive set function $\nu : \Sigma \times \Sigma \to \mathbb{R}$ for which*

$$\omega(A, B) \leq \nu(A, B) - \nu(B, A) \leq \rho(A, B) \;\textit{and}\; \sigma(A, B) \leq \nu(A, B) \leq \tau(A, B)$$

for all $A, B \in \Sigma$ if and only if the conditions

$$\omega(A, B) + \omega(B, A) \leq 0 \leq \rho(A, B) + \rho(B, A) \quad \textit{and}$$

$$\omega(A, B)\,, -\rho(B, A) \leq \tau(A, B) - \sigma(B, A)$$

hold for all $A, B \in \Sigma$.

Proof. The necessity of the latter conditions is again immediate. To verify the sufficiency, we consider the vector lattice F of all real-valued $\Sigma \circ \Sigma$-measurable simple functions on $S \times S$ and the linear operator $T : F \to F$, given by $(Tw)(s, t) := w(s, t) - w(t, s)$ for all $f \in F$ and $s, t \in S$. Then it is clear that $T\chi_{A \times B} = \chi_{A \times B} - \chi_{B \times A}$ for all $A, B \in \Sigma$. With the canonical identification of the spaces F^* and $\mathfrak{B}(\Sigma \times \Sigma)$, it follows that the algebraic adjoint of T is given by $(T^*\nu)(A, B) = \nu(A, B) - \nu(B, A)$ for all $A, B \in \Sigma$.

Now let $\varphi_1, \varphi_2, \psi_1, \psi_2 \in F^*$ denote the linear functionals on F which correspond to the set functions $\omega, \rho, \sigma, \tau \in \mathfrak{B}(\Sigma \times \Sigma)$ via integration. In this context, assertion (a) of Theorem 2 means precisely the existence of a set function $\nu \in \mathfrak{B}(\Sigma \times \Sigma)$ with the desired properties. Hence it remains to be seen that the feasibility conditions of Theorem 11 imply, in the present setting, condition (b) of Theorem 2, namely

$$\varphi_1(w^+) - \varphi_2(w^-) \leq \psi_2\left((Tw)^+\right) - \psi_1\left((Tw)^-\right) \quad \text{for all } w \in F. \tag{4}$$

In the particular case of a function of the form $w = \alpha\,\chi_{A\times A}$ with $\alpha \in \mathbb{R}$ and $A \in \Sigma$, the inequality (4) follows immediately from the fact that $Tw = 0$ and $\omega(A, A) \leq 0 \leq \rho(A, A)$.

We next consider the case $w = \alpha\,\chi_{A\times B} + \beta\,\chi_{B\times A}$, where $\alpha, \beta \in \mathbb{R}$ are arbitrary real numbers and $A, B \in \Sigma$ are disjoint sets. By symmetry, we may and do assume that $\alpha \geq \beta$. In this case, the inequality (4) is equivalent to

$$\alpha^+\omega(A,B) + \beta^+\omega(B,A) - \alpha^-\rho(A,B) - \beta^-\rho(B,A) \leq$$
$$\leq (\alpha - \beta)\,(\tau(A,B) - \sigma(B,A))\,.$$

If $\alpha \geq \beta \geq 0$, we obtain from our assumptions that

$$\begin{aligned}\alpha\,\omega(A,B) + \beta\,\omega(B,A) &= (\alpha-\beta)\,\omega(A,B) + \beta\,(\omega(A,B)+\omega(B,A))\\ &\leq (\alpha-\beta)\,\omega(A,B) \leq (\alpha-\beta)\,(\tau(A,B)-\sigma(B,A))\,,\end{aligned}$$

as desired. Similarly, if $0 \geq \alpha \geq \beta$, we conclude that

$$\begin{aligned}\alpha\,\rho(A,B) + \beta\,\rho(B,A) &= \alpha\,(\rho(A,B)+\rho(B,A)) + (\beta-\alpha)\,\rho(B,A)\\ &\leq (\beta-\alpha)\,\rho(B,A) \leq (\alpha-\beta)\,(\tau(A,B)-\sigma(B,A))\,,\end{aligned}$$

again as desired. Finally, if $\alpha \geq 0 \geq \beta$, we obtain from our assumptions that

$$\alpha\,\omega(A,B) + \beta\,\rho(B,A) \leq (\alpha-\beta)\,(\tau(A,B)-\sigma(B,A))\,,$$

which completes the proof of (4) for functions of the form $w = \alpha\,\chi_{A\times B} + \beta\,\chi_{B\times A}$.

To handle the general case, we now consider an arbitrary $w \in F$, and represent this function in the form

$$w = \sum_{j=1}^{n}\sum_{k=1}^{n} \alpha_{jk}\,\chi_{A_j\times A_k}$$

with real numbers α_{jk} and pairwise disjoint sets $A_j \in \Sigma$. Then the preceding estimates apply separately to each of the terms in the representation

$$w = \sum_{k=1}^{n} \alpha_{kk}\,\chi_{A_k\times A_k} + \sum_{j<k}\left(\alpha_{jk}\,\chi_{A_j\times A_k} + \alpha_{kj}\,\chi_{A_k\times A_j}\right),$$

and may be added to ensure that (4) holds for any $w \in F$. Hence the assertion follows indeed from Theorem 2. □

Theorem 12 *There is a biadditive set function $\nu : \Sigma \times \Sigma \to \mathbb{R}$ for which*

$$\omega(A,B) \le \nu(A,B) + \nu(B,A) \le \rho(A,B) \text{ and } \sigma(A,B) \le \nu(A,B) \le \tau(A,B)$$

for all $A, B \in \Sigma$ if and only if the conditions

$$\omega(A,B) \le \tau(A,B) + \tau(B,A)\ ,\ \sigma(A,B) + \sigma(B,A) \le \rho(A,B)$$
$$\text{and}\quad \omega(A,B) \le \rho(B,A)$$

hold for all $A, B \in \Sigma$.

Proof. The assertion is again a consequence of Theorem 2, which has to be applied to the same functionals as those considered in the preceding proof and now to the operator $T : F \to F$, given by $(Tw)(s,t) := w(s,t) + w(t,s)$ for all $w \in F$ and $s, t \in S$. Note that this operator is not a lattice homomorphism, but fortunately it may be handled exactly as the corresponding operator from the proof of Theorem 11.

Indeed, a simple calculation shows that the feasibility conditions of Theorem 12 imply that the estimate $\varphi_1(w^+) - \varphi_2(w^-) \le \psi_2((Tw)^+) - \psi_1((Tw)^-)$ holds for all functions of the form $w = \alpha\,\chi_{A\times A}$ and $w = \alpha\,\chi_{A\times B} + \beta\,\chi_{B\times A}$ for arbitrary $\alpha, \beta \in \mathbb{R}$ and disjoint sets $A, B \in \Sigma$. As before, from this it easily follows that condition (b) of Theorem 2 is fulfilled. □

7. Marginal Problems

In this short section, we consider two non-empty sets S_1 and S_2, endowed with algebras of subsets Σ_1 and Σ_2, respectively, as well as biadditive set functions $\sigma\,,\,\tau : \Sigma_1 \times \Sigma_2 \to \mathbb{R}$ such that $\sigma \le \tau$ on $\Sigma_1 \times \Sigma_2$, and additive set functions $\lambda_k\,,\,\mu_k : \Sigma_k \to \mathbb{R}$ such that $\lambda_k \le \mu_k$ on Σ_k for $k = 1, 2$.

The following two theorems provide solutions to the corresponding one- and two-sided marginal problems for biadditive set functions. Results of this kind date back to Kellerer [7], [8] in a measure-theoretic context and are usually developed with completely different methods.

Theorem 13 *There exists a biadditive set function $\nu : \Sigma_1 \times \Sigma_2 \to \mathbb{R}$ such that*

$$\lambda_1(A) \le \nu(A, S_2) \le \mu_1(A) \quad \text{and} \quad \sigma(A,B) \le \nu(A,B) \le \tau(A,B)$$

for all $A \in \Sigma_1\,,\ B \in \Sigma_2$ if and only if

$$\lambda_1(A) \le \tau(A, S_2) \quad \text{and} \quad \sigma(A, S_2) \le \mu_1(A)$$

for all $A \in \Sigma_1$.

Proof. Let X and Y denote the vector lattices of all real-valued measurable simple functions on S_1 and $S_1 \times S_2$, respectively, and consider the linear functionals $\varphi_1, \varphi_2 \in X^*$ and $\psi_1, \psi_2 \in Y^*$ which correspond, via integration, with the set functions $\lambda_1, \mu_1 \in \mathfrak{A}(\Sigma_1)$ and $\sigma, \tau \in \mathfrak{B}(\Sigma_1 \times \Sigma_2)$, respectively. Since the definition $(Tf)(s,t) := f(s)$ for all $(s,t) \in S_1 \times S_2$ yields a lattice homomorphism $T : X \to Y$, the assertion follows from Theorem 2 via Remark 3 (iii). □

Theorem 14 *There exists a biadditive set function* $\nu : \Sigma_1 \times \Sigma_2 \to \mathbb{R}$ *such that*

$$\lambda_1(A) \leq \nu(A, S_2) \leq \mu_1(A) \ , \ \lambda_2(B) \leq \nu(S_1, B) \leq \mu_2(B)$$
$$\textit{and} \quad \sigma(A,B) \leq \nu(A,B) \leq \tau(A,B)\,,$$

for all $A \in \Sigma_1\,,\ B \in \Sigma_2$ *if and only if*

$$\lambda_1(A) - \mu_2(B)\,,\ \lambda_2(B^c) - \mu_1(A^c) \leq \tau(A, B^c) - \sigma(A^c, B)$$

for all $A \in \Sigma_1\,,\ B \in \Sigma_2$.

Proof. We may assume that S_1 and S_2 are disjoint, and endow $S := S_1 \cup S_2$ with the canonical algebra $\Sigma := \{A \subseteq S : A \cap S_1 \in \Sigma_1 \text{ and } A \cap S_2 \in \Sigma_2\}$. Moreover, let $\lambda, \mu \in \mathfrak{A}(\Sigma)$ be given by $\lambda(A) := \lambda_1(A \cap S_1) + \lambda_2(A \cap S_2)$ and $\mu(A) := \mu_1(A \cap S_1) + \mu_2(A \cap S_2)$ for all $A \in \Sigma$. Similarly, let $\sigma, \tau \in \mathfrak{B}(\Sigma \times \Sigma)$ denote the natural extensions of the set functions σ and τ from $\Sigma_1 \times \Sigma_2$ to $\Sigma \times \Sigma$, so that $\sigma(A,B) := \sigma(A \cap S_1, B \cap S_2)$ and $\tau(A,B) := \tau(A \cap S_1, B \cap S_2)$ for all $A \in \Sigma_1$, $B \in \Sigma_2$. The assertion is then a direct consequence of Theorem 8. □

We note that the preceding result can also be obtained from Theorem 5; see also [15]. Of particular interest in Theorem 14 is the classical case that $\lambda_1 = \mu_1$ and $\lambda_2 = \mu_2$, in which the marginal functions for ν are completely described. It is easily seen that, in this case, the feasibility conditions of Theorem 14 are equivalent to $\lambda_1(S_1) = \lambda_2(S_2)$ and $\lambda_1(A) - \lambda_2(B) \leq \tau(A, B^c) - \sigma(A^c, B)$ for all $A \in \Sigma_1\,,\ B \in \Sigma_2$. These conditions were introduced by Kellerer [7] in his investigation of marginal problems for functions and measures.

Let us finally mention that a similar approach may be used in the case of product spaces with more than two factors. Indeed, it is not hard to verify that, in this case, part (ii) of Remark 3 leads to an extension of a corresponding result due to Kellerer [8] for bounded measures on product spaces with an arbitrary number of prescribed marginals.

8. Concluding Remarks

We close with a brief discussion of some extensions and applications of the preceding results. First observe that the Hahn - Banach and sandwich theorems used in this paper actually hold in the more general context of operators from a real vector space into a Dedekind complete ordered real vector space.

Consequently, the algebraic versions of the results of this paper remain valid for operators and set functions with values in a Dedekind complete ordered vector space. A typical example of such a space is the vector lattice $L_0(\Omega, \mathfrak{S}, \mathrm{P})$ consisting of the equivalence classes of all real-valued measurable functions on a given complete probability space $(\Omega, \mathfrak{S}, \mathrm{P})$, endowed with the usual order structure induced by the cone of almost everywhere non-negative measurable functions. In this setting, the vector-valued extensions of our results lead, in an obvious way, to probabilistic versions of our feasibility theorems for flows and potentials.

Moreover, it is easily seen that our results on bounded flows and potentials remain valid for additive and biadditive set functions with values in a normed Dedekind complete vector lattice.

As intimated earlier, the theory of biadditive set functions subsumes the case of bounded measures on product spaces. More precisely, if Σ_1 and Σ_2 are σ-algebras of subsets of the non-empty sets S_1 and S_2, respectively, and if $\sigma, \tau : \Sigma_1 \otimes \Sigma_2 \to \mathbb{R}$ are finite real measures such that the estimate $\sigma \leq \tau$ holds on the product σ-algebra $\Sigma_1 \otimes \Sigma_2$, then for every biadditive set function $\widehat{\nu} : \Sigma_1 \times \Sigma_2 \to \mathbb{R}$ with the property that $\sigma(A \times B) \leq \widehat{\nu}(A, B) \leq \tau(A \times B)$ for all $A \in \Sigma_1$ and $B \in \Sigma_2$, there exists exactly one real measure $\nu : \Sigma_1 \otimes \Sigma_2 \to \mathbb{R}$ such that $\sigma \leq \nu \leq \tau$ on $\Sigma_1 \otimes \Sigma_2$ and $\nu(A \times B) = \widehat{\nu}(A, B)$ for all $A \in \Sigma_1$ and $B \in \Sigma_2$. This result follows easily from the natural identification of the biadditive set functions on $\Sigma_1 \times \Sigma_2$ with the additive set functions on the algebra generated by $\Sigma_1 \times \Sigma_2$, together with a standard extension theorem from measure theory; see also Lemma 7.3 of [10] for details.

Consequently, in the setting of the preceding paragraph, the biadditive set functions on $\Sigma_1 \times \Sigma_2$ correspond canonically with the real measures on $\Sigma_1 \otimes \Sigma_2$. We conclude that our feasibility theorems for biadditive set functions lead to similar results for bounded measures on product spaces. Moreover, an obvious combination with the Radon - Nikodým theorem then covers the case of integrable functions. In this sense, the results of Section 7 subsume and extend the measure-theoretic results due to Kellerer [7] , [8].

In the same vein, we observe that the abstract theory of Section 2 have immediate applications to topological measure theory. In fact, our general results apply to operators between spaces of continuous functions on locally compact Hausdorff spaces and provide, via the Riesz representation theorem, the existence of regular Borel measures with certain properties. Let us note in passing that biadditive set functions are also very suitable in the context of Radon bimeasures in the sense of Berg - Christensen - Ressel [1].

Concerning further applications of the theory of flows in infinite networks, we finally mention Ford - Fulkerson type theorems for maximal flows and minimal cuts as well as certain results on supply and demand problems from mathematical economics. For details we refer to [4], [9], [14], [15], [17], [18], and again to the monographs [5] and [10].

Acknowledgment

The results of this paper were presented by the first author in the Workshop on Minimax Theory and Applications at the Ettore Majorana Center for Scientific Culture in Erice, Sicily, in October 1996. The warm hospitality, inspiring atmosphere, and excellent organization of this meeting are very much appreciated. Our special thanks go to the directors of the workshop, Biagio Ricceri and Stephen Simons.

References

1. C. Berg, J.P.R. Christensen and P. Ressel, *Harmonic Analysis on Semigroups,* Graduate Texts in Mathematics, Vol. 100, Springer-Verlag, Berlin, 1984.
2. L.R. Ford and D.R. Fulkerson, *Flows in Networks,* Princeton University Press, Princeton, NJ, 1962.
3. B. Fuchssteiner, *An abstract disintegration theorem,* Pacific J. Math. **94** (1981), 303-309.
4. B. Fuchssteiner, *Disintegration methods in mathematical economics,* in O. Moeschlin and D. Pallaschke (eds.), Game Theory and Mathematical Economics, North-Holland Publishing Company, Amsterdam, 1981, 193-204.
5. B. Fuchssteiner and W. Lusky, *Convex Cones,* North-Holland Math. Studies, Vol. 56, North-Holland Publishing Company, Amsterdam, 1981.
6. S. Goldberg, *Unbounded Linear Operators,* Dover Publications, New York, 1985.
7. H.G. Kellerer, *Funktionen auf Produkträumen mit vorgegebenen Marginal-Funktionen,* Math. Ann. **144** (1961), 323-344.
8. H.G. Kellerer, *Maßtheoretische Marginalprobleme,* Math. Ann. **153** (1964), 168-198.
9. H. König, *On some basic theorems in convex analysis,* in B. Korte (ed.), Modern Applied Mathematics, North-Holland Publishing Company, Amsterdam, 1982, pp. 107-144.
10. H. König and M.M. Neumann, *Mathematische Wirtschaftstheorie mit einer Einführung in die konvexe Analysis,* Math. Systems in Economics Vol. 100, Hain Verlag bei Athenäum, Königstein, 1986.
11. P. Meyer-Nieberg, *Banach Lattices,* Springer-Verlag, Berlin, 1991.
12. S. Mazur and W. Orlicz, *Sur les espaces métriques linéaires II,* Studia Math. **13** (1953), 137-179.
13. M.M. Neumann, *Bemerkungen zum von Neumannschen Minimaxtheorem,* Arch. Math. (Basel) **29** (1977), 96-105.
14. M.M. Neumann, *A Ford-Fulkerson type theorem concerning vector-valued flows in infinite networks,* Czech. Math. J. **34** (1984), 156-162.
15. M.M. Neumann, *The theorem of Gale for infinite networks and applications,* in E.J. Anderson and A.B. Philpott (eds.), Infinite Programming, Lecture Notes in Economics and Math. Systems, Vol. 259, Springer-Verlag, Berlin, 1985, 154-171.

16. M.M. Neumann, *Sandwich theorems for operators of convex type,* J. Math. Anal. Appl. **188** (1994), 759-773.
17. W. Oettli and M. Yamasaki, *On Gale's feasibility theorem for certain infinite networks,* Arch. Math. **62** (1994), 378-384.
18. W. Oettli and M. Yamasaki, *Existence of feasible potentials on infinite networks,* J. Convex Anal. **3** (1996), 71-81.
19. V. Pták, *On a theorem of Mazur and Orlicz,* Studia Math. **15** (1956), 365-366.
20. R.T. Rockafellar, *Network Flows and Monotropic Optimization,* John Wiley and Sons, New York, 1984.
21. S. Simons, *Extended and sandwich versions of the Hahn-Banach theorem,* J. Math. Anal. Appl. **21** (1968), 112-122.

ON A TOPOLOGICAL MINIMAX THEOREM AND ITS APPLICATIONS

BIAGIO RICCERI
Department of Mathematics
University of Catania
95125 Catania, Italy
e-mail: ricceri@dipmat.unict.it

1. Introduction

Let X, Y be two non-empty sets and let f be a real function defined on $X \times Y$. The classical minimax problem is to find suitable conditions under which the equality

$$\sup_{y \in Y} \inf_{x \in X} f(x,y) = \inf_{x \in X} \sup_{y \in Y} f(x,y) \tag{1.1}$$

does hold.

We refer to the excellent survey given in [18] for a general account on the subject.

The aim of the present paper, which groups together the contents of a series of three lectures given at the workshop, is mainly to give an overview of some applications of Theorem 1.1 below which do not seem to be directly obtainable from other known minimax theorems.

Theorem 1.1 ([12], Theorem 1.1) *Let X, Y be topological spaces, with Y connected and admitting a continuous bijection onto $[0,1]$. Assume that, for each $\lambda \in \mathbf{R}$, $x_0 \in X$, $y_0 \in Y$, the sets*

$$\{x \in X : f(x, y_0) \leq \lambda\}$$

and

$$\{y \in Y : f(x_0, y) > \lambda\}$$

B. Ricceri and S. Simons (eds.), Minimax Theory and Applications. 191–216.

are connected. In addition, assume that at least one of the following three sets of conditions is satisfied:
(h_1) *$f(x,\cdot)$ is upper semicontinuous in Y for each $x \in X$, and $f(\cdot,y)$ is lower semicontinuous in X for each $y \in Y$;*
(h_2) *Y is compact, and f is upper semicontinuous in $X \times Y$;*
(h_3) *X is compact, and f is lower semicontinuous in $X \times Y$.*
Under such hypotheses, equality (1.1) does hold.

To appreciate the novelties contained in Theorem 1.1, it is better to recall the statement of a result of H. König ([6]) which arrived at the end of a long refining process related to Wu's minimax theorem (see [2],[3],[19],[20]).

Theorem 1.2 ([6], Theorem 1.2) *Let X,Y be topological spaces, with X compact. Assume that, for each $x \in X$, the function $f(x,\cdot)$ is upper semicontinuous in Y and that, for each $y \in Y$, the function $f(\cdot,y)$ is lower semicontinuous in X. Further, assume that:*
(i_1) *for each $\lambda \in \mathbf{R}$ and each non-empty finite set $H \subseteq Y$, the set*

$$\bigcap_{y\in H} \{x \in X : f(x,y) \leq \lambda\}$$

is connected;
(i_2) *for each $\lambda \in \mathbf{R}$ and each non-empty set $H \subseteq X$, the set*

$$\bigcap_{x\in H} \{y \in Y : f(x,y) > \lambda\}$$

is connected.
Then, equality (1.1) does hold.

The comparison between Theorems 1.1 and 1.2 is then clear. In Theorem 1.2, the space Y is arbitrary, but we are in the presence of conditions (i_1), (i_2) whose checking is, in general, very difficult out of the (classical) case where, X,Y being convex sets in topological vector spaces, the sets $\{x \in X : f(x,y) \leq \lambda\}$ and $\{y \in Y : f(x,y) > \lambda\}$ are convex. In Theorem 1.1, on the contrary, the space Y is of very special type, but X is arbitrary and we are able to assume simply the connectedness of those sets, an optimal condition from a topological point of view.

The following example shows, at the same time, that Theorems 1.1 and 1.2 are no longer true if, respectively, Y does not admit any continuous bijection onto $[0,1]$ and condition (i_1) is replaced by requiring that, for each $\lambda \in \mathbf{R}, y \in Y$, the set $\{x \in X : f(x,y) \leq \lambda\}$ is connected.

Example 1.1 Take :

$$X = \{(t,u) \in \mathbf{R}^2 : t^2 + u^2 = 1\}$$

$$Y = \{(v, z) \in \mathbf{R}^2 : v^2 + z^2 \leq 1\}$$

and, for each $(t, u) \in X, (v, z) \in Y$,

$$f(t, u, v, z) = tv + uz.$$

So, for fixed $(t, u) \in X$, we have

$$\sup_{(v,z) \in Y} f(t, u, v, z) = \sqrt{t^2 + u^2} = 1.$$

Moreover, for fixed $(v, z) \in Y$, we have

$$\inf_{(t,u) \in X} f(t, u, v, z) = -\sqrt{v^2 + z^2}.$$

Hence, it follows that

$$0 = \sup_{(v,z) \in Y} \inf_{(t,u) \in X} f(t, u, v, z) < \inf_{(t,u) \in X} \sup_{(v,z) \in Y} f(t, u, v, z) = 1.$$

2. Preliminaries

In this section, for the reader's convenience, we recall a series of various results which will be used to prove the ones in the successive sections.

We first recall that if X, Y are two topological spaces, a multifunction F from X into Y is said to be lower (resp. upper) semicontinuous provided that, for each open (resp. closed) set $\Omega \subseteq Y$, the set $\{x \in X : F(x) \cap \Omega \neq \emptyset\}$ is open (resp. closed) in X. If $A \subseteq X$ we also put $F(A) = \cup_{x \in A} F(x)$.

Proposition 2.1 ([5], Theorem 7.4.4) *Let X, Y be two topological spaces, with X connected, and let F, G be two multifunctions from X into Y, with non-empty connected values. Assume that both F, G are either upper semicontinuous with compact values, or lower semicontinuous.*

Then, the set $\cup_{x \in X}(F(x) \times G(x))$ is connected.

Given a set $S \subseteq X \times Y$, for every $x \in X$ and $y \in Y$, we put

$$S_x = \{v \in Y : (x, v) \in S\}$$

and

$$S^y = \{u \in X : (u, y) \in S\}.$$

Proposition 2.2 ([12], Theorem 2.4) *Let X, Y be two topological spaces, with Y connected, and let S be a subset of $X \times Y$. Assume that at least one of the following two sets of conditions is satisfied:*

(a_1) S^y is non-empty and connected for each $y \in Y$, and S_x is open for each $x \in X$;
(a_2) X is compact, S is closed, and S^y is non-empty and connected for each $y \in Y$.

Then, S is connected.

Let E be a real vector space and X a non-empty subset of E. We denote by $\mathrm{aff}(X)$ the affine hull of X. Moreover, we put

$$B_X = \{x \in X : \exists y \in \mathrm{aff}(X) \setminus \{x\} : [x, y] \cap X = \{x\}\},$$

where $[x, y]$ denotes the line segment joining x and y.

When E is endowed with a vector topology, we denote, respectively, by $\mathrm{int}_r(X)$ and $\partial_r X$ the interior and the boundary of X in the relative topology of $\mathrm{aff}(X)$. The subscript r is, of course, deleted if $\mathrm{aff}(X) = E$. We have

Proposition 2.3 ([7], Proposition 2.1) *Let E be a real topological vector space and X a convex subset of E. Assume that $\partial_r X \subseteq X$ and $\mathrm{int}_r(X) \neq \emptyset$.*

Then, $B_X = \partial_r X$.

As usual, we say that X is finitely compact (resp. finitely closed), if for every finite-dimensional affine subspace S of E, the set $X \cap S$ is compact (resp. closed) in the Euclidean topology of S. Also, $\partial_S(X \cap S)$ will denote the boundary of $X \cap S$ in such a topology.

We denote by $\mathcal{S}(X)$ the family of all affine subspaces S of E such that $\dim(S) = \dim(X \cap S) = 2$ (if $Y \subseteq E$, $\dim(Y)$ denotes the algebraic dimension of Y). We have

Proposition 2.4 ([7], Proposition 2.2) *Let X be a non-empty, convex and finitely closed subset of a real vector space, with $\dim(X) \geq 2$.*

Then, one has

$$B_X = \bigcup_{S \in \mathcal{S}(X)} \partial_S(X \cap S).$$

Let us also recall that, when X is convex, a function $f : X \to \mathbf{R}$ is said to be quasi-concave if, for every $\lambda \in \mathbf{R}$, the set $\{x \in X : f(x) \geq \lambda\}$ is convex.

Two other facts we will use are the following

Proposition 2.5 ([7], Lemma 3.1) *Let X be a compact and convex subset of $\mathbf{R}^2$, with $\mathrm{int}(X) \neq \emptyset$, and H a closed half-plane of $\mathbf{R}^2$.*

Then, the set $H \cap \partial X$ is connected.

Proposition 2.6 ([1], Theorem 6, p. 318) *Let X be a compact topological space, Y a non-empty set, and f a real function on $X \times Y$. Assume that, for each $y \in Y$, the function $f(\cdot, y)$ is lower semicontinuous in X.*

Then, if $\mathcal{H}$ denotes the family of all finite non-empty subsets of Y, one has

$$\sup_{H \in \mathcal{H}} \inf_{x \in X} \sup_{y \in H} f(x,y) = \inf_{x \in X} \sup_{y \in Y} f(x,y).$$

Let us recall that the Hausdorff distance between two non-empty sets A, B in a metric space (denoted by $d_H(A,B)$) is given by

$$d_H(A,B) = \max \left\{ \sup_{x \in A} \operatorname{dist}(x,B), \sup_{x \in B} \operatorname{dist}(x,A) \right\}.$$

Proposition 2.7 ([11], Théorème 2) *Let X, Y be two real Banach spaces, let $\Phi : X \to Y$ be a continuous linear surjective operator, and let $\Psi : X \to Y$ be a Lipschitzian operator, with Lipschitz constant $L < \frac{1}{\alpha_\Phi}$, where $\alpha_\Phi = \sup_{\|y\| \le 1} \operatorname{dist}(0, \Phi^{-1}(y))$.*

Then, for each $y \in Y$, the set $(\Phi + \Psi)^{-1}(y)$ is a (non-empty) retract of X, and, for each $y, z \in Y$, one has

$$d_H((\Phi+\Psi)^{-1}(y), (\Phi+\Psi)^{-1}(z)) \le \frac{\alpha_\Phi}{1 - L\alpha_\Phi} \parallel y - z \parallel .$$

Let $(T, \mathcal{F}, \mu)$ be a σ-finite non-atomic measure space, E a real Banach space ($E \neq \{0\}$), and p a real number greater than or equal to 1.

As usual, $L^p(T,E)$ denotes the space of all (equivalence classes of) strongly μ-measurable functions $u : T \to E$ such that $\int_T \parallel u(t) \parallel^p d\mu < +\infty$, equipped with the norm $\parallel u \parallel_{L^p(T,E)} = (\int_T \parallel u(t) \parallel^p d\mu)^{\frac{1}{p}}$.

A set $D \subseteq L^p(T,E)$ is said to be decomposable if, for every $u, v \in D$ and every $S \in \mathcal{F}$, the function $t \to \chi_S(t)u(t) + (1 - \chi_S(t))v(t)$ belongs to D, where χ_S denotes the characteristic function of S.

A function $f : T \times E \to \mathbf{R}$ is said to be sup-measurable if for every strongly μ-measurable function $u : T \to E$, the function $t \to f(t, u(t))$ is μ-measurable.

Proposition 2.8 ([17], Théorème 7) *Let $f : T \times E \to \mathbf{R}$ be a sup-measurable function, and let $D \subseteq L^p(T,E)$ be a decomposable set.*

Then, if we put

$$Y = \{u \in D : f(\cdot, u(\cdot)) \in L^1(T)\},$$

for each $\lambda \in \mathbf{R}$, *the set*

$$\left\{u \in Y : \int_T f(t, u(t))d\mu \leq \lambda\right\}$$

is arcwise connected.

Proposition 2.9 ([5], p. 187) *Let* $f_1, ..., f_n$ *be* n *sup-measurable real functions on* $T \times E$, *and let* $D \subseteq L^p(T, E)$ *be a decomposable set. Assume that, for each* $u \in D$ *and each* $i = 1, ..., n$, *the function* $t \to f_i(t, u(t))$ *belongs to* $L^1(T)$.

Then, the set

$$\left\{\left(\int_T f_1(t, u(t))d\mu, ..., \int_T f_n(t, u(t))d\mu\right) : u \in D\right\}$$

is convex.

3. Proof of Theorem 1.1

The proof of Theorem 1.1 is based on the following principle:

Theorem 3.1 ([12], Theorem 2.3) *Let* X, Y *be two topological spaces, with* Y *admitting a continuous bijection onto* $[0, 1]$, *and let* S, T *be two subsets of* $X \times Y$, *with* S *connected and* $S^y \neq \emptyset$ *for all* $y \in Y$. *Assume that at least one of the following two sets of conditions is satisfied:*
(α) T_x *is non-empty and connected for each* $x \in X$, *and* T^y *is open for each* $y \in Y$;
(β) Y *is compact,* T *is closed, and* T_x *is non-empty and connected for each* $x \in X$.

Then, one has $S \cap T \neq \emptyset$.

Proof. Let φ be any continuous bijection from Y onto $[0, 1]$. Arguing by contradiction, assume that $S \cap T = \emptyset$. For each $(x, y) \in S$, put

$$H(x, y) = \varphi(y) \times \varphi(T_x).$$

Observe that the multifunction $x \to T_x$ is clearly lower semicontinuous when (α) holds, while is upper semicontinuous when (β) holds ([5], Theorem 7.1.16). Consequently, thanks to Proposition 2.1, the set $H(S)$ is connected. Now, put

$$A = \{(t, u) \in [0, 1] \times [0, 1] : t < u\}$$

and

$$B = \{(t, u) \in [0, 1] \times [0, 1] : t > u\}.$$

Of course, we have

$$H(S) \subseteq A \cup B. \tag{3.1}$$

Pick $x_0, x_1 \in X$ in such a way that $(x_0, \varphi^{-1}(0)), (x_1, \varphi^{-1}(1)) \in S$. Moreover, choose $y_0 \in T_{x_0}$ and $y_1 \in T_{x_1}$. Of course, $\varphi(y_0) > 0$, since, otherwise, we would have $(x_0, y_0) \in S \cap T$. Likewise, we have $\varphi(y_1) < 1$. Consequently,

$$(0, \varphi(y_0)) \in A \cap H(S) \tag{3.2}$$

and

$$(1, \varphi(y_1)) \in B \cap H(S). \tag{3.3}$$

Clearly, since A, B are disjoint and open in $[0,1] \times [0,1]$, the simultaneous occurrence of (3.1), (3.2), (3.3) is against the fact that $H(S)$ is connected. So, the proof is complete. △

Proof of Theorem 1.1. Arguing by contradiction, assume that equality (1.1) does not hold. Then, we can choose a number λ in such a way that $\sup_{y\in Y} \inf_{x\in X} f(x,y) < \lambda < \inf_{x\in X} \sup_{y\in Y} f(x,y)$.

Let (h_1) hold. Put

$$S = \{(x,y) \in X \times Y : f(x,y) < \lambda\}$$

and

$$T = \{(x,y) \in X \times Y : f(x,y) > \lambda\}.$$

By the choice of λ, we have $S^y \neq \emptyset$ for all $y \in Y$. Moreover, since $S^y = \cup_{\mu<\lambda}\{x \in X : f(x,y) \leq \mu\}$, S^y is connected. Then, taking into account that S_x is open for all $x \in X$, it follows from Proposition 2.2 (case (a_1)) that S is connected. On the other hand, for all $x \in X$, T_x is connected and non-empty (again by the choice of λ), and T^y is open for all $y \in Y$. Consequently, by Theorem 3.1 (case (α)), we should have $S \cap T \neq \emptyset$ which is clearly false.

Now, let (h_2) hold. This time, put

$$S = \{(x,y) \in X \times Y : f(x,y) < \lambda\}$$

and

$$T = \{(x,y) \in X \times Y : f(x,y) \geq \lambda\}.$$

Reasoning as before, we get a contradiction by case (β) of Theorem 3.1.

Finally, let (h_3) hold. In this case, put

$$S = \{(x,y) \in X \times Y : f(x,y) \leq \lambda\}$$

and

$$T = \{(x,y) \in X \times Y : f(x,y) > \lambda\}.$$

This time the contradiction is obtained applying case (a_2) of Proposition 2.2 jointly with case (α) of Theorem 3.1. △

4. An Application of Theorem 1.1 to the Problem $\inf_X f = \inf_{\partial X} f$

Let E be a real normed space and X a closed, bounded, convex subset of E, with $\mathrm{int}(X) \neq \emptyset$. In this setting, a well-known class of functions $f : X \to \mathbf{R}$ satisfying the equality

$$\inf_{x \in X} f(x) = \inf_{x \in \partial X} f(x) \tag{4.1}$$

is that of bounded below, concave functions ([16], p.127). Theorem 1.1 turns out to be particularly useful in realizing another broad class of functions f satisfying (4.1). In this section, we report the main results on this matter obtained in [7].

We first need the following consequence of Theorem 1.1:

Theorem 4.1 ([7], Theorem 3.1) *Let X be a compact topological space, I a real interval, and f a real function defined on $X \times I$. For each $\lambda \in \mathbf{R}$, $x_0 \in X$, $t_0 \in I$, assume that the sets*

$$\{x \in X : f(x, t_0) \leq \lambda\}$$

and

$$\{t \in I : f(x_0, t) > \lambda\}$$

are connected, the function $f(x_0, \cdot)$ is upper semicontinuous in I, and the function $f(\cdot, t_0)$ is lower semicontinuous in X.

Then, one has

$$\sup_{t \in I} \inf_{x \in X} f(x, t) = \inf_{x \in X} \sup_{t \in I} f(x, t).$$

Proof. Fix any non-decreasing sequence $\{I_n\}$ of compact intervals such that $I = \bigcup_{n \in \mathbf{N}} I_n$. For each $n \in \mathbf{N}$, we can apply Theorem 1.1 to the restriction of f to $X \times I_n$. So, we have

$$\sup_{t \in I_n} \inf_{x \in X} f(x, t) = \inf_{x \in X} \sup_{t \in I_n} f(x, t).$$

On the other hand, if $\mathcal{H}$ denotes the family of all finite non-empty subsets of I, by Proposition 2.6, we have

$$\sup_{H \in \mathcal{H}} \inf_{x \in X} \sup_{t \in H} f(x, t) = \inf_{x \in X} \sup_{t \in I} f(x, t).$$

Then, since each $H \in \mathcal{H}$ is contained in some I_n, we get

$$\sup_{t \in I} \inf_{x \in X} f(x, t) = \sup_{n \in \mathbf{N}} \inf_{x \in X} \sup_{t \in I_n} f(x, t) \geq \sup_{H \in \mathcal{H}} \inf_{x \in X} \sup_{t \in H} f(x, t)$$

and hence

$$\sup_{t\in I}\inf_{x\in X} f(x,t) \geq \inf_{x\in X}\sup_{t\in I} f(x,t)$$

that yields, of course, the conclusion. △

Let E be a real vector space. Then, we denote by $\mathcal{Z}_E$ the family of all functions $\varphi : E \to \mathbf{R}$ for which there exist a linear functional ψ on E and a lower semicontinuous and monotone function $h : \mathbf{R} \to \mathbf{R}$ such that $\varphi(x) = h(\psi(x))$ for every $x \in E$.

The key lemma for the main result of this section is as follows:

Lemma 4.1 ([7], Lemma 3.2) *Let X be a compact and convex subset of $\mathbf{R}^2$, with $\operatorname{int}(X) \neq \emptyset$, I a real interval, and φ a real function defined on $\mathbf{R}^2 \times I$. Suppose that:*
(i) *for each $t \in I$, $\varphi(\cdot,t) \in \mathcal{Z}_{\mathbf{R}^2}$;*
(ii) *for each $x \in X$, $\varphi(x,\cdot)$ is upper semicontinuous and quasi-concave in I.*

Then, the function $f : X \to \overline{\mathbf{R}}$ defined by

$$f(x) = \sup_{t\in I}\varphi(x,t)$$

satisfies the equality

$$\inf_{x\in X} f(x) = \inf_{x\in\partial X} f(x).$$

Proof. By (i), for each $t \in I$, there exist a lower semicontinuous and monotone function $g_t : \mathbf{R} \to \mathbf{R}$, and a linear function $h_t : \mathbf{R}^2 \to \mathbf{R}$ such that $\varphi(x,t) = g_t(h_t(x))$ for every $x \in \mathbf{R}^2$. Observe that, for each $\lambda \in \mathbf{R}$, the set $\{s \in \mathbf{R} : g_t(s) \leq \lambda\}$ is either empty or a closed unbounded interval. Consequently, the set $\{x \in \mathbf{R}^2 : \varphi(x,t) \leq \lambda\}$ is either empty, or a closed half-plane, or the whole $\mathbf{R}^2$. So, in any case, we infer that the set $\{x \in X : \varphi(x,t) \leq \lambda\}$ is closed and convex, and that, by Proposition 2.5, the set $\{x \in \partial X : \varphi(x,t) \leq \lambda\}$ is closed and connected. Therefore, we are allowed to apply Theorem 4.1 to both φ and $\varphi_{|\partial X\times I}$. Hence, we have

$$\sup_{t\in I}\inf_{x\in X}\varphi(x,t) = \inf_{x\in X}\sup_{t\in I}\varphi(x,t) \tag{4.2}$$

as well as

$$\sup_{t\in I}\inf_{x\in\partial X}\varphi(x,t) = \inf_{x\in\partial X}\sup_{t\in I}\varphi(x,t). \tag{4.3}$$

For each $t \in I$, applying to h_t the known result recalled at the beginning of this section ([16], p.127), we get

$$\inf_{x\in X} g_t(h_t(x)) = g_t\left(\inf_{x\in X} h_t(x)\right) = g_t\left(\inf_{x\in\partial X} h_t(x)\right) = \inf_{x\in\partial X} g_t(h_t(x))$$

or

$$\inf_{x\in X} g_t(h_t(x)) = g_t\left(\sup_{x\in X} h_t(x)\right) = g_t\left(\sup_{x\in\partial X} h_t(x)\right) = \inf_{x\in\partial X} g_t(h_t(x))$$

according to whether g_t is non-decreasing or non-increasing. Now, our conclusion follows directly from (4.2) and (4.3), once observed that the left-hand sides are equal. △

Now, we can prove the main result of this section:

Theorem 4.2 ([7] Theorem 3.2) *Let E be a real vector space, X a convex and finitely compact subset of E, with $\dim(X) \geq 2$, I a real interval, $\mathcal{F}$ a family of real functions defined on $E \times I$. Suppose that every function $\varphi \in \mathcal{F}$ verifies the following conditions:*
(j) *for each $t \in I$, $\varphi(\cdot, t) \in \mathcal{Z}_E$;*
(jj) *for each $x \in X$, $\varphi(x, \cdot)$ is upper semicontinuous and quasi-concave in I.*

Then, the function $f : X \to \overline{\mathbf{R}}$ defined by

$$f(x) = \inf_{\varphi\in\mathcal{F}} \sup_{t\in I} \varphi(x,t)$$

satisfies the equality

$$\inf_{x\in X} f(x) = \inf_{x\in B_X} f(x).$$

Proof. Fix any $\varphi \in \mathcal{F}$ and any $S \in \mathcal{S}(X)$. Let us show that

$$\inf_{x\in X\cap S} \sup_{t\in I} \varphi(x,t) = \inf_{x\in\partial_S(X\cap S)} \sup_{t\in I} \varphi(x,t). \tag{4.4}$$

Let $x_0 \in X \cap S$. Then, $S = F + x_0$, where F is the linear hull of $(X\cap S) - x_0$. Let $\omega : \mathbf{R}^2 \to F$ be any surjective linear function. By (j), for each $t \in I$, there exist a monotone and lower semicontinuous function $g_t : \mathbf{R} \to \mathbf{R}$, and a linear function $h_t : E \to \mathbf{R}$ such that $\varphi(x,t) = g_t(h_t(x))$ for every $x \in E$. So, the function $\xi \to \varphi(\omega(\xi) + x_0, t)$ belongs to the class $\mathcal{Z}_{\mathbf{R}^2}$, since the function $s \to g_t(s + h_t(x_0))$ is monotone and lower semicontinuous, and the function $\xi \to h_t(\omega(\xi))$ is linear. Observing that $\xi \to \omega(\xi) + x_0$ is a homeomorphism with respect to the Euclidean topology on S, thanks to Lemma 4.1, we have

$$\inf_{x\in X\cap S} \sup_{t\in I} \varphi(x,t) = \inf_{\xi\in V_S} \sup_{t\in I} \varphi(\omega(\xi) + x_0, t) =$$

$$= \inf_{\xi\in\partial V_S} \sup_{t\in I} \varphi(\omega(\xi) + x_0, t) = \inf_{x\in\partial_S(X\cap S)} \sup_{t\in I} \varphi(x,t),$$

where $V_S = \{\xi \in \mathbf{R}^2 : \omega(\xi) + x_0 \in X \cap S\}$. So, (4.4) is proved. Now, taking into account Proposition 2.4 and the obvious fact that $X = \bigcup_{S \in \mathcal{S}(X)} X \cap S$, we get

$$\inf_{x \in X} \sup_{t \in I} \varphi(x,t) = \inf_{S \in \mathcal{S}(X)} \inf_{x \in X \cap S} \sup_{t \in I} \varphi(x,t) =$$

$$= \inf_{S \in \mathcal{S}(X)} \inf_{x \in \partial_S (X \cap S)} \sup_{t \in I} \varphi(x,t) = \inf_{x \in B_X} \sup_{t \in I} \varphi(x,t).$$

From this, we finally infer

$$\inf_{x \in X} f(x) = \inf_{x \in X} \inf_{\varphi \in \mathcal{F}} \sup_{t \in I} \varphi(x,t) = \inf_{\varphi \in \mathcal{F}} \inf_{x \in B_X} \sup_{t \in I} \varphi(x,t) =$$

$$= \inf_{x \in B_X} \inf_{\varphi \in \mathcal{F}} \sup_{t \in I} \varphi(x,t) = \inf_{x \in B_X} f(x)$$

as desired. △

As an example of application of Theorem 4.2, we now establish Theorem 4.3 below. As usual, a set X in a real topological vector space E is said to be bounded if for every neighbourhood U of the origin of E, there exists $\lambda > 0$ such that $X \subseteq \lambda U$.

Theorem 4.3 ([7], Theorem 3.3) *Let E be a real Hausdorff topological vector space, $\alpha_1, ..., \alpha_n$ n linear functionals on E, $\varphi, \varphi_0, \varphi_1, ..., \varphi_n$ $n+2$ concave real functions on an open interval $I \subseteq \mathbf{R}$, ψ a non-decreasing concave real function on $\mathbf{R}$. Then, if for each $i = 1, ..., n$, we put*

$$V_i = \begin{cases} \{x \in E : \alpha_i(x) \geq 0\} & \text{if } \varphi_i \text{ is not affine} \\ E & \text{if } \varphi_i \text{ is affine,} \end{cases}$$

for every closed, bounded and convex set $X \subseteq \bigcap_{i=1}^n V_i$, with $\dim(X) \geq 2$, the function $f : X \to \overline{\mathbf{R}}$ defined by

$$f(x) = \sup_{t \in I} \left(\psi \left(\sum_{i=1}^n \varphi_i(t)\alpha_i(x) + \varphi_0(t) \right) + \varphi(t) \right)$$

satisfies the equality

$$\inf_{x \in X} f(x) = \inf_{x \in B_X} f(x).$$

Proof. Let X be as in the statement. Since E is Hausdorff, on any finite-dimensional affine subspace of E, the relative topology agrees with the Euclidean one. Hence, X is finitely compact. Consider the function $\hat{\varphi} : E \times I \to \mathbf{R}$ defined by

$$\hat{\varphi}(x,t) = \psi\left(\sum_{i=1}^{n} \varphi_i(t)\alpha_i(x) + \varphi_0(t)\right) + \varphi(t).$$

Clearly, for each $t \in I$, $\hat{\varphi}(\cdot,t)$ belongs to $\mathcal{Z}_E$, since it is the composition of the linear functional $x \to \sum_{i=1}^{n} \varphi_i(t)\alpha_i(x)$ and the continuous non-decreasing function $s \to \psi(s + \varphi_0(t)) + \varphi(t)$. Moreover, our assumptions imply that, for each $x \in X$, $\hat{\varphi}(x,\cdot)$ is concave (and hence continuous) on I. Now, applying Theorem 4.2 to $\hat{\varphi}$, we directly get our conclusion. △

In particular, applying Theorem 4.3, we can get the following

Theorem 4.4 ([7], Theorem 3.4) *Let E be a real Hausdorff topological vector space, p a real number greater than 1, and $\beta_1, \beta_2, \beta_3$ three affine functionals on E, with $\beta_3(0) \geq 0$.*

Then, for every closed, bounded and convex set $X \subseteq \beta_3^{-1}(]0,+\infty[) \cap \beta_3^{-1}([\beta_3(0),+\infty[)$, with $\dim(X) \geq 2$, one has

$$\inf_{x\in X}\left(\beta_1(x) + \left(\frac{|\beta_2(x)|^p}{\beta_3(x)}\right)^{\frac{1}{p-1}}\right) = \inf_{x\in B_X}\left(\beta_1(x) + \left(\frac{|\beta_2(x)|^p}{\beta_3(x)}\right)^{\frac{1}{p-1}}\right).$$

Proof. Apply Theorem 4.3 taking $I = \mathbf{R}$, $\alpha_i(x) = \beta_i(x) - \beta_i(0)$ $(i = 1,2,3)$, $\varphi(t) = \beta_1(0) + \beta_2(0)t - \frac{(p-1)^{p-1}}{p^p}\beta_3(0)|t|^p$, $\varphi_0(t) = 0$, $\varphi_1(t) = 1$, $\varphi_2(t) = t$, $\varphi_3(t) = -\frac{(p-1)^{p-1}}{p^p}|t|^p$, $\psi = \mathrm{id}_{\mathbf{R}}$. Now, to conclude, it suffices to observe that, for each $x \in X$, since $\beta_3(x) > 0$, one has

$$\sup_{t\in\mathbf{R}}\left(\beta_1(x) + \beta_2(x)t - \frac{(p-1)^{p-1}}{p^p}\beta_3(x)|t|^p\right) = \beta_1(x) + \left(\frac{|\beta_2(x)|^p}{\beta_3(x)}\right)^{\frac{1}{p-1}},$$

as a simple calculation shows. △

We conclude observing that, in Theorem 4.4, the boundedness of X is essential. To see this, take $E = \mathbf{R}^2$, $p = 2$, $\beta_1(u,v) = v$, $\beta_2(u,v) = u$, $\beta_3(u,v) = 1$, $X = \{(u,v) \in \mathbf{R}^2 : u^2 + v \leq 0\}$. Clearly, we have $\inf_{(u,v)\in X}(v + u^2) = -\infty$, while $\inf_{(u,v)\in B_X}(v + u^2) = 0$ (take into account that, by Proposition 2.3, $B_X = \partial X$).

5. A Variational Property of Integral Functionals

Throughout this section, $(T, \mathcal{F}, \mu)$ is a σ-finite non-atomic measure space ($\mu(T) > 0$), $(E, \|\cdot\|)$ is a real Banach space ($E \neq \{0\}$), and p is a number in $[1, +\infty[$.

Let X be a linear subspace of $L^p(T, E)$. We denote by $\mathcal{V}(X)$ the family of all sets $V \subseteq X$ of the following type:

$$V = \left\{ u \in X : \Psi(u) = \int_T g(t, u(t)) d\mu \right\}$$

where Ψ is a continuous linear functional on X, and $g : T \times E \to \mathbf{R}$ is such that the integral functional $u \to \int_T g(t, u(t)) d\mu$ is (well-defined and) Lipschitzian in X, with Lipschitz constant strictly less than $\|\Psi\|_{X^*}$. Also, we denote by $\mathcal{I}(X)$ the family of all closed hyperplanes of X, that is the family of all sets V of the type $V = \{u \in X : \Psi(u) = r\}$, where $\Psi \in X^* \setminus \{0\}$, $r \in \mathbf{R}$. Clearly, $\mathcal{I}(X) \subseteq \mathcal{V}(X)$.

Lemma 5.1 ([13], Lemma 1) *Let X be a closed decomposable linear subspace of $L^p(T, E)$, Ψ be a continuous linear functional on X and f, g two real functions on $T \times E$. Assume that:*

(i) *the functional $u \to \int_T f(t, u(t)) d\mu$ is (well defined) bounded below and uniformly continuous in X;*

(ii) *the functional $u \to \int_T g(t, u(t)) d\mu$ is (well defined and) Lipschitzian in X, with Lipschitz constant c strictly less than $\|\Psi\|_{X^*}$;*

(iii) *there exist two sequences $\{u_n\}$, $\{v_n\}$ in $X \setminus \{0\}$ such that*

$$\lim_{n \to +\infty} \|u_n\|_X = \lim_{n \to +\infty} \|v_n\|_X = +\infty,$$

$$\inf_{n \in \mathbf{N}} \min \left\{ \Psi\left(\frac{u_n}{\|u_n\|_X} \right), \Psi\left(\frac{v_n}{\|v_n\|_X} \right) \right\} > c$$

and

$$\lim_{n \to +\infty} \frac{\int_T f(t, -u_n(t)) d\mu}{\|u_n\|_X} = \lim_{n \to +\infty} \frac{\int_T f(t, v_n(t)) d\mu}{\|v_n\|_X} = 0.$$

Then, one has

$$\inf_{u \in V} \int_T f(t, u(t)) d\mu = \inf_{u \in X} \int_T f(t, u(t)) d\mu$$

where

$$V = \left\{ u \in X : \Psi(u) = \int_T g(t, u(t)) d\mu \right\}.$$

When $c = 0$, the closedness of X can be omitted.

Proof. First, assume $c > 0$. Observe that, since X is complete and (by the Ascoli lemma)

$$\sup_{|s|\leq 1} \operatorname{dist}(0, \Psi^{-1}(s)) = \frac{1}{\|\Psi\|_{X^*}}$$

and, by condition (ii), $c < \|\Psi\|_{X^*}$, we can apply Proposition 2.7 to Ψ and Φ_g. So, in particular, $V \neq \emptyset$. Now, arguing by contradiction, suppose that $\inf_X \Phi_f < \inf_V \Phi_f$. Fix any $\epsilon \in]0, \inf_V \Phi_f - \inf_X \Phi_f[$. Since Φ_f is assumed to be uniformly continuous, there is some $\delta > 0$ such that $|\Phi_f(u) - \Phi_f(v)| < \epsilon$ for each $u, v \in X$ satisfying $\|u - v\|_X < \delta$. Put

$$r = \frac{2(\inf_V \Phi_f - \inf_X \Phi_f)}{(\|\Psi\|_{X^*} - c)\delta} .$$

Let $u \in X$. By Proposition 2.7 again, there exists some $w \in V$ such that

$$\|u - w\|_X < \frac{|\Psi(u) - \Phi_g(u)|}{\|\Psi\|_{X^*} - c} + \frac{\delta}{2} . \tag{5.1}$$

If $|\Psi(u) - \Phi_g(u)| \geq (\|\Psi\|_{X^*} - c)\delta/2$, we have

$$\Phi_f(u) + r|\Psi(u) - \Phi_g(u)| \geq \inf_X \Phi_f + r\frac{(\|\Psi\|_{X^*} - c)\delta}{2} = \inf_V \Phi_f.$$

If, otherwise, $|\Psi(u) - \Phi_g(u)| < (\|\Psi\|_{X^*} - c)\delta/2$, then, by (5.1), we get $\|u - w\|_X < \delta$, and hence

$$\Phi_f(u) + r|\Psi(u) - \Phi_g(u)| \geq \Phi_f(u) > \Phi_f(w) - \epsilon \geq \inf_V \Phi_f - \epsilon.$$

It remains so proved that

$$\inf_{u\in X} (\Phi_f(u) + r|\Psi(u) - \Phi_g(u)|) \geq \inf_V \Phi_f - \epsilon > \inf_X \Phi_f. \tag{5.2}$$

Now, observe that, by the Hahn-Banach theorem, Ψ is the restriction to X of a continuous linear functional on $L^p(T, E)$. So, by a well-known representation theorem (see [4], pp.94-99), there exists a mapping $A \to E^*$ such that $\Psi(u) = \int_T A(t)(u(t))d\mu$ for all $u \in X$. Then, by Proposition 2.8, for each $\xi, \lambda \in \mathbf{R}$, the set

$$\{u \in X : \Phi_f(u) + \xi(\Psi(u) - \Phi_g(u)) \leq \lambda\}$$

is connected in X. With this in mind, we see that it is possible to apply Theorem 1.1 to the function

$$(u,\xi) \to \Phi_f(u) + \xi(\Psi(u) - \Phi_g(u)) \quad ((u,\xi) \in X \times [-r,r]).$$

So, by that result, taking into account that

$$\inf_{u\in X}\sup_{\xi\in[-r,r]}(\Phi_f(u) + \xi(\Psi(u) - \Phi_g(u))) = \inf_{u\in X}(\Phi_f(u) + r|\Psi(u) - \Phi_g(u)|),$$

we have

$$\sup_{\xi\in[-r,r]}\inf_{u\in X}(\Phi_f(u) + \xi(\Psi(u) - \Phi_g(u))) =$$
$$\inf_{u\in X}(\Phi_f(u) + r|\Psi(u) - \Phi_g(u)|). \tag{5.3}$$

Let $\xi \in [-r,r]\setminus\{0\}$. Recalling condition (iii), for each $n \in \mathbf{N}$, put

$$s_n = \begin{cases} \xi\|u_n\|_X & \text{if } \xi > 0 \\ -\xi\|v_n\|_X & \text{if } \xi < 0 \end{cases}$$

as well as

$$w_n = \begin{cases} \dfrac{u_n}{\|u_n\|_X} & \text{if } \xi > 0 \\ \dfrac{v_n}{\|v_n\|_X} & \text{if } \xi < 0. \end{cases}$$

Now, pick $\eta \in]0, |\xi|(\inf_{n\in\mathbf{N}}\Psi(w_n) - c)[$ and choose $\nu \in \mathbf{N}$ in such a way that

$$\Phi_f\left(-\frac{s_n}{\xi}w_n\right) < \eta\frac{s_n}{|\xi|}$$

for all $n > \nu$. So, for each $n > \nu$, we have

$$\Phi_f\left(-\frac{s_n}{\xi}w_n\right) + \xi\left(\Psi\left(-\frac{s_n}{\xi}w_n\right) - \Phi_g\left(-\frac{s_n}{\xi}w_n\right)\right) \le$$
$$\eta\frac{s_n}{|\xi|} - s_n\Psi(w_n) + |\xi|\left|\Phi_g\left(-\frac{s_n}{\xi}w_n\right)\right| \le$$
$$|\xi\Phi_g(0)| + \left(\frac{\eta}{|\xi|} - \inf_{m\in\mathbf{N}}\Psi(w_m) + c\right)s_n.$$

From this, since $\lim_{n\to+\infty} s_n = +\infty$, it follows that

$$\inf_{u\in X}(\Phi_f(u) + \xi(\Psi(u) - \Phi_g(u))) = -\infty.$$

Consequently, one has

$$\sup_{\xi\in[-r,r]}\inf_{u\in X}(\Phi_f(u)+\xi(\Psi(u)-\Phi_g(u)))=\inf_X\Phi_f.$$

Hence, from (5.3) we get

$$\inf_{u\in X}(\Phi_f(u)+r|\Psi(u)-\Phi_g(u)|)=\inf_X\Phi_f.$$

This is against (5.2), and so (in the case $c>0$) our result is proved. Finally, when $c=0$, we proceed exactly as before, taking into account that Proposition 2.7, when $L=0$, holds for any normed space X. △

Now, we can prove

Theorem 5.1 ([15], Theorem 1) *Let $f: T\times E\to \mathbf{R}$ be a sup-measurable function. Assume that there exist $\alpha\in L^1(T)$, $\gamma_i\in]0,1[$ and $\beta_i\in L^{\frac{p}{p-\gamma_i}}(T)$ $(i=1,...,k)$ such that*

$$-\alpha(t)\le f(t,x)\le \alpha(t)+\sum_{i=1}^{k}\beta_i(t)\|x\|^{\gamma_i} \tag{5.4}$$

for almost every $t\in T$ and for every $x\in E$.

Then, for every closed decomposable linear subspace X of $L^p(T,E)$ and every $V\in\mathcal{V}(X)$, one has

$$\inf_{u\in V}\int_T f(t,u(t))d\mu=\inf_{u\in X}\int_T f(t,u(t))d\mu \tag{5.5}$$

Equality (5.5) *also holds for every decomposable linear subspace X of $L^p(T,E)$ and every $V\in\mathcal{I}(X)$.*

Proof. Let X be a closed decomposable linear subspace of $L^p(T,E)$ and $V\in\mathcal{V}(X)$. Let $V=\{u\in X:\Psi(u)=\int_T g(t,u(t))d\mu\}$, with the obvious meaning of Ψ and g, by definition. By Proposition 2.7, the range of the functional $u\to\Psi(u)-\int_T g(t,u(t))d\mu$ $(u\in X)$ is the whole $\mathbf{R}$. For each $r\in\mathbf{R}$, put

$$V_r=\left\{u\in X:\Psi(u)-\int_T g(t,u(t))d\mu=r\right\}$$

as well as

$$\varphi_f(r)=\inf_{u\in V_r}\int_T f(t,u(t))d\mu.$$

Observe that

$$\varphi_f(r)=\inf\{s\in\mathbf{R}:(s,r)\in\Gamma\} \tag{5.6}$$

where

$$\Gamma = \left\{ \left(\int_T f(t, u(t))d\mu, \Psi(u) - \int_T g(t, u(t))d\mu \right) : u \in X \right\}.$$

By the already quoted representation theorem ([4], pp.94-99), there exists a mapping $A : T \to E^*$ such that $\Psi(u) = \int_T A(t)(u(t))d\mu$ for all $u \in X$. With this in mind, one can apply Proposition 2.9 to realize that Γ is convex. Of course, thanks to (5.6), this implies that the function φ_f is convex in $\mathbf{R}$. Now, for each $r \in \mathbf{R}$, we show that

$$\inf_{u \in V_r} \sum_{i=1}^{k} \int_T \beta_i(t) \|u(t)\|^{\gamma_i} d\mu = 0. \tag{5.7}$$

To this end, we intend to apply Lemma 5.1. First, observe that, for each $i = 1, ..., k$, the functional

$$u \to \int_T \beta_i(t) \|u(t)\|^{\gamma_i} d\mu$$

is Hölder continuous in X, with exponent γ_i. Consequently, the functional

$$u \to \sum_{i=1}^{k} \int_T \beta_i(t) \|u(t)\|^{\gamma_i} d\mu$$

is uniformly continuous in X. Let c be the Lipschitz constant of the functional $u \to \int_T g(t, u(t))d\mu$ ($u \in X$). Choose any $u_0 \in X$, with $\|u_0\|_X = 1$, such that $\Psi(u_0) > c$. Then, for each $n \in \mathbf{N}$, we have

$$\sum_{i=1}^{k} \int_T \beta_i(t) \| \pm n u_0(t) \|^{\gamma_i} d\mu \le \sum_{i=1}^{k} n^{\gamma_i} \|\beta_i\|_{L^{p/(p-\gamma_i)}(T)}.$$

Consequently

$$\lim_{n \to +\infty} \frac{\sum_{i=1}^{k} \int_T \beta_i(t) \| \pm n u_0(t) \|^{\gamma_i} d\mu}{n} = 0.$$

At this point, we realize that we can apply Lemma 5.1 to Ψ and to the functions

$$(t, x) \to \sum_{i=1}^{k} \beta_i(t) \|x\|^{\gamma_i}$$

and

$$(t, x) \to g(t, x) + \frac{r}{\mu(S)} \chi_S(t),$$

where S is any member of $\mathcal{F}$ satisfying $0 < \mu(S) < +\infty$. Actually, to satisfy (iii), we can take

$$u_n = v_n = nu_0$$

for all $n \in \mathbf{N}$. Hence, we directly get (5.7) from that lemma, since, of course,

$$\inf_{u \in X} \sum_{i=1}^{k} \int_T \beta_i(t) \|u(t)\|^{\gamma_i} d\mu = 0.$$

Finally, observe that, thanks to (5.4) and (5.7), we have

$$\varphi_f(r) \leq \|\alpha\|_{L^1(T)}$$

for all $r \in \mathbf{R}$. Summarizing: the real function φ_f is convex and bounded above in $\mathbf{R}$, and so it is necessarily constant there. This, of course, implies (5.5). The validity of (5.5) for every decomposable linear subspace X of $L^p(T, E)$ and every $V \in \mathcal{I}(X)$ follows, from the above proof, on the basis of the last assertion in Lemma 5.1. △

Let us now observe a consequence of Theorem 5.1 which extends the classical fact that, for $\gamma \in]0, 1[$, the topological dual of $L^\gamma(T, E)$ reduces to zero. Precisely, we denote by $\mathcal{M}$ the set of all metrics d on $L^p(T, E)$ of the following type:

$$d(u, v) = \sum_{i=1}^{k} \int_T \beta_i(t) \|u(t) - v(t)\|^{\gamma_i} d\mu$$

where $u, v \in L^p(T, E)$, $\gamma_i \in]0, 1[$, $\beta_i \in L^{\frac{p}{p-\gamma_i}}(T)$, $\beta_i > 0$ in T $(i = 1, ..., k)$. Note that each $d \in \mathcal{M}$ is a metric inducing a vector topology which is weaker than the $\|\cdot\|_{L^p(T,E)}$-topology.

Theorem 5.2 ([15], Theorem 2) *For every $d \in \mathcal{M}$ and every decomposable linear subspace X of $L^p(T, E)$, the topological dual of (X, d) reduces to zero.*

Proof. Let X, d be as in the statement. Hence,

$$d(u, v) = \sum_{i=1}^{k} \int_T \beta_i(t) \|u(t) - v(t)\|^{\gamma_i} d\mu,$$

for suitable γ_i, β_i as in the definition above. Arguing by contradiction, assume that there is some non-null d-continuous linear functional Ψ on X. So, Ψ is also $\|\cdot\|_{L^p(T,E)}$-continuous. Put

$$V = \{u \in X : \Psi(u) = 1\}.$$

Then, thanks to Theorem 5.1, we have

$$\inf_{u\in V}\sum_{i=1}^{k}\int_T \beta_i(t)\|u(t)\|^{\gamma_i}d\mu = 0.$$

This means that 0 belongs to the d-closure of V. But V is d-closed, and so $0 \in V$, which is clearly absurd. △

When we take $X = L^p(T,E)$, the conclusion of Theorem 5.1 can be extended to a class of functions f with a more general growth. This is shown by Theorem 5.3 below. So, in it and in Theorem 5.4 as well, we assume $X = L^p(T,E)$.

Theorem 5.3 ([14], Theorem 2) *Let $f : T \times E \to [0,+\infty[$ be such that $f(\cdot,x)$ is μ-measurable for each $x \in E$ and $f(t,\cdot)$ is Lipschitzian with Lipschitz constant $M(t)$ for almost every $t \in T$, where $M \in L^{\frac{p}{p-1}}(T)$. Assume that $f(\cdot,0) \in L^1(T)$ and that there exists a sequence $\{\lambda_n\}$ in $]0,+\infty[$, with $\lim_{n\to+\infty}\lambda_n = +\infty$, such that, for almost every $t \in T$ and for every $x \in E$, one has*

$$\lim_{n\to+\infty}\frac{f(t,\lambda_n x)}{\lambda_n} = 0.$$

Then, for every $V \in \mathcal{V}(X)$, one has

$$\inf_{u\in V}\int_T f(t,u(t))d\mu = \inf_{u\in X}\int_T f(t,u(t))d\mu.$$

Proof. By a standard result, for each $u \in X$, the real function $t \to f(t,u(t))$ is μ-measurable, and so it belongs to $L^1(T)$ in view of the inequality

$$f(t,u(t)) \le M(t)\|u(t)\| + f(t,0).$$

Also, note that (by the Hölder inequality if $p > 1$) the integral functional $u \to \int_T f(t,u(t))d\mu$ is Lipschitzian in X. Now, fix $V \in \mathcal{V}(X)$. Then, by definition, there are $\Psi \in X^*$ and $g : T \times E \to \mathbf{R}$ such that

$$V = \left\{u \in X : \Psi(u) = \int_T g(t,u(t))d\mu\right\},$$

the integral functional $u \to \int_T g(t,u(t))d\mu$ being Lipschitzian in X, with Lipschitz constant $c < \|\Psi\|_{X^*}$. Pick any simple function $u_0 \in X \setminus \{0\}$ in such a way that

$$\frac{\Psi(u_0)}{\|u_0\|_X} > c.$$

This can be done taking into account the definiton of $\|\Psi\|_{X^*}$, the continuity of the function $u \longrightarrow \frac{\Psi(u)}{\|u\|_X}$ in $X \setminus \{0\}$, and the density in X of the set of all simple functions belonging to X. Let $T_1, ..., T_k$ be a measurable partition of T and $x_1, ..., x_k$ k vectors in E such that $u_0(t) = x_i$ if $t \in T_i$ $(i = 1, ..., k)$. Fix i. We want to show that

$$\lim_{n \to +\infty} \frac{\int_{T_i} f(t, \pm\lambda_n x_i) d\mu}{\lambda_n} = 0. \tag{5.8}$$

If $\mu(T_i) = +\infty$, then $x_i = 0$, and so (5.8) follows at once. Therefore, assume that $\mu(T_i) < +\infty$. By assumption, for almost every $t \in T$, we have

$$\lim_{n \to +\infty} \frac{f(t, \pm\lambda_n x_i)}{\lambda_n} = 0.$$

On the other hand, we also have

$$\frac{f(t, \pm\lambda_n x_i)}{\lambda_n} \leq M(t)\|x_i\| + f(t, 0)$$

for every $t \in T_i$ and n large enough.
Since $\mu(T_i) < +\infty$, we have $\int_{T_i} M(t) d\mu < +\infty$, and so (5.8) follows by the dominated convergence theorem. Next, observe that

$$\lim_{n \to +\infty} \frac{\int_T f(t, \pm\lambda_n u_0(t)) d\mu}{\|\lambda_n u_0\|_X} = \lim_{n \to +\infty} \sum_{i=1}^{k} \frac{\int_{T_i} f(t, \pm\lambda_n x_i) d\mu}{\lambda_n \|u_0\|_X} = 0.$$

At this point, our conclusion follows directly from Lemma 5.1 (in fact we can take $u_n = v_n = \lambda_n u_0$ to satisfy condition (iii) there). △

Here is an application of Theorem 5.3. Let us first recall that a multifunction $F : T \to 2^E$ is said to be measurable if, for every open set $\Omega \subseteq E$, one has $\{t \in T : F(t) \cap \Omega \neq \emptyset\} \in \mathcal{F}$. A function $u : T \to E$ is a selection of the multifunction $F : T \to 2^E$ if $u(t) \in F(t)$ for all $t \in T$. We denote by $\mathcal{S}_F$ the set of all selections of F belonging to $L^1(T, E)$.

Theorem 5.4 ([14], Theorem 1) *Let E be separable, let $p = 1$ and let $F : T \to 2^E$ be a measurable multifunction, with non-empty closed values. Assume that* $\operatorname{dist}(0, F(\cdot)) \in L^1(T)$ *and that there exists a sequence $\{\lambda_n\}$ in $]0, +\infty[$, with $\lim_{n \to +\infty} \lambda_n = +\infty$, such that, for almost every $t \in T$ and for every $x \in E$, one has*

$$\lim_{n \to +\infty} \frac{\operatorname{dist}(\lambda_n x, F(t))}{\lambda_n} = 0.$$

Then, $\mathcal{S}_F$ intersects each member of the family $\mathcal{V}(X)$.

Proof. Fix $V \in \mathcal{V}(X)$. So, we have

$$V = \left\{ u \in X : \Psi(u) = \int_T g(t, u(t)) d\mu \right\},$$

where $\Psi \in X^*$ and the integral functional $u \to \int_T g(t, u(t)) d\mu$ is Lipschitzian in X, with Lipschitz constant $c < \|\Psi\|_{X^*}$. Now, put

$$A = \left\{ u \in X : \Psi(u) > \int_T g(t, u(t)) d\mu \right\}$$

and

$$B = \left\{ u \in X : \Psi(u) < \int_T g(t, u(t)) d\mu \right\}.$$

Also, consider the sets

$$V_1 = \left\{ u \in X : \Psi(u) = \int_T g(t, u(t)) d\mu + 1 \right\}$$

and

$$V_2 = \left\{ u \in X : \Psi(u) = \int_T g(t, u(t)) d\mu - 1 \right\}.$$

Of course, $V_1, V_2 \in \mathcal{V}(X)$, $V_1 \subseteq A$, and $V_2 \subseteq B$. Moreover, observe that $\mathcal{S}_F \neq \emptyset$ (see, for instance, [10]). Plainly, the function $(t, x) \to \mathrm{dist}(x, F(t))$ satisfies the assumptions of Theorem 5.3. Consequently, for each $U \in \mathcal{V}(X)$, we have $\inf_{u \in U} \int_T \mathrm{dist}(u(t), F(t)) d\mu = \inf_{u \in X} \int_T \mathrm{dist}(u(t), F(t)) d\mu$. But, the fact that $\mathcal{S}_F \neq \emptyset$ imply that $\inf_{u \in X} \int_T \mathrm{dist}(u(t), F(t)) d\mu = 0$, and so we can find $u_1 \in V_1$, $u_2 \in V_2$ in such a way that

$$\max\left\{ \int_T \mathrm{dist}(u_1(t), F(t)) d\mu, \int_T \mathrm{dist}(u_2(t), F(t)) d\mu \right\} < \frac{1}{\|\Psi\|_{X^*} + c}.$$

Now, fix any strictly positive function ϵ in $L^1(T)$ such that

$$\|\epsilon\|_{L^1(T)} < \frac{1}{\|\Psi\|_{X^*} + c} - \max\left\{ \int_T d(u_1(t), F(t)) d\mu, \int_T d(u_2(t), F(t)) d\mu \right\}.$$

Applying Lemmas 2.1 and 2.2 of [10] jointly with the Kuratowski and Ryll-Nardzewski theorem ([5], Theorem 14.2.1) we then can get $v_1, v_2 \in \mathcal{S}_F$ such that

$$\|u_1(t) - v_1(t)\| \leq d(u_1(t), F(t)) + \epsilon(t)$$

and

$$\|u_2(t) - v_2(t)\| \leq d(u_2(t), F(t)) + \epsilon(t)$$

for all $t \in T$. So, we have

$$\max\{\|u_1 - v_1\|_X, \|u_2 - v_2\|_X\} < \frac{1}{\|\Psi\|_{X^*} + c}.$$

Now, recalling that $u_1 \in V_1$, we have

$$1 - \Psi(v_1) + \int_T g(t, v_1(t))d\mu =$$

$$= \Psi(u_1) - \int_T g(t, u_1(t))d\mu - \Psi(v_1) + \int_T g(t, v_1(t))d\mu \leq$$

$$\leq (\|\Psi\|_{X^*} + c)\|u_1 - v_1\|_X < 1.$$

Hence, $v_1 \in A$. Likewise, since $u_2 \in V_2$, we have

$$1 + \Psi(v_2) - \int_T g(t, v_2(t))d\mu =$$

$$\int_T g(t, u_2(t))d\mu - \Psi(u_2) + \Psi(v_2) - \int_T g(t, v_2(t))d\mu \leq$$

$$\leq (\|\Psi\|_{X^*} + c)\|u_2 - v_2\|_X < 1.$$

Hence, $v_2 \in B$. By Proposition 2.8, $\mathcal{S}_F$, being decomposable, is connected. Then, since A, B are open and disjoint, $A \cup B = X \setminus V$, and (for what seen above) $\mathcal{S}_F \cap A \neq \emptyset$, $\mathcal{S}_F \cap B \neq \emptyset$, we must have $\mathcal{S}_F \cap V \neq \emptyset$, as claimed. △

When $E = \mathbf{R}$, Theorem 5.4 can be established in an optimal form. Indeed, we have

Theorem 5.5 ([8], Theorem 2) *Let μ be complete, and let $F : T \to 2^{\mathbf{R}}$ be a measurable multifunction with non-empty closed values.*

Then, the following are equivalent:

(α) *$\mathcal{S}_F$ intersects each member of the family $\mathcal{V}(L^1(T))$.*

(β) *$\mathcal{S}_F$ intersects each closed hyperplane of $L^1(T)$.*

(γ) *The real function $t \to \mathrm{dist}(0, F(t))$ belongs to $L^1(T)$ and, for almost every $t \in T$, one has $\sup F(t) = +\infty$ and $\inf F(t) = -\infty$.*

Proof. On the basis of a previous remark, the implication (α) $\Rightarrow$ (β) is obvious.

Suppose that (β) is true. So, in particular, $\mathcal{S}_F \neq \emptyset$. Then, the integrability of the function $t \to \mathrm{dist}(0, F(t))$ follows from its measurability and from the inequality $\mathrm{dist}(0, F(t)) \leq w(t)$, where $w \in \mathcal{S}_F$. Now, let us put

$$A = \{t \in T : \sup F(t) < +\infty\}.$$

The function $t \to \sup F(t)$ is measurable, and so $A \in \mathcal{F}$. Arguing by contradiction, suppose that $\mu(A) > 0$. Then, we can choose $M > 0$ and $B \in \mathcal{F}$, $B \subseteq A$, so that $0 < \mu(B) < +\infty$ and $\sup F(t) \leq M$ for every $t \in B$. Let $r > M\mu(B)$. By (β), since $u \to \int_B u(t)d\mu$ is a non-null continuous linear functional on $L^1(T)$, there exists $v \in \mathcal{S}_F$ such that $\int_B v(t)d\mu = r$. But $\int_B v(t)d\mu \leq M\mu(B) < r$, that is absurde. So, $\mu(A) = 0$. In an analogous way, it is proved that $\mu(\{t \in T : \inf F(t) > -\infty\}) = 0$. So, (γ) holds.

Finally, let us prove that (γ) implies (α). To this end, fix $V \in \mathcal{V}(L^1(T))$. So, $V = \{u \in L^1(T) : \Psi(u) = \int_T g(t, u(t))d\mu\}$, with obvious meaning of Ψ and g. For each $(t, x) \in T \times \mathbf{R}$, put

$$f(t, x) = \operatorname{dist}(x, F(t)).$$

We want to apply Lemma 5.1 to f, g and Ψ. In connection with such a lemma, let us construct two sequences $\{u_n\}$, $\{v_n\}$ with the properties expressed in (iii). To this purpose, let $\varphi \in L^\infty(T)$ be such that $\Psi(u) = \int_T \varphi(t)u(t)d\mu$ for every $u \in L^1(T)$. Since $\|\varphi\|_{L^\infty(T)} = \|\Psi\|_{(L^1(T))^*}$, we have $\|\varphi\|_{L^\infty(T)} > c$, where c is the Lipschitz constant of the functional $u \to \int_T g(t, u(t))d\mu$. Consequently, we have $\mu(\{t \in T : |\varphi(t)| > c\}) > 0$. Let $c' > c$ be such that the set $\{t \in T : |\varphi(t)| > c'\}$ has a positive measure. Also, put

$$S_1 = \{t \in T : \varphi(t) > c'\}$$

and

$$S_2 = \{t \in T : \varphi(t) < -c'\}.$$

By (γ), we have $F(t) \cap [0, +\infty[\neq \emptyset$ and $F(t) \cap]-\infty, 0] \neq \emptyset$ a.e. in T. So, by the Kuratowski and Ryll-Nardzewski theorem, we get two measurable functions $h_1 : T \to [0, +\infty[$, $h_2 : T \to]-\infty, 0]$ such that $\{h_1(t), h_2(t)\} \subseteq F(t)$ for all $t \in T \setminus T_0$, where $T_0 \in \mathcal{F}$ and $\mu(T_0) = 0$.

We now distinguish two cases. First, assume that $\mu(S_1) > 0$. Fix $S_3 \in \mathcal{F}$, with $S_3 \subseteq S_1 \setminus T_0$ and $\mu(S_3) > 0$, so that $\int_{S_3}(h_1(t) - h_2(t))d\mu < +\infty$. We then are allowed to apply Theorem 3.2′ of [9] to the restrictions of the multifunctions $t \to F(t) \cap [0, +\infty[$ and $t \to F(t) \cap]-\infty, 0]$ to S_3. Denote them by F_1, F_2, respectively. So, we have

$$\sup\left\{\int_{S_3} \psi(t)d\mu : \psi \in \mathcal{S}_{F_1}\right\} = \int_{S_3} \sup F_1(t)d\mu = +\infty$$

and

$$\sup\left\{-\int_{S_3} \psi(t)d\mu : \psi \in \mathcal{S}_{F_2}\right\} = \int_{S_3} \sup -F_2(t)d\mu = +\infty.$$

Consequently, we can choose a sequence $\{\psi_n\}$ in $\mathcal{S}_{F_1} \setminus \{0\}$ and a sequence $\{\xi_n\}$ in $\mathcal{S}_{F_2} \setminus \{0\}$ so that

$$\lim_{n\to+\infty} \int_{S_3} \psi_n(t)d\mu = \lim_{n\to+\infty} \int_{S_3} |\xi_n(t)|d\mu = +\infty.$$

Now, for each $n \in \mathbf{N}$, define $u_n, v_n : T \to \mathbf{R}$ by putting

$$u_n(t) = \begin{cases} -\xi_n(t) & \text{if } t \in S_3 \\ 0 & \text{if } t \in T \setminus S_3 \end{cases}$$

and

$$v_n(t) = \begin{cases} \psi_n(t) & \text{if } t \in S_3 \\ 0 & \text{if } t \in T \setminus S_3. \end{cases}$$

The sequences $\{u_n\}, \{v_n\}$ have the required properties. In fact, for each $n \in \mathbf{N}$, we have

$$\Psi\left(\frac{u_n}{\|u_n\|_{L^1(T)}}\right) = \frac{1}{\|u_n\|_{L^1(T)}} \int_{S_3} -\varphi(t)\xi_n(t)d\mu >$$
$$> \frac{1}{\|u_n\|_{L^1(T)}} \int_{S_3} -c'\xi_n(t)d\mu = c'.$$

Hence

$$\inf_{n\in\mathbf{N}} \Psi\left(\frac{u_n}{\|u_n\|_{L^1(T)}}\right) > c.$$

Moreover, we have

$$\lim_{n\to+\infty} \frac{\int_T f(t, -u_n(t))d\mu}{\|u_n\|_{L^1(T)}} = \lim_{n\to+\infty} \frac{\int_{T\setminus S_3} d(0, F(t))d\mu}{\|u_n\|_{L^1(T)}} = 0.$$

The properties of $\{v_n\}$ are checked in an analogous way.

Now, assume that $\mu(S_1) = 0$. Hence, $\mu(S_2) > 0$. Proceeding as before, we find $S_4 \in \mathcal{F}$, with $S_4 \subseteq S_2 \setminus T_0$ and $\mu(S_4) > 0$, and two sequences $\{\bar{\psi}_n\}$, $\{\bar{\xi}_n\}$ in $L^1(S_4) \setminus \{0\}$, with $\bar{\psi}_n(t) \in F(t) \cap [0, +\infty[$, $\bar{\xi}_n(t) \in F(t) \cap]-\infty, 0]$ for all $t \in S_4$, $n \in \mathbf{N}$, such that

$$\lim_{n\to+\infty} \int_{S_4} \bar{\psi}_n(t)d\mu = \lim_{n\to+\infty} \int_{S_4} |\bar{\xi}_n(t)|d\mu = +\infty.$$

In this case, the desired sequences $\{u_n\}$, $\{v_n\}$ are obtained by putting

$$u_n(t) = \begin{cases} -\bar{\psi}_n(t) & \text{if } t \in S_4 \\ 0 & \text{if } t \in T \setminus S_4 \end{cases}$$

and

$$v_n(t) = \begin{cases} \bar{\xi}_n(t) & \text{if } t \in S_4 \\ 0 & \text{if } t \in T \setminus S_4. \end{cases}$$

Then, Lemma 5.1 assures that

$$\inf_{u \in V} \int_T \operatorname{dist}(u(t), F(t)) d\mu = \inf_{u \in L^1(T)} \int_T \operatorname{dist}(u(t), F(t)) d\mu.$$

At this point, the proof goes on exactly as that of Theorem 5.3. △

Finally, as an application of Theorem 5.5, one gets the following

Theorem 5.6 ([8], Theorem 1) *Let X be a real topological vector space, let Φ be a linear homeomorphism from X onto $L^1(T)$, and let $f : T \times \mathbf{R} \to \mathbf{R}$ be a Carathéodory function.*

Then, the following are equivalent:
(a) *The set*

$$\{u \in X : f(t, \Phi(u)(t)) = 0 \quad a.e. \text{ } in \text{ } T\}$$

is arcwise connected and intersects each closed hyperplane of X.
(b) *The function*

$$t \to \inf\{|x| : x \in \mathbf{R},\ f(t, x) = 0\}$$

belongs to $L^1(T)$ and, for almost every $t \in T$, one has

$$\sup\{x \in \mathbf{R} : f(t, x) = 0\} = +\infty$$

and

$$\inf\{x \in \mathbf{R} : f(t, x) = 0\} = -\infty.$$

Proof. For every $t \in T$, put

$$F(t) = \{x \in \mathbf{R} : f(t, x) = 0\}.$$

Note that the multifunction F is measurable (and closed-valued, of course). Also, put

$$\Gamma = \{u \in X : f(t, \Phi(u)(t)) = 0 \quad \text{a.e. in } T\},$$

and observe that

$$\Gamma = \Phi^{-1}(\mathcal{S}_F). \tag{5.9}$$

Suppose that (a) holds. Let $\psi \in (L^1(T))^* \setminus \{0\}$ and $r \in \mathbf{R}$. If we put $\varphi(x) = \psi(\Phi(x))$, $x \in X$, we have $\varphi \in X^* \setminus \{0\}$, and so, by (a), there exists $u \in \Gamma$ such that $\varphi(u) = r$. Then, thanks to (5.9), $\Phi(u)$ belongs to

$\mathcal{S}_F \cap \psi^{-1}(r)$, and so ($\beta$) of Theorem 5.5 holds. Hence, (γ) of the same theorem assures (b).

Now, let us prove that (*b*) implies (*a*). Since $\mathcal{S}_F$ is decomposable, Proposition 2.8 assures that it is arcwise connected. Observing that Φ^{-1} is a continuous operator, (5.9) implies at once that Γ is arcwise connected. Now, let $\varphi \in X^* \setminus \{0\}$ and $r \in \mathbf{R}$. If we put $\psi(v) = \varphi(\Phi^{-1}(v)), v \in L^1(T)$, we have $\psi \in (L^1(T))^* \setminus \{0\}$. By Theorem 5.5 again (implication $(\gamma) \Rightarrow (\beta)$), there exists $v \in \mathcal{S}_F$ such that $\psi(v) = r$. Then, $\Phi^{-1}(v)$ belongs to $\Gamma \cap \varphi^{-1}(r)$, and the proof is complete. △

References

1. J.-P. Aubin and I. Ekeland, *Applied Nonlinear Analysis,* John Wiley & Sons, 1984.
2. M. A. Geraghty and B.-L. Lin, *Topological minimax theorems,* Proc. Amer. Math. Soc.**91** (1984), 377-380.
3. C. Horvath, *Quelques théorèmes en théorie des mini-max,* C. R. Acad. Sci. Paris, Série I, **310** (1990), 269-2727.
4. A. Ionescu Tulcea and C. Ionescu Tulcea, *Topics in the theory of lifting,* Springer-Verlag, 1969.
5. E. Klein and A. C. Thompson, *Theory of correspondences,* John Wiley & Sons, 1984.
6. H. König, *A general minimax theorem based on connectedness,* Arch. Math. **59** (1992), 55-64.
7. O. Naselli, *On a class of functions with equal infima over a domain and its boundary,* J. Optim. Theory Appl. **91** (1996), 81-90.
8. O. Naselli, *On the solution set of an equation of the type* $f(t,\Phi(u)(t))=0$, Set-Valued Anal. **4** (1996), 399-405.
9. P. Pucci and G. Vitillaro, *A representation theorem for Aumann integrals,* J. Math. Anal. Appl. **102** (1984), 86-101.
10. B. Ricceri, *Sur l'approximation des sélections mesurables,* C. R. Acad. Sci. Paris, Série I, **295** (1982), 527-530.
11. B. Ricceri, *Structure, approximation et dépendance continue des solutions de certaines équations non linéaires,* C. R. Acad. Sci. Paris, Série I, **305** (1987), 45-47.
12. B. Ricceri, *Some topological mini-max theorems via an alternative principle for multifunctions,* Arch. Math. **60** (1993), 367-377.
13. B. Ricceri, *A variational property of integral functionals on* L^p*-spaces of vector-valued functions,* C. R. Acad. Sci. Paris, Série I, **318** (1994), 337-342.
14. B. Ricceri, *On the integrable selections of certain multifunctions,* Set-Valued Anal. **4** (1996), 91-99.
15. B. Ricceri, *More on a variational property of integral functionals,* J. Optim. Theory Appl. **94** (1997), 757-763.
16. A. W. Roberts and D. E. Varberg, *Convex Functions,* Academic Press, 1973.
17. J. Saint Raymond, *Connexité des sous-niveaux des fonctionnelles intégrales,* Rend. Circ. Mat. Palermo, **44** (1995), 162-168.
18. S. Simons, *Minimax theorems and their proofs,* in Minimax and Applications, D.-Z. Du and P. M. Pardalos (eds.), Kluwer Academic Publishers, 1995, 1-23.
19. H. Tuy, *On a general minimax theorem,* Soviet Math. Dokl. **15** (1974), 1689-1693.
20. Wu Wen-Tsün, *A remark on the fundamental theorem in the theory of games,* Sci. Rec. (New Ser.), **3** (1959), 229-233.

THREE LECTURES ON MINIMAX AND MONOTONICITY

STEPHEN SIMONS
Department of Mathematics
University of California
Santa Barbara, CA 93106-3080
e-mail: simons@math.ucsb.edu

0. Introduction

This is an edited version of three lectures that were given at the workshop on minimax theory and applications that was held at the "Ettore Majorana" international center for scientific culture in Erice, Sicily, September 30 – October 6, 1996. The purpose of these lectures is to show how a standard minimax theorem can be used to give some new results, and very simplified proofs of some old results, about monotone multifunctions on a Banach space.

We shall show how to use the minimax theorem to transform problems on the existence of elements of the dual space into problems on the existence of a real constant. We will use this technique three times (in Lemmas 4.1, 6.8 and 8.1). As should be clear from the later analysis (see, for instance, the discussion in the introduction to Section 8) this technique has the advantage that in many cases one can works backwards from a conjecture, find whether it reasonable and, if it is, obtain a proof.

We suppose throughout that E is a nonzero real Banach space with dual E^*. We define multifunctions and monotonicity formally in Section 1. For some problems, it is convenient to think of a multifunction from E into 2^{E^*} as a subset of $E \times E^*$. This is pursued in Section 2, where we introduce a *convexification* of any nonempty subset of $E \times E^*$. Using this, we show in Section 3 how to associate a closed convex subset $\Pi(G)$ of the first quadrant of the plane with every nonempty subset G of $E \times E^*$. The main result here is Lemma 3.3, in which we prove that

$$M \text{ a nonempty monotone subset of } E \times E^* \quad \Longrightarrow \quad \Pi(M) \neq \emptyset.$$

Section 4 is about reflexive spaces. In Theorem 4.5, we prove that *if M is a monotone subset of $E \times E^*$ then*

$$M \text{ is maximal monotone} \iff M + G(-J) = E \times E^*,$$

where J is the duality map, and we deduce in Theorem 4.6 that

$$S\colon\ E \to 2^{E^*} \text{ maximal monotone} \quad \Longrightarrow \quad R(S + J) = E^*.$$

This is one direction of Rockafellar's "surjectivity theorem", except that our results hold for any reflexive space, not only ones where the norms of E and E^* are strictly convex.

Many problems about monotone multifunctions can be reduced to ones about convex functions by the device explained in Section 5, where we show how to associate a convex lower semicontinuous function χ_S with every nontrivial multifunction $S\colon\ E \to 2^{E^*}$.

We start Section 6 by collecting together various results on convex functions that depend ultimately on Baire's theorem. The least known of these is probably Lemma 6.1, a close relative of the closed graph theorem. We then return to our discussion of multifunctions in a general Banach space. Theorem 6.6 is a general local boundedness result for (not necessarily monotone) multifunctions, while in the "six sets theorem", Theorem 6.10, and the "nine sets theorem", Theorem 6.11, we settle a conjecture and improve some known results about $D(S)$ in the case when S is maximal monotone.

In Sections 7 and 8 we obtain sufficient conditions ("constraint qualifications") under which the sum of maximal monotone multifunctions on a reflexive space is maximal monotone. The main result here is Corollary 8.4. In some sense, our analysis actually contains a necessary and sufficient condition (still, for reflexive spaces) for the sum of maximal monotone multifunctions to be maximal monotone, but this is hard to write down. We refer the reader to the introduction to Section 8 for more explanation of this rather cryptic comment. We end Section 8 with a brief discussion of other constraint qualifications.

We shall use the minimax theorem below that follows from a result of Fan (see [8]). It is important to note that the set A has no topological structure.

Theorem 0.1. *Let A be a nonempty convex subset of a vector space, B be a nonempty convex subset of a vector space and B also be a compact*

Hausdorff topological space. Let $h : A \times B \to \mathbb{R}$ be convex on A, and concave and upper semicontinuous on B. Then

$$\inf_A \max_B h = \max_B \inf_A h.$$

1. Multifunctions and Monotonicity

We now introduce some general notation for "multifunctions" or "set-valued maps". We write 2^{E^*} for the set of all subsets of E^*. If S: $E \to 2^{E^*}$, we write

$$G(S) := \{(x, x^*)\colon\ x \in E,\ x^* \in Sx\}.$$

$G(S)$ is the "graph" of S. We shall always suppose that $G(S) \neq \emptyset$ — we shall emphasize this by saying that S is *nontrivial.* We write

$$D(S) := \{x \in X\colon\ Sx \neq \emptyset\}.$$

$D(S)$ is the "domain" of S. We write

$$R(S) := \textstyle\bigcup_{x \in E} Sx.$$

$R(S)$ is the "range" of S. If S_1: $E \to 2^{E^*}$ and S_2: $E \to 2^{E^*}$ are nontrivial, we define $S_1 + S_2$: $E \to 2^{E^*}$ (with $D(S_1 + S_2) = D(S_1) \cap D(S_2)$) by

$$(S_1 + S_2)x := S_1x + S_2x \quad (x \in E), \tag{1.0.1}$$

where $S_1x + S_2x$ is the "Minkowski sum" $\{x_1^* + x_2^*\colon\ x_1^* \in S_1x,\ x_2^* \in S_2x\}$.
Let S: $E \to 2^{E^*}$. S is said to be *monotone* if

$$(x, x^*),\ (y, y^*) \in G(S) \quad \Longrightarrow \quad \langle x - y, x^* - y^*\rangle \geq 0.$$

S is said to be *maximal monotone* if S is monotone, and S has no proper monotone extension. This is equivalent to the statement:

$$(z, z^*) \in E \times E^* \text{ and } \inf_{(s,s^*)\in G(S)} \langle s - z, s^* - z^*\rangle \geq 0 \quad \Longrightarrow \quad (z, z^*) \in G(S).$$

It will be convenient in Section 8 to have the more quantitative version of this that is provided by Lemma 1.1. (a) follows by taking $(s, s^*) := (z, z^*)$, (b) is immediate from the definition of maximal monotonicity, and (c) follows from (a) and (b).

Lemma 1.1. *Let $S : E \to 2^{E^*}$ be maximal monotone.*
(a) If $(z, z^*) \in G(S)$ then $\inf_{(s,s^*)\in G(S)}\langle s - z, s^* - z^*\rangle = 0$.
(b) If $(z, z^*) \in E \times E^* \setminus G(S)$ then $\inf_{(s,s^*)\in G(S)}\langle s - z, s^* - z^*\rangle < 0$.
(c) For all $(z, z^*) \in E \times E^*$, $\inf_{(s,s^*)\in G(S)}\langle s - z, s^* - z^*\rangle \leq 0$.

We point to [2], [10] and [11] as general references on multifunctions and monotonicity. We now give some examples of maximal monotone multifunctions.

The first one we consider is that of *positive linear maps.* Let $T\colon\ E \to E^*$ be linear and

$$x \in E \quad \Longrightarrow \quad \langle x, Tx \rangle \geq 0.$$

Then T *is a (single valued) maximal monotone map.* More precisely, *the multifunction* S *defined by* $Sx := \{Tx\}$ *is maximal monotone.* As a special case of the above, we mention *skew linear maps.* These are linear maps $T\colon\ E \to E^*$ such that

$$x \in E \quad \Longrightarrow \quad \langle x, Tx \rangle = 0.$$

The second example that we consider is that of *subdifferentials.* Suppose that $f : E \to \mathbb{R} \cup \{\infty\}$ is convex and lower semicontinuous, and $\operatorname{dom} f \neq \emptyset$. If $x \in E$, the *subdifferential* of f at x is defined by

$$\partial f(x) := \{z^* \in E^*\colon \quad y \in E \Longrightarrow f(x) + \langle y - x, z^* \rangle \leq f(y)\}.$$

Then $\partial f\colon\ E \to 2^{E^*}$ *is maximal monotone.* The maximality is not easy, and is *Rockafeller's maximal monotonicity theorem.* (See [16] for a proof of this, and other properties of subdifferentials.) It is easy to see in this situation that $D(\partial f) \subset \operatorname{dom} f$, however this inclusion may be proper: let $E := \mathbb{R}$ and $f\colon\ \mathbb{R} \to \mathbb{R} \cup \{\infty\}$ be defined by

$$f(x) := \begin{cases} -\sqrt{1 - x^2} & (x \in [-1, 1]) \\ \qquad \infty & (x \in \mathbb{R} \setminus [-1, 1]); \end{cases}$$

then $D(\partial f) = (-1, 1)$ but $\operatorname{dom} f = [-1, 1]$. Incidentally, if T is linear, skew and nonzero then T is not a subdifferential so, provided that E has dimension > 1, there always exist maximal monotone multifunctions that are not subdifferentials. (If $E = \mathbb{R}$ then every maximal monotone multifunction on E is a subdifferential.)

The final example that we mention here is that of the *normality multifunction.* Let C be a nonempty closed convex subset of E and $N_C\colon\ E \to 2^{E^*}$ be defined by

$$(x, x^*) \in G(N_C) \iff x \in C \text{ and } \langle x, x^* \rangle = \max_C x^*.$$

Then N_C *is maximal monotone.* The maximality can be seen in two ways. It can be deduced from the result of Rockafellar mentioned above using the "indicator function" of C. Alternatively, one can proceed directly from the definition of N_C and use the consequence of the Bishop-Phelps theorem in [10], Proposition 3.20, p. 49 that *C is the intersection of the closed half-spaces defined by its supporting hyperplanes.*

2. A Convexification of $E \times E^*$ and the Three Affine Maps

We write $\mathcal{C}(E \times E^*)$ for the set of functions μ: $E \times E^* \to [0, \infty)$ such that

$$\{(s, s^*) \in E \times E^*\colon \mu(s, s^*) > 0\} \text{ is finite} \quad \text{and} \quad \sum_{(s,s^*)\in E\times E^*} \mu(s, s^*) = 1.$$

$\mathcal{C}(E \times E^*)$ is a convex set. If $(y, y^*) \in E \times E^*$ then $\delta_{(y,y^*)} \in \mathcal{C}(E \times E^*)$, where $\delta_{(y,y^*)}$ is defined by

$$\delta_{(y,y^*)}(s, s^*) := \begin{cases} 1 & ((s, s^*) = (y, y^*)) \\ 0 & ((s, s^*) \neq (y, y^*)), \end{cases}$$

thus $\mathcal{C}(E \times E^*)$ can be thought of as a "convexification" of $E \times E^*$. We can also think of $\mathcal{C}(E \times E^*)$ as the set of probability measures on $E \times E^*$ with finite support. Then $\delta_{(y,y^*)}$ corresponds to the point mass at (y, y^*). Any nonempty subset of $E \times E^*$ also has a "convexification": let $\emptyset \neq G \subset E \times E^*$, and write $\mathcal{C}(G)$ for the set of those elements μ of $\mathcal{C}(E \times E^*)$ such that

$$\mu(s, s^*) > 0 \quad \Longrightarrow \quad (s, s^*) \in G.$$

Clearly, $\mathcal{C}(G)$ is convex.

We now introduce the three affine maps p: $\mathcal{C}(E \times E^*) \to E$, q: $\mathcal{C}(E \times E^*) \to E^*$ and $r\colon \mathcal{C}(E \times E^*) \to \mathbb{R}$, defined by

$$p(\mu) := \textstyle\sum_{(s,s^*)\in E\times E^*} \mu(s, s^*)s,$$
$$q(\mu) := \textstyle\sum_{(s,s^*)\in E\times E^*} \mu(s, s^*)s^*$$

and

$$r(\mu) := \textstyle\sum_{(s,s^*)\in E\times E^*} \mu(s, s^*)\langle s, s^*\rangle.$$

We shall have frequent occasion to use the fact that

$$(y, y^*) \in E \times E^* \quad \Longrightarrow \quad p(\delta_{(y,y^*)}) = y,\ q(\delta_{(y,y^*)}) = y^* \text{ and } r(\delta_{(y,y^*)}) = \langle y, y^*\rangle.$$

3. Monotone Subsets and their "Pictures"

A subset M of $E \times E^*$ is *monotone* if

$$(x, x^*) \text{ and } (y, y^*) \in M \quad \Longrightarrow \quad \langle x - y, x^* - y^* \rangle \geq 0.$$

The *pqr*-lemma, Lemma 3.1, is going to be basic in everything that follows. In it, we give a characterization of the monotone subsets of $E \times E^*$ in terms of p, q and r. The manipulations contained in Lemma 3.1 are part of the folklore of monotonicity.

Lemma 3.1. *Let M be a nonempty subset of $E \times E^*$. Then M is monotone* $\Longleftrightarrow$

$$\mu \in \mathcal{C}(M) \quad \Longrightarrow \quad r(\mu) \geq \langle p(\mu), q(\mu) \rangle.$$

Proof. Suppose first that M is a monotone subset of $E \times E^*$.
Let $(s_1, s_1^*), \ldots, (s_m, s_m^*)$ be an enumeration of those elements (s, s^*) of M for which $\mu(s, s^*) > 0$, and write α_i for $\mu(s_i, s_i^*)$. Then, with the summations going from 1 to m,

$$\begin{aligned} r(\mu) - \langle p(\mu), q(\mu) \rangle &= \textstyle\sum_i \alpha_i \langle s_i, s_i^* \rangle - \langle \sum_i \alpha_i s_i, \sum_i \alpha_i s_i^* \rangle \\ &= \textstyle\sum_{i,j} \alpha_i \alpha_j \langle s_i, s_i^* \rangle - \sum_{i,j} \alpha_i \alpha_j \langle s_i, s_j^* \rangle \\ &= \textstyle\sum_{i,j} \alpha_i \alpha_j \langle s_i, s_i^* - s_j^* \rangle \\ &= \textstyle\sum_{i<j} \alpha_i \alpha_j \langle s_i, s_i^* - s_j^* \rangle + \sum_{j<i} \alpha_i \alpha_j \langle s_i, s_i^* - s_j^* \rangle \\ &= \textstyle\sum_{i<j} \alpha_i \alpha_j \langle s_i, s_i^* - s_j^* \rangle + \sum_{i<j} \alpha_i \alpha_j \langle s_j, s_j^* - s_i^* \rangle \\ &= \textstyle\sum_{i<j} \alpha_i \alpha_j \langle s_i - s_j, s_i^* - s_j^* \rangle \geq 0. \end{aligned}$$

We leave the proof of the converse to the reader. ▮

We define $h_1\colon \mathcal{C}(E \times E^*) \times \mathbb{R}^2 \to \mathbb{R}$ by

$$h_1(\mu, (\alpha, \beta)) := 2r(\mu) + 2\alpha\|p(\mu)\| + 2\beta\|q(\mu)\| - \alpha^2 - \beta^2.$$

If $\emptyset \neq G \subset E \times E^*$, we define the "picture" of G by

$$\Pi(G) := \{(\alpha, \beta)\colon \quad \alpha \geq 0,\ \beta \geq 0, \quad \mu \in \mathcal{C}(M) \Longrightarrow h_1(\mu, (\alpha, \beta)) \leq 0\} \subset \mathbb{R}^2.$$

$\Pi(G)$ is the intersection of the first quadrant in $\mathbb{R}^2$ with a family of closed disks and is, consequently, a compact convex subset of the plane. However, $\Pi(G)$ may be empty. The main result of this section is Lemma 3.3, in which we shall prove that *if M is a nonempty monotone subset of $E \times E^*$ then* $\Pi(M) \neq \emptyset$. The analysis in the rest of this section and the next section first appeared in [15]. We first deal with the special case of *bounded* monotone sets.

Lemma 3.2. *Let N be a nonempty bounded* monotone *subset of $E \times E^*$. Then* $\Pi(N) \neq \emptyset$.

Proof. Let $A_1 := \mathcal{C}(N)$, $K := \sup_{(s,s^*)\in N} [\|s\| \vee \|s^*\|]$ and $B_1 := [0,K] \times [0,K]$. If $\mu \in A_1$ then $(\|p(\mu)\|, \|q(\mu)\|) \in B_1$ and, using the pqr-Lemma, Lemma 3.1,

$$\begin{aligned} h_1(\mu, (\|p(\mu)\|, \|q(\mu)\|)) &= \|p(\mu)\|^2 + \|q(\mu)\|^2 + 2r(\mu) \\ &\geq \|p(\mu)\|^2 + \|q(\mu)\|^2 + 2\langle p(\mu), q(\mu)\rangle \\ &\geq \|p(\mu)\|^2 + \|q(\mu)\|^2 - 2\|p(\mu)\|\|q(\mu)\| \\ &= (\|p(\mu)\| - \|q(\mu)\|)^2 \geq 0. \end{aligned}$$

Thus

$$\inf_{A_1} \max_{B_1} h_1 \geq 0.$$

Since h_1 is convex on A_1, and concave and continuous on B_1, from the minimax theorem, Theorem 0.1,

$$\max_{B_1} \inf_{A_1} h_1 \geq 0.$$

This gives the required result. ▌

We now bootstrap Lemma 3.2 into Lemma 3.3, which is the main result of this section. Theorem 3.4 is simply a rephrasal of Lemma 3.3.

Lemma 3.3. *Let M be a nonempty monotone subset of $E \times E^*$. Then $\Pi(M) \neq \emptyset$.*

Proof. For all $n \geq 1$, let $M_n := \{(s,s^*) \in M\colon\ \|s\| \leq n,\ \|s^*\| \leq n\}$. Fix $n_0 \geq 1$ so that $M_{n_0} \neq \emptyset$. From Lemma 3.2, for all $n \geq n_0$, $\Pi(M_n) \neq \emptyset$. Since the sets $\Pi(M_n)$ decrease as n increases, from compactness, there exists $(\alpha, \beta) \in \bigcap_{n\geq n_0} \Pi(M_n)$. Using the fact that

$$\mu \in \mathcal{C}(M) \quad \Longrightarrow \quad \text{there exists } n \geq n_0 \text{ such that } \mu \in \mathcal{C}(M_n),$$

it follows that $(\alpha, \beta) \in \Pi(M)$. ▌

Theorem 3.4. *Let M be a nonempty monotone subset of $E \times E^*$. Then there exist $\alpha, \beta \geq 0$ such that*

$$\mu \in \mathcal{C}(M) \quad \Longrightarrow \quad 2r(\mu) + 2\alpha\|p(\mu)\| + 2\beta\|q(\mu)\| - \alpha^2 - \beta^2 \geq 0. \quad (3.4.1)$$

4. For Reflexive Spaces Only

In this section we assume that E is reflexive. We start off by proving an equivalence for *arbitrary* nonempty subsets of $E \times E^*$. We note that (4.1.1) is identical with (3.4.1), except that M has been replaced by G.

Lemma 4.1. *Let E be reflexive and $\emptyset \neq G \subset E \times E^*$. Then the conditions (a) and (b) are equivalent.*
(a) There exist $\alpha, \beta \geq 0$ such that

$$\mu \in \mathcal{C}(G) \quad \Longrightarrow \quad 2r(\mu) + 2\alpha\|p(\mu)\| + 2\beta\|q(\mu)\| - \alpha^2 - \beta^2 \geq 0. \quad (4.1.1)$$

(b) There exists $(x, x^*) \in E \times E^*$ such that

$$(s, s^*) \in G \quad \Longrightarrow \quad 2\langle s - x, s^* - x^* \rangle \geq \|x^*\|^2 + \|x\|^2 + 2\langle x, x^* \rangle. \quad (4.1.2)$$

Proof. We shall prove that (a) $\Longrightarrow$ (b), and leave the converse to the reader. Define h_2: $\mathcal{C}(G) \times (E \times E^*) \to \mathbb{R}$ by

$$h_2(\mu, (x, x^*)) := 2r(\mu) - 2\langle p(\mu), x^* \rangle - 2\langle x, q(\mu) \rangle - \|x\|^2 - \|x^*\|^2.$$

Let α, β be as in (a), $A_2 := \mathcal{C}(G)$ and

$$B_2 := \{(x, x^*) \in E \times E^*\colon \ \|x\| \leq \beta, \ \|x^*\| \leq \alpha\},$$

with the product of the topologies $w(E, E^*)$ and $w(E^*, E)$. If $\mu \in A_2$ then, from the reflexivity of E and the Hahn-Banach Theorem, we can find $x \in E$ such that

$$\|x\| \leq \beta \quad \text{and} \quad \langle x, q(\mu) \rangle = -\beta\|q(\mu)\|$$

and $x^* \in E^*$ such that

$$\|x^*\| \leq \alpha \quad \text{and} \quad \langle p(\mu), x^* \rangle = -\alpha\|p(\mu)\|.$$

So $(x, x^*) \in B_2$. Further, from (a),

$$h_2(\mu, (x, x^*)) \geq 2r(\mu) + 2\alpha\|p(\mu)\| + 2\beta\|q(\mu)\| - \alpha^2 - \beta^2 \geq 0.$$

Thus

$$\inf_{A_2} \max_{B_2} h_2 \geq 0.$$

From the reflexivity of E and the Banach-Alaoglu theorem, B_2 is compact. Since A_2 and B_2 are convex h_2 is affine on A_2, and concave and upper semicontinuous on B_2, from the minimax theorem, Theorem 0.1,

$$\max_{B_2} \inf_{A_2} h_2 \geq 0,$$

hence there exists $(x, x^*) \in B_2 \subset E \times E^*$ such that

$$\mu \in A_2 \quad \Longrightarrow \quad 2r(\mu) - 2\langle p(\mu), x^* \rangle - 2\langle x, q(\mu) \rangle - \|x\|^2 - \|x^*\|^2 \geq 0.$$

Allowing μ to run through the values $\delta_{(s,s^*)}$ $\quad ((s, s^*) \in G)$,

$$(s, s^*) \in G \quad \Longrightarrow \quad 2\langle s, s^* \rangle - 2\langle s, x^* \rangle - 2\langle x, s^* \rangle - \|x\|^2 - \|x^*\|^2 \geq 0.$$

(b) follows by rearranging and adding $2\langle x, x^* \rangle$ to each side. ▌

Note that we introduce the term $2\langle x, x^* \rangle$ *after* the minimax theorem has been used in the proof of ((a) $\Longrightarrow$ (b)) above, since this term is not generally semicontinuous on $E \times E^*$ with respect to the product of the topologies $w(E, E^*)$ and $w(E^*, E)$.

Theorem 4.2. *Let E be reflexive and M be a nonempty monotone subset of $E \times E^*$. Then:*
(a) There exists $(x, x^*) \in E \times E^*$ such that

$$(s, s^*) \in M \quad \Longrightarrow \quad 2\langle s - x, s^* - x^* \rangle \geq \|x^*\|^2 + \|x\|^2 + 2\langle x, x^* \rangle. \quad (4.2.1)$$

(b) Suppose now that M is a maximal monotone *subset of $E \times E^*$. Then there exists $(x, x^*) \in M$ such that*

$$\|x\|^2 + \|x^*\|^2 + 2\langle x, x^* \rangle = 0.$$

Proof. We note that (4.2.1) is identical with (4.1.2), except that G has been replaced by M. Thus (a) immediate from Theorem 3.4 and Lemma 4.1. Now let $(x, x^*) \in E \times E^*$ be as in (a). Since

$$\|x\|^2 + \|x^*\|^2 + 2\langle x, x^* \rangle \geq \|x\|^2 + \|x^*\|^2 - 2\|x\|\|x^*\| = (\|x\| - \|x^*\|)^2 \geq 0,$$

it follows that

$$(s, s^*) \in M \quad \Longrightarrow \quad 2\langle s - x, s^* - x^* \rangle \geq 0.$$

Since M is *maximal* monotone, $(x, x^*) \in M$, and (b) then follows by substituting $(s, s^*) := (x, x^*)$ in (4.2.1). ▌

Theorem 4.3. *Let E be reflexive and M be a monotone subset of $E \times E^*$. Then M is maximal monotone $\Longleftrightarrow$ for all $(w, w^*) \in E \times E^*$, there exists $(x, x^*) \in M$ such that*

$$\|x - w\|^2 + \|x^* - w^*\|^2 + 2\langle x - w, x^* - w^* \rangle = 0 \qquad (4.3.1)$$

Proof. ($\Longrightarrow$) We apply Theorem 4.2(b), with M replaced by $M-(w,w^*) \subset E \times E^*$, which is also maximal monotone.
($\Longleftarrow$) Let $(w,w^*) \in E \times E^*$ and

$$(x,x^*) \in M \quad \Longrightarrow \quad \langle x-w, x^*-w^* \rangle \geq 0.$$

Choose $(x,x^*) \in M$ as in (4.3.1). Then $\|x-w\|^2 + \|x^*-w^*\|^2 \leq 0$, and so $(w,w^*) = (x,x^*) \in M$. Thus M is maximal monotone. ▌

The *duality map* $J : E \to 2^{E^*}$ is defined by:

$$x^* \in Jx \iff \langle x, x^* \rangle = \|x\|^2 = \|x^*\|^2.$$

Further, $-J : E \to 2^{E^*}$ is defined by:

$$(-J)x := -Jx.$$

We leave the proof of the following lemma to the reader.

Lemma 4.4. *If* $(x,x^*) \in E \times E^*$ *then*

$$(x,x^*) \in G(-J) \iff \|x\|^2 + \|x^*\|^2 + 2\langle x, x^* \rangle = 0.$$

Theorem 4.5. *Let* E *be reflexive and* M *be a monotone subset of* $E \times E^*$. *Then*

$$M \textit{ is maximal monotone} \iff M + G(-J) = E \times E^*.$$

Proof. From Lemma 4.4 and Theorem 4.3, M is maximal monotone $\iff$ for all $(w,w^*) \in E \times E^*$, there exists $(x,x^*) \in M$ such that $(x-w, x^*-w^*) \in G(-J)$. But this last is equivalent to $(w-x, w^*-x^*) \in G(-J)$, i.e.,

$$(w,w^*) \in (x,x^*) + G(-J). \ ▌$$

There is an obvious one-to-one correspondence between multifunctions from E into 2^{E^*} and subsets of $E \times E^*$: if S: $E \to 2^{E^*}$ then $G(S)$ is the corresponding subset of $E \times E^*$, while if $G \subset E \times E^*$ then the corresponding multifunction is defined by

$$Sx := \{x^* \in E^* \colon (x,x^*) \in G\}.$$

Further, monotone multifunctions correspond to monotone subsets and maximal monotone multifunctions correspond to maximal monotone subsets. For the rest of these lectures, we return to the "multifunction" notation rather than the "subset of $E \times E^*$" notation that we have been using since the beginning of Section 3. Now, it was proved by Minty that *if E is a Hilbert space and S: $E \to 2^{E^*}$ is monotone then S is maximal monotone $\iff$ $R(S+J) = E^*$.* Rockafellar showed that Minty's result can be extended to the case where E is reflexive and the norms of E and E^* are "strictly convex". Further, it was proved by Asplund that any reflexive Banach space can be renormed so that this property holds. This renorming theorem is not easy. We shall show in Theorem 4.6 that the implication ($\Longrightarrow$), known as "Rockafellar's surjectivity theorem", is true even without the renorming. It was pointed out by Simon Fitzpatrick that if ($\Longleftarrow$) holds then J is single–valued.

Theorem 4.6. *Let E be reflexive and S: $E \to 2^{E^*}$ be maximal monotone. Then*

$$R(S+J) = E^*.$$

Proof. Let $w^* \in E^*$. From Theorem 4.5,

$$(0, w^*) \in G(S) + G(-J).$$

Thus there exist $x \in E$, $x^* \in Sx$ and $y^* \in (-J)(-x)$ such that $x^* + y^* = w^*$. But then $y^* \in Jx$, hence

$$w^* = x^* + y^* \in Sx + Jx \subset R(S+J). \blacksquare$$

5. The Convex Function Determined by a Multifunction

We shall abuse our notation a little, and use $\mathcal{C}(S)$ as an abbreviation for $\mathcal{C}(G(S))$. We suppose throughout this section that $S : E \to 2^{E^*}$ is nontrivial. We define the function $\chi_S : E \to \mathbb{R} \cup \{\infty\}$ by

$$\chi_S(x) := \sup_{\mu \in \mathcal{C}(S)} \frac{\langle x, q(\mu) \rangle - r(\mu)}{1 + \|p(\mu)\|} \quad (x \in E).$$

Since χ_S is the supremum of a family of continuous affine functions, χ_S is convex and lower semicontinuous. The function χ_S was first defined, and many of its properties were established in [7]. Lemma 5.1 shows that if S is monotone then $\operatorname{dom}\chi_S \neq \emptyset$. We use the notation "co" to stand for "convex hull of".

Lemma 5.1. *Let $S : E \to 2^{E^*}$ be nontrivial and monotone. Then*

$$D(S) \subset \operatorname{co} D(S) \subset \operatorname{dom}\chi_S.$$

Proof. Since $\operatorname{dom}\chi_S$ is convex, it suffices to prove that

$$D(S) \subset \operatorname{dom}\chi_S. \tag{5.1.1}$$

To this end, let $w \in D(S)$. Pick $w^* \in Sw$, and define $\beta := \langle w, w^*\rangle \vee \|w^*\|$. If $(s, s^*) \in G(S)$ then, since S is monotone,

$$\langle w, s^*\rangle - \langle s, s^*\rangle = \langle w - s, s^*\rangle \leq \langle w - s, w^*\rangle = \langle w, w^*\rangle - \langle s, w^*\rangle.$$

Thus, for all $\mu \in \mathcal{C}(S)$ and $(s, s^*) \in G(S)$,

$$\langle w, \mu_{(s,s^*)}s^*\rangle - \mu_{(s,s^*)}\langle s, s^*\rangle \leq \mu_{(s,s^*)}\langle w, w^*\rangle - \langle \mu_{(s,s^*)}s, w^*\rangle.$$

Summing over $(s, s^*) \in G(S)$,

$$\begin{aligned}\langle w, q(\mu)\rangle - r(\mu) &\leq \langle w, w^*\rangle - \langle p(\mu), w^*\rangle \leq \\ &\leq \langle w, w^*\rangle + \|p(\mu)\|\|w^*\| \leq \beta(1 + \|p(\mu)\|).\end{aligned}$$

Dividing by $1 + \|p(\mu)\|$, and taking the supremum over $\mu \in \mathcal{C}(S)$, we see that $\chi_S(w) \leq \beta$, which implies that $w \in \operatorname{dom}\chi_S$. This completes the proof of (5.1.1), and hence that of Lemma 5.1. ∎

6. Surrounding Sets and the Dom-Dom Lemma

We start this section by collecting together some results on convex lower semicontinuous functions that we shall use later.

Let $x \in E$ and $A \subset E$. A is said to be *absorbing* if $\bigcup_{\lambda>0} \lambda A = E$. Any neighborhood of 0 is absorbing. We write "$x \in \operatorname{sur} A$" and say that "$A$ surrounds x" if, for each $w \in E\backslash\{0\}$, there exists $\delta > 0$ such that $x + \delta w \in A$. The statement "$x \in \operatorname{sur} A$" is related to x being an "absorbing point" of A (see [10], Definition 2.27(b), p. 28), but differs in that we do not require that $x \in A$. We also note that

$$\text{if } A \text{ is convex then} \qquad (0 \in \operatorname{sur} A \iff A \text{ is absorbing}). \tag{6.0.1}$$

Lemma 6.1, the "dom-dom lemma", which is related to the open mapping theorem can, in, fact be generalized considerably. (In this connection, we refer the reader to Robinson [12], Ursescu [18] and Borwein [3]). Here we confine our attention to what we will need in these lectures. More details can be found in [17].

We shall use the notation $\{g \vee \| \; \| \le n\}$ as a shorthand for $\{x \in E\colon\ g(x) \le n,\ \|x\| \le n\}$.

Lemma 6.1. *Let* $g_1,\ g_2 : E \to \mathbb{R} \cup \{\infty\}$ *be convex, lower semicontinuous functions and* $\operatorname{dom} g_1 - \operatorname{dom} g_2$ *surround* 0. *Then there exists* $n \ge 1$ *such that*

$$\{g_1 \vee \| \; \| \le n\} - \{g_2 \vee \| \; \| \le n\} \quad \textit{is a neighborhood of } 0. \tag{6.1.1}$$

Lemma 6.2 is the "dom lemma". It follows from Lemma 6.1, with $g_1 := f$ and g_2 the "indicator function" of $\{0\}$, i.e.,

$$g_2(x) := \begin{cases} 0 & (x = 0) \\ \infty & (x \ne 0). \end{cases}$$

It can also be proved directly from Baire's theorem. The "dom corollary", Corollary 6.3, is an easy consequence of Lemma 6.2.

Lemma 6.2. *Let* $f : E \to \mathbb{R} \cup \{\infty\}$ *be a convex, lower semicontinuous function and* $\operatorname{dom} f$ *surround* 0. *Then there exist* $\eta > 0$ *and* $n \ge 1$ *such that*

$$w \in E \text{ and } \|w\| \le \eta \quad \Longrightarrow \quad f(w) \le n. \tag{6.2.1}$$

Corollary 6.3. *Let* $f : E \to \mathbb{R} \cup \{\infty\}$ *be convex and lower semicontinuous. Then*

$$\operatorname{sur}(\operatorname{dom} f) = \operatorname{int}(\operatorname{dom} f).$$

Remark 6.4. The classical "uniform boundedness theorem" can easily be deduced from the dom lemma. Thus, in some sense, Lemma 6.1 is a "second order" generalization of the uniform boundedness theorem. Lemma 6.2 also implies the result that a convex lower semicontinuous function is locally bounded on the interior of its domain. (See, for instance, [10], Proposition 3.3, p. 39.)

We now come to the "χ local boundedness theorem", Theorem 6.6. We start off with an equivalence valid for any nontrivial multifunction on a Banach space. We will use Lemma 6.5 again later in Lemma 6.9.

Lemma 6.5. *Let* $S\colon E \to 2^{E^*}$ *be nontrivial. Then* $\operatorname{dom} \chi_S$ *surrounds* $0 \iff$ *there exist* $\eta > 0$ *and* $n \ge 1$ *such that*

$$\mu \in \mathcal{C}(S) \quad \Longrightarrow \quad r(\mu) + n(1 + \|p(\mu)\|) \ge \eta \|q(\mu)\|. \tag{6.5.1}$$

Proof. ($\Longrightarrow$) From the dom lemma, Lemma 6.2, there exist $\eta > 0$ and $n \geq 1$ such that

$$x \in E \text{ and } \|x\| \leq \eta \quad \Longrightarrow \quad \chi_S(x) \leq n.$$

Thus, from the definition of χ_S,

$$x \in E,\ \|x\| \leq \eta \quad \text{and} \quad \mu \in \mathcal{C}(S) \quad \Longrightarrow \quad \langle x, q(\mu)\rangle - r(\mu) \leq n(1 + \|p(\mu)\|),$$

that is to say,

$$\mu \in \mathcal{C}(S),\ x \in E \quad \text{and} \quad \|x\| \leq \eta \quad \Longrightarrow \quad r(\mu) + n(1 + \|p(\mu)\|) \geq \langle x, q(\mu)\rangle.$$

We now obtain (6.5.1) by taking the supremum of the right-hand expression over all $x \in E$ such that $\|x\| \leq \eta$.

($\Longleftarrow$) We leave the proof of this to the reader. ▌

Theorem 6.6. *Let S: $E \to 2^{E^*}$ be nontrivial, $x \in E$ and* $\operatorname{dom}\chi_S$ *surround x. Then there exist θ, $Q > 0$ such that*

$$(s, s^*) \in G(S) \text{ and } \|s - x\| < \delta \quad \Longrightarrow \quad \|s^*\| \leq Q.$$

Proof From a translation argument, we can and will assume that $x = 0$. Let $\eta > 0$ and $n \geq 1$ be as in Lemma 6.5. Applying (6.5.1) with $\mu := \delta_{(s,s^*)}$ $((s, s^*) \in G(S))$,

$$(s, s^*) \in G(S) \quad \Longrightarrow \quad \langle s, s^*\rangle + n(1 + \|s\|) \geq \eta\|s^*\|,$$

hence

$$(s, s^*) \in G(S) \quad \Longrightarrow \quad \eta\|s^*\| - \|s\|\|s^*\| \leq n(1 + \|s\|).$$

Let $\theta := \eta/2$. Then

$$(s, s^*) \in G(S) \text{ and } \|s\| \leq \theta \quad \Longrightarrow \quad \theta\|s^*\| \leq n(1 + \theta).$$

Thus the required result follows with $Q := n(1 + \theta)/\theta$. ▌

In Corollary 6.7, we give a special case of Theorem 6.6 that is phrased in terms of the more classical concept of $D(S)$. Corollary 6.7 follows easily from Theorem 6.6 and Lemma 5.1. In turn, Corollary 6.7 generalizes the following result of Borwein and Fitzpatrick (see [4]): *Let S be nontrivial and monotone, and x be an absorbing point of $D(S)$. Then S is locally bounded at x.*

Corollary 6.7. *Let S: $E \to 2^{E^*}$ be nontrivial and monotone and* co$D(S)$ *surround x. Then there exist θ, $Q > 0$ such that*

$$(s, s^*) \in G(S) \text{ and } \|s - x\| < \theta \quad \Longrightarrow \quad \|s^*\| \le Q.$$

We now use the minimax theorem to obtain an equivalence for general (i.e., possibly nonreflexive) Banach spaces that is similar in spirit to Lemma 4.1.

Lemma 6.8. *Let S: $E \to 2^{E^*}$ be nontrivial. Then* (6.8.1) $\Longleftrightarrow$ (6.8.2).

$$\text{There exists } K > 0 \text{ such that } \quad \mu \in \mathcal{C}(S) \Longrightarrow r(\mu) + K\|p(\mu)\| \ge 0. \quad (6.8.1)$$

$$\text{There exists } x^* \in E^* \text{ such that } (s, s^*) \in G(S) \Longrightarrow \langle s, s^* - x^* \rangle \ge 0. \quad (6.8.2)$$

Proof. ($\Longrightarrow$) Define h_3: $\mathcal{C}(E \times E^*) \times E^* \to \mathbb{R}$ by

$$h_3(\mu, x^*) := r(\mu) - \langle p(\mu), x^* \rangle. \tag{6.8.3}$$

Let K satisfy (6.8.1), $A_3 := \mathcal{C}(S)$ and $B_3 := \{x^* \in E^*\colon \|x^*\| \le K\}$, with the topology $w(E^*, E)$. If $\mu \in A_3$ then

$$\max_{x^* \in B_3} h_3(\mu, x^*) = r(\mu) + K\|p(\mu)\|,$$

hence

$$\inf_{A_3} \max_{B_3} h_3 \ge 0.$$

A_3 and B_3 are convex. From the Banach-Alaoglu theorem, B_3 is compact. Since h_3 is affine in each variable and continuous on B_3, from the minimax theorem, Theorem 0.1,

$$\max_{B_3} \inf_{A_3} h_3 \ge 0,$$

hence there exists $x^* \in B_3 \subset E^*$ such that

$$\mu \in A_3 \quad \Longrightarrow \quad r(\mu) - \langle p(\mu), x^* \rangle \ge 0.$$

(6.8.2) now follows by allowing μ to run through the values $\delta_{(s,s^*)}$ ($(s, s^*) \in G(S)$).

($\Longleftarrow$) We leave the proof of this to the reader. ∎

Lemma 6.9 will lead easily to our main results on maximal monotone multifunctions on general Banach spaces.

Lemma 6.9. *Let* $S\colon E \to 2^{E^*}$ *be maximal monotone. Then*

$$\operatorname{sur}(\operatorname{dom}\chi_S) \subset \operatorname{int} D(S).$$

Proof. We first prove that

$$\operatorname{dom}\chi_S \text{ surrounds } 0 \quad\Longrightarrow\quad 0 \in D(S). \tag{6.9.1}$$

If $\operatorname{dom}\chi_S$ surrounds 0 then, from Lemma 6.5 there exist $\eta > 0$ and $n \geq 1$ such that

$$\mu \in \mathcal{C}(S) \quad\Longrightarrow\quad r(\mu) + n(1 + \|p(\mu)\|) \geq \eta\|q(\mu)\|. \tag{6.5.1}$$

By decreasing η if necessary, we can and will suppose that $\eta \in (0, 1]$. We now write $K := n/\eta$, and we shall prove that

$$\mu \in \mathcal{C}(S) \quad\Longrightarrow\quad r(\mu) + K\|p(\mu)\| \geq 0. \tag{6.9.2}$$

To this end, let $\mu \in \mathcal{C}(S)$. If $\|q(\mu)\| \leq K$ then, from the pqr-lemma, Lemma 3.1,

$$\begin{aligned} r(\mu) + K\|p(\mu)\| &\geq \langle p(\mu), q(\mu)\rangle + K\|p(\mu)\| \\ &\geq K\|p(\mu)\| - \|p(\mu)\|\|q(\mu)\| \\ &= (K - \|q(\mu)\|)\|p(\mu)\| \geq 0, \end{aligned}$$

and (6.9.2) follows. If, on the other hand, $\|q(\mu)\| > K$ then, since $K = n/\eta \geq n$,

$$r(\mu) + K\|p(\mu)\| \geq r(\mu) + n\|p(\mu)\|,$$

from (6.5.1),

$$\begin{aligned} &\geq \eta\|q(\mu)\| - n \\ &= \eta(\|q(\mu)\| - K) > 0, \end{aligned}$$

and (6.9.2) follows again. From Lemma 6.8,

$$\text{there exists } x^* \in E^* \text{ such that} \quad (s, s^*) \in G(S) \Longrightarrow \langle s - 0, s^* - x^*\rangle \geq 0. \tag{6.8.2}$$

Since S is maximal monotone,

$$(0, x^*) \in G(S).$$

This gives (6.9.1). From (6.9.1) and a translation argument,

$$\operatorname{sur}(\operatorname{dom}\chi_S) \subset D(S).$$

The result follows since, from the dom corollary, Corollary 6.3, $\operatorname{sur}(\operatorname{dom}\chi_S)$ is open. ▍

The "six sets theorem", Theorem 6.10, is our main result on maximal monotone multifunctions on general Banach spaces. Theorem 6.10 answers in the affirmative a question raised by Phelps (see [10], p. 29 and [11], p. 8), namely whether an absorbing point of $D(S)$ is necessarily an interior point. If $S\colon E \to 2^{E^*}$ is maximal monotone then, for all $x^* \in E^*$, $\{x \in E\colon Sx \ni x^*\}$ is closed and convex. On the other hand, it is not in general true that $D(S)$ is convex. In fact, we shall now give an example of a convex lower semicontinuous function $f\colon \mathbb{R}^2 \to \mathbb{R} \cup \{\infty\}$ such that $D(\partial f)$ is not convex. Define $f\colon \mathbb{R}^2 \to \mathbb{R} \cup \{\infty\}$ by

$$f(x_1, x_2) := \begin{cases} |x_2| \vee (1 - \sqrt{1 - {x_1}^2}) & (|x_1| \vee |x_2| \leq 1) \\ \infty & (|x_1| \vee |x_2| > 1). \end{cases}$$

However, Rockafellar proved in [13], Theorem 1, p. 398 (see also [11], Theorem 1.9, p. 6) that $D(S)$ is not far from being convex in the following sense: *Let S be maximal monotone and* $\operatorname{int}(\operatorname{co} D(S)) \neq \emptyset$. *Then* $\operatorname{int} D(S)$ *is convex and* $\overline{D(S)} = \overline{\operatorname{int} D(S)}$, *so* $\overline{D(S)}$ *is also convex.* These result are extended by Theorem 6.10 and the "nine sets theorem", Theorem 6.11.

Theorem 6.10. *Let $S\colon E \to 2^{E^*}$ be maximal monotone. Then*

$$\begin{aligned} \operatorname{int} D(S) &= \operatorname{int}(\operatorname{co} D(S)) = \operatorname{int}(\operatorname{dom} \chi_S) \\ = \operatorname{sur} D(S) &= \operatorname{sur}(\operatorname{co} D(S)) = \operatorname{sur}(\operatorname{dom} \chi_S). \end{aligned}$$

Proof. From Lemma 5.1,

$$\operatorname{int} D(S) \subset \operatorname{int}(\operatorname{co} D(S)) \subset \operatorname{int}(\operatorname{dom} \chi_S)$$

and

$$\operatorname{sur} D(S) \subset \operatorname{sur}(\operatorname{co} D(S)) \subset \operatorname{sur}(\operatorname{dom} \chi_S).$$

Obviously $\operatorname{int}(\ldots) \subset \operatorname{sur}(\ldots)$, and the result follows from Lemma 6.9. ∎

Theorem 6.11. *Let S be maximal monotone and* $\operatorname{sur}(\operatorname{dom} \chi_S) \neq \emptyset$. *Then*

$$\begin{aligned} \overline{D(S)} &= \overline{\operatorname{co} D(S)} = \overline{\operatorname{dom} \chi_S} \\ = \overline{\operatorname{int} D(S)} &= \overline{\operatorname{int}(\operatorname{co} D(S))} = \overline{\operatorname{int}(\operatorname{dom} \chi_S)} \\ = \overline{\operatorname{sur} D(S)} &= \overline{\operatorname{sur}(\operatorname{co} D(S))} = \overline{\operatorname{sur}(\operatorname{dom} \chi_S)}. \end{aligned}$$

Proof. Obviously, $\overline{\operatorname{int} D(S)} \subset \overline{D(S)}$ and, from Lemma 5.1,

$$\overline{D(S)} \subset \overline{\operatorname{co} D(S)} \subset \overline{\operatorname{dom} \chi_S}.$$

From the dom corollary, Corollary 6.3, $\operatorname{int}(\operatorname{dom}\chi_S) = \operatorname{sur}(\operatorname{dom}\chi_S) \neq \emptyset$, hence (see, for instance, [9], 13.1(i), p. 100-111), $\overline{\operatorname{dom}\chi_S} = \overline{\operatorname{int}(\operatorname{dom}\chi_S)}$. Thus we have

$$\overline{\operatorname{int} D(S)} \subset \overline{D(S)} \subset \overline{\operatorname{co} D(S)} \subset \overline{\operatorname{dom}\chi_S} = \overline{\operatorname{int}(\operatorname{dom}\chi_S)}.$$

The result now follows by combining this with Theorem 6.10. ▌

7. The "Dom-Dom Constraint Qualification"

If $S_1\colon E \to 2^{E^*}$ and $S_2\colon E \to 2^{E^*}$ are nontrivial and monotone and $D(S_1) \cap D(S_2) \neq \emptyset$ then $S_1 + S_2$, defined in (1.0.1), is obviously nontrivial and monotone. On the other hand, it does *not* follow that if S_1 and S_2 are maximal monotone and $D(S_1) \cap D(S_2) \neq \emptyset$ then $S_1 + S_2$ is maximal monotone. As an example of this, let C_1 and C_2 be two closed disks in the plane that touch at the point p. Then N_{C_1} and N_{C_2} (see Section 1) are maximal monotone, but $N_{C_1} + N_{C_2}$ is not — $G(N_{C_1} + N_{C_2})$ is a proper subset of $G(N_{\{p\}})$. (See p. 54 of [10]). Determining conditions on S_1 and S_2 (normally called "constraint qualifications") that ensure that $S_1 + S_2$ is maximal monotone is one of the fundamental questions in the theory of monotone multifunctions. We shall say that S_1 and S_2 satisfy the "dom-dom constraint qualification" if

$$\operatorname{dom}\chi_{S_1} - \operatorname{dom}\chi_{S_2} \quad \text{is absorbing,} \tag{7.0.1}$$

and we shall prove in Corollary 8.4 that if E is reflexive and S_1 and S_2 are maximal monotone and satisfy the dom-dom constraint qualification then $S_1 + S_2$ is maximal monotone.

Lemma 7.1, which assumes neither the reflexivity of E nor the maximality of S_1 or S_2, will be our first step towards Corollary 8.4. In some senses, Lemma 7.1 shares some structure with Lemma 6.5($\Longrightarrow$). It is, however, much harder since it uses the dom-dom lemma, Lemma 6.1 rather than the dom lemma, Lemma 6.2.

Lemma 7.1. *Let $S_1\colon E \to 2^{E^*}$ and $S_2\colon E \to 2^{E^*}$ be nontrivial and monotone, and satisfy the dom-dom constraint qualification. Let $h_3(\mu, x^*) := r(\mu) - \langle p(\mu), x^*\rangle$ be as in (6.8.3) and $A_4 := C(S_1) \times C(S_2)$. Then*

$$\left.\begin{array}{l} \text{there exists } \gamma > 0 \text{ such that, for all } (\mu_1, \mu_2) \in A_4, \\ \text{there exist } \lambda \in [0, 2\gamma] \quad \text{and} \quad (x_1^*, x_2^*) \in (E^*)^2 \\ \text{such that} \quad \|x_1^*\| \vee \|x_2^*\| \leq \gamma \quad \text{and} \\ 2h_3(\mu_1, x_1^*) + 2h_3(\mu_2, x_2^*) + 2\lambda\|q(\mu_1) + q(\mu_2)\| - \\ \qquad - \lambda^2 - \|x_1^* + x_2^*\|^2 \geq 0. \end{array}\right\} \tag{7.1.1}$$

Proof. To simplify the expressions in what follows, we write χ_i instead of χ_{S_i}. From the dom-dom lemma, Lemma 6.1, there exist $n \geq 1$ and $\eta \in (0,1]$ such that

$$w \in E \text{ and } \|w\| \leq \eta \quad \Longrightarrow \quad w \in \{\chi_1 \vee \|\ \| \leq n\} - \{\chi_2 \vee \|\ \| \leq n\}. \quad (7.1.2)$$

We write $\gamma := 5n^2/\eta$, and we shall show that γ has the required property. So let $(\mu_1, \mu_2) \in A_4$. We first consider the case that $\|q(\mu_1)\| \vee \|q(\mu_2)\| \leq \gamma$. Here we write $(x_1^*, x_2^*) = (q(\mu_1), q(\mu_2))$ and $\lambda := \|q(\mu_1) + q(\mu_2)\|$. From the *pqr*-lemma, Lemma 3.1,

$$h_3(\mu_k, x_k^*) = r(\mu_k) - \langle p(\mu_k), q(\mu_k)\rangle \geq 0 \quad (k = 1,\ 2),$$

and (7.1.1) follows immediately. We now consider the other alternative, namely that $\|q(\mu_1)\| \vee \|q(\mu_2)\| > \gamma$, and we shall prove that

$$(7.1.1) \text{ is satisfied with } \quad \lambda = n \quad \text{ and } \quad \|x_1^*\| \vee \|x_2^*\| \leq n. \quad (7.1.3)$$

We fix $j \in \{1,\ 2\}$ such that $\|q(\mu_j)\| > \gamma$, choose an element v of E such that

$$\|v\| \leq \eta \quad \text{and} \quad \langle v, q(\mu_j)\rangle \geq \gamma\eta = 5n^2, \quad (7.1.4)$$

and write k for the element of $\{1,\ 2\}$ different from j. From (7.1.2) applied to $w := \pm v$ (depending on whether $j = 1$ or 2), there exist $x,\ y \in E$ such that

$$\chi_j(x) \leq n, \quad \chi_k(y) \leq n, \quad \|y\| \leq n \quad \text{and} \quad x - y = v.$$

From the Hahn-Banach theorem, there exist $x_1^*,\ x_2^* \in E^*$ such that, for $i = 1,\ 2$,

$$\|x_i^*\| \leq n, \quad \text{and} \quad \langle p(\mu_i), x_i^*\rangle = -n\|p(\mu_i)\|.$$

Since $\gamma \geq n$, $\|x_1^*\| \leq \gamma$ and $\|x_2^*\| \leq \gamma$ as required. Further, since $\chi_j(x) \leq n$ and $x = v + y$,

$$\begin{aligned} h_3(\mu_j, x_j^*) &:= r(\mu_j) - \langle p(\mu_j), x_j^*\rangle \\ &= r(\mu_j) + n\|p(\mu_j)\| \geq \langle x, q(\mu_j)\rangle - n \\ &= \langle v + y, q(\mu_j)\rangle - n = \langle v, q(\mu_j)\rangle + \langle y, q(\mu_j)\rangle - n. \end{aligned}$$

Thus, from (7.1.4),

$$h_3(\mu_j, x_j^*) \geq 5n^2 + \langle y, q(\mu_j)\rangle - n \geq 4n^2 + \langle y, q(\mu_j)\rangle. \quad (7.1.5)$$

Since $\chi_k(y) \leq n$,

$$\begin{aligned} h_3(\mu_k, x_k^*) &:= r(\mu_k) - \langle p(\mu_k), x_k^*\rangle = \\ &= r(\mu_k) + n\|p(\mu_k)\| \geq \langle y, q(\mu_k)\rangle - n. \quad (7.1.6) \end{aligned}$$

We now add (7.1.5) and (7.1.6). Since $\|y\| \le n$, we derive

$$h_3(\mu_1, x_1^*) + h_3(\mu_2, x_2^*) \ge 4n^2 + \langle y, q(\mu_1) + q(\mu_2)\rangle - n \ge$$
$$\ge 3n^2 - n\|q(\mu_1) + q(\mu_2)\|.$$

Thus

$$2h_3(\mu_1, x_1^*) + 2h_3(\mu_2, x_2^*) + 2n\|q(\mu_1) + q(\mu_2)\| - n^2 - \|x_1^* + x_2^*\|^2$$
$$\ge 6n^2 - n^2 - \|x_1^* + x_2^*\|^2,$$

since $\|x_1^*\| \le n$ and $\|x_2^*\| \le n$,

$$\ge 5n^2 - (2n)^2 > 0.$$

Thus (7.1.3) is satisfied. This completes the proof of (7.1.1). ▌

8. A "Sum Theorem" for Reflexive Spaces

We point out that the equivalence proved in our next result, Lemma 8.1 is not "pulled out of a hat". Remember that our ultimate aim is to find conditions that ensure that S_1+S_2 is maximal monotone. Suppose we know that this is the case. From Theorem 4.2(b), there exists $(z, z^*) \in G(S_1+S_2)$ such that (8.2.2) is satisfied. Let $(z, x_1^*) \in G(S_1)$, $(z, x_2^*) \in G(S_2)$ and $x_1^* + x_2^* = z^*$. Then, for all $(s_1, s_1^*) \in G(S_1)$ and $(s_2, s_2^*) \in G(S_2)$, (8.2.1) is satisfied, which implies that (8.1.2) is satisfied. From Lemma 8.1, (8.1.1) is also satisfied. Put another way, there is no real loss between (8.1.1), which is identical to (7.1.1), and Corollary 8.4. The problem is, of course, to find a value of γ satisfying this condition. This is where we have used the dom-dom lemma, which depends ultimately on the Baire category theorem.

Lemma 8.1. *Let E be reflexive and S_1: $E \to 2^{E^*}$ and S_2: $E \to 2^{E^*}$ be nontrivial. Let $h_3(\mu, x^*) := r(\mu) - \langle p(\mu), x^*\rangle$ be as in* (6.8.3) *and $A_4 := C(S_1) \times C(S_2)$ be as in Lemma 7.1. Then* (8.1.1) $\iff$ (8.1.2).

$$\left.\begin{array}{l} \textit{there exists } \gamma > 0 \textit{ such that, for all } (\mu_1, \mu_2) \in A_4, \\ \textit{there exist } \lambda \in [0, 2\gamma] \quad \textit{and} \quad (x_1^*, x_2^*) \in (E^*)^2 \\ \textit{such that} \quad \|x_1^*\| \vee \|x_2^*\| \le \gamma \quad \textit{and} \\ 2[h_3(\mu_1, x_1^*) + 2h_3(\mu_2, x_2^*) + 2\lambda\|q(\mu_1) + q(\mu_2)\| - \\ \qquad - \lambda^2 - \|x_1^* + x_2^*\|^2 \ge 0; \end{array}\right\} \quad (8.1.1)$$

$$\left.\begin{array}{l} \textit{there exist } (x_1^*, x_2^*) \in (E^*)^2 \textit{ and } z \in E \textit{ such that,} \\ \textit{for all } (s_1, s_1^*) \in G(S_1) \textit{ and } (s_2, s_2^*) \in G(S_2), \\ 2\langle s_1 - z, s_1^* - x_1^*\rangle + 2\langle s_2 - z, s_2^* - x_2^*\rangle \geq \\ \qquad \geq \|z\|^2 + \|x_1^* + x_2^*\|^2 + 2\langle z, x_1^* + x_2^*\rangle. \end{array}\right\} \qquad (8.1.2)$$

Proof. ($\Longrightarrow$) Let γ be as in (8.1.1), and

$$B_4 := \{(x_1^*, x_2^*, z)\colon \ \|x_k^*\| \leq \gamma, \ \ \|z\| \leq 2\gamma\} \subset E^* \times E^* \times E$$

with the product of the weak topologies. Define $h_4\colon A_4 \times B_4 \to \mathbb{R}$ by

$$h_4\big((\mu_1, \mu_2), (x_1^*, x_2^*, z)\big) := \\ 2h_3(\mu_1, x_1^*) + 2h_3(\mu_2, x_2^*) - 2\langle z, q(\mu_1) + q(\mu_2)\rangle - \|z\|^2 - \|x_1^* + x_2^*\|^2.$$

If now $(\mu_1, \mu_2) \in A_4$, then we choose (x_1^*, x_2^*, λ) as in (8.1.1). Since E is reflexive, there exists $z \in E$ such that $\|z\| \leq \lambda$ and

$$\langle z, q(\mu_1) + q(\mu_2)\rangle = -\lambda\|q(\mu_1) + q(\mu_2)\|.$$

So $(x_1^*, x_2^*, z) \in B_4$ and, further,

$$h_4\big((\mu_1, \mu_2), (x_1^*, x_2^*, z)\big) \\ \geq 2h_3(\mu_1, x_1^*) + 2h_3(\mu_2, x_2^*) + 2\lambda\|q(\mu_1) + 2q(\mu_2)\| - \lambda^2 - \|x_1^* + x_2^*\|^2 \geq 0.$$

Thus

$$\inf_{A_4} \max_{B_4} h_4 \geq 0.$$

A_4 and B_4 are convex. From the Banach-Alaoglu theorem, B_4 is compact. h_4 is affine on A_4, and concave and upper semicontinuous on B_4. Thus, from the minimax theorem, Theorem 0.1,

$$\max_{B_4} \inf_{A_4} h_4 \geq 0,$$

hence there exists $(x_1^*, x_2^*, z) \in E^* \times E^* \times E$ such that, for all $(\mu_1, \mu_2) \in A_4$,

$$2h_3(\mu_1, x_1^*) + 2h_3(\mu_2, x_2^*) - 2\langle z, q(\mu_1) + q(\mu_2)\rangle - \|z\|^2 - \|x_1^* + x_2^*\|^2 \geq 0.$$

Now let $(s_1, s_1^*) \in G(S_1)$ and $(s_2, s_2^*) \in G(S_2)$. If we let $\mu_1 := \delta_{(s_1, s_1^*)}$ and $\mu_2 := \delta_{(s_2, s_2^*)}$ in the above inequality, we obtain

$$2\langle s_1, s_1^*\rangle - 2\langle s_1, x^*\rangle + 2\langle s_2, s_2^*\rangle - 2\langle s_2, x^*\rangle - 2\langle z, s_1^* + s_2^*\rangle - \|z\|^2 - \|x_1^* + x_2^*\|^2 \geq 0.$$

(8.1.2) now follows by rearranging the terms and adding $2\langle z, x_1^* + x_2^*\rangle$ to both sides.

($\Longleftarrow$) We leave the proof of this to the reader. ▌

The technique used to prove Lemma 8.2(b) is derived from a technique due to Minty and Browder (see [5], Lemma 6, p. 99).

Lemma 8.2. *Let E be reflexive and S_1: $E \to 2^{E^*}$ and S_2: $E \to 2^{E^*}$ be nontrivial and monotone and satisfy the dom-dom constraint qualification. Then:*

(a) There exist $(x_1^*, x_2^*) \in (E^*)^2$ and $z \in E$ such that, for all $(s_1, s_1^*) \in G(S_1)$ and $(s_2, s_2^*) \in G(S_2)$,

$$2\langle s_1 - z, s_1^* - x_1^*\rangle + 2\langle s_2 - z, s_2^* - x_2^*\rangle \geq \|z\|^2 + \|x_1^* + x_2^*\|^2 + 2\langle z, x_1^* + x_2^*\rangle. \quad (8.2.1)$$

(b) If, in addition, S_1 and S_2 are maximal monotone then there exists $(z, z^*) \in G(S_1 + S_2)$ such that

$$\|z\|^2 + \|z^*\|^2 + 2\langle z, z^*\rangle = 0. \quad (8.2.2)$$

Proof. (a) is immediate from Lemmas 8.1 and 7.1.

(b) Let $(x_1^*, x_2^*) \in (E^*)^2$ and $z \in E$ be as in (a). Putting $z^* := x_1^* + x_2^*$, (8.2.1) can be rewritten

$$2\langle s_1 - z, s_1^* - x_1^*\rangle + 2\langle s_2 - z, s_2^* - x_2^*\rangle \geq \|z\|^2 + \|z^*\|^2 + 2\langle z, z^*\rangle.$$

Taking the infimum over all $(s_1, s_1^*) \in G(S_1)$ and $(s_2, s_2^*) \in G(S_2)$,

$$2 \inf_{(s,s^*) \in G(S_1)} \langle s - z, s^* - x_1^*\rangle + 2 \inf_{(s,s^*) \in G(S_2)} \langle s - z, s^* - x_2^*\rangle \geq \|z\|^2 + \|z^*\|^2 + 2\langle z, z^*\rangle.$$

Since $\|z\|^2 + \|z^*\|^2 + 2\langle z, z^*\rangle \geq \|z\|^2 + \|z^*\|^2 - 2\|z\|\|z^*\| \geq 0$, it follows from Lemma 1.1(c) that (8.2.2) is satisfied,

$$\inf_{(s,s^*) \in G(S_1)} \langle s - z, s^* - x_1^*\rangle = 0 \quad \text{and} \quad \inf_{(s,s^*) \in G(S_2)} \langle s - z, s^* - x_2^*\rangle = 0.$$

We now deduce from Lemma 1.1(b) that $(z, x_1^*) \in G(S_1)$ and $(z, x_2^*) \in G(S_2)$. Consequently,

$$(z, z^*) = (z, x_1^* + x_2^*) \in G(S_1 + S_2).$$

This completes the proof of Lemma 8.2(b).

Theorem 8.3. *Let E be reflexive, S_1: $E \to 2^{E^*}$ and S_2: $E \to 2^{E^*}$ be maximal monotone and satisfy the dom-dom constraint qualification. Suppose also that $w \in E$ and $w^* \in E^*$. Then there exists $(\zeta, \zeta^*) \in G(S_1 + S_2)$ such that*

$$\|\zeta - w\|^2 + \|\zeta^* - w^*\|^2 + 2\langle \zeta - w, \zeta^* - w^*\rangle = 0. \quad (8.3.1)$$

Proof. This follows from Lemma 8.2(b) and translation arguments. ▌

Corollary 8.4. *Let E be reflexive, S_1: $E \to 2^{E^*}$ and S_2: $E \to 2^{E^*}$ be maximal monotone and satisfy the dom-dom constraint qualification. Then $S_1 + S_2$ is maximal monotone.*

Proof. This is immediate from Theorems 8.3 and 4.3. ▌

We end these lectures by discussing various constraint qualifications that have been shown to imply the maximal monotonicity of $S_1 + S_2$ when E is reflexive and S_1 and S_2 are maximal monotone. The original result due to Rockafellar (see [14], Theorem 1, p. 76) was that this is the case if

$$D(S_1) \cap \mathrm{int} D(S_2) \neq \emptyset. \tag{8.4.1}$$

Attouch-Riahi-Théra proved (see [1], Théorème 4) that this is also the case if

$$D(S_1) - D(S_2) \quad \text{is absorbing,} \tag{8.4.2}$$

and Chu proved (this is contained in [6], Corollary 3.5) that this is also the case if

$$\mathrm{co}D(S_1) - \mathrm{co}D(S_2) \quad \text{is a neighborhood of } 0. \tag{8.4.3}$$

Finally, we have proved in Corollary 8.4 that this is also the case if

$$\mathrm{dom}\,\chi_{S_1} - \mathrm{dom}\,\chi_{S_2} \quad \text{is absorbing.} \tag{8.4.4}$$

It is clear from Lemma 5.1 that Corollory 8.4 is a formal generalization of the preceding three results. One can, in fact, show that (8.4.2), (8.4.3) and (8.4.4) are equivalent. This equivalence is not trivial, since it relies on Lemma 8.2(b). We point out that, unlike what we have presented here, the analysis in [14], [1] and [6] uses the renorming of E (see the remarks preceding Theorem 4.6).

References

1. H. Attouch, H. Riahi and M. Théra, *Somme Ponctuelle d'Opérateurs Maximaux Monotones,* Serdica Math. J. **22** (1996), 267-292.
2. J.-P. Aubin and H. Frankowska, *Set-Valued Analysis,* Birkhäuser, Boston – Basel – Berlin (1990).
3. J. M. Borwein, *A Lagrange multiplier theorem and a sandwich theorem for convex relations,* Math. Scand. **48** (1981), 198-204.
4. J. M. Borwein and S. Fitzpatrick, *Local boundedness of monotone operators under minimal hypotheses,* Bull. Australian Math. Soc. **39** (1988), 439-441.
5. F. E. Browder, *Nonlinear maximal monotone operators in Banach spaces,* Math. Annalen **175** (1968), 89-113.
6. L.-J. Chu, *On the sum of monotone operators,* Michigan Math. J. **43**(1996), 273-289.
7. M. Coodey and S. Simons, *The convex function determined by a multifunction,* Bull. Austral. Math. Soc. **54** (1996), 87-97.
8. K. Fan, *Minimax theorems,* Proc. Nat. Acad. Sci. U.S.A. **39** (1953), 42-47.
9. J. L. Kelley, I. Namioka and co-authors, *Linear Topological Spaces,* D. Van Nostrand Co., Inc., Princeton-Toronto-London-Melbourne (1963).
10. R. R. Phelps, *Convex Functions, Monotone Operators and Differentiability,* Lecture Notes in Mathematics **1364** (1993), Springer-Verlag (Second Edition).

11. R. R. Phelps, *Lectures on Maximal Monotone Operators,* 2nd Summer School on Banach Spaces, Related Areas and Applications, Prague and Paseky, August 15–28, 1993. (Preprint, 30 pages.) TeX file: Banach space bulletin board archive: <ftp://ftp:@math.okstate.edu/pub/banach/phelpsmaxmonop.tex>. Posted Nov. 1993.
12. S. M. Robinson, *Regularity and stability for convex multivalued functions,* Math. Oper. Res. **1** (1976), 130–143.
13. R. T. Rockafellar, *Local boundedness of Nonlinear, Monotone Operators,* Michigan Math. J. **16** (1969), 397–407.
14. R. T. Rockafellar, *On the Maximality of Sums of Nonlinear Monotone Operators,* Trans. Amer. Math. Soc. **149** (1970),75-88.
15. S. Simons, *Pictures of monotone operators,* Set-Valued Analysis **4** (1996), 271-282.
16. S. Simons, *Subdifferentials of convex functions,* Contemporary Mathematics, in press.
17. S. Simons, *Sum theorems for monotone operators and convex functions,* Trans. Amer. Math. Soc., in press.
18. C. Ursescu, *Multifunctions with convex closed graph,* Czechoslovak Math. J. **25** (1975), 438-441.

FAN'S EXISTENCE THEOREM FOR INEQUALITIES CONCERNING CONVEX FUNCTIONS AND ITS APPLICATIONS

WATARU TAKAHASHI
Tokyo Institute of Technology
Department of Mathematical and Computing Sciences
Ohokayama, Meguro-ku, Tokyo 152, JAPAN
e-mail: wataru@is.titech.ac.jp

1. Introduction

In [5], Fan proved the following useful existence theorem for inequalities concerning convex functions in topological vector spaces, by using his minimax theorem [3].

Theorem A (Fan's system theorem) *Let X be a nonempty compact convex subset of a topological vector space and let $\{f_1, f_2, \dots, f_n\}$ be a finite family of lower semicontinuous convex functions on X with values in $(-\infty, \infty)$. Then the following are equivalent:*

(1) *The system of convex inequalities*

$$f_i(x) \le 0 \quad \text{for all } i \in \{1, 2, \dots, n\} \tag{$*$}$$

is consistent on X, i.e., there exists an $x \in X$ satisfying $(*)$;

(2) *for any n nonnegative numbers α_i with $\sum_{i=1}^{n} \alpha_i = 1$, there exists a $y \in X$ such that*

$$\sum_{i=1}^{n} \alpha_i f_i(y) \le 0.$$

We also know that the theorem is proved by using Fan-Browder's fixed point theorem [1]; see [19].

In this article, we first prove some results concerning minimax theorems which are obtained by Fan's system theorem (Theorem A); see Section 2. In Section 3, using Fan's system theorem, we obtain basic results in

B. Ricceri and S. Simons (eds.), Minimax Theory and Applications, 241–260.

Functional Analysis. In particular, we prove the Mazur-Orlicz theorem by using Fan's system theorem and Markov-Kakutani's fixed point theorem [2]. The Mazur-Orlicz theorem is used to obtain another Fan's system theorem (Theorem 3.3) concerning linear inequalities. In Section 4, using this another Fan's system, we discuss cores of games. Finally, we prove a fixed point theorem for noncommutative semigroups of nonexpansive mappings. This theorem answers affirmatively a problem posed during the International Conference on Fixed Point Theory and Applications held at Marseille, 1989. Our proof of the theorem depends on Fan's system theorem (Theorem A).

2. Generalizations of Fan's System Theorem

In this section, we prove the first minimax theorem by using Fan's system theorem. Let X and Y be nonempty sets and let F be a real valued function on $X \times Y$. Then, F is said to be *concavelike* in its second variable if for any $y_1, y_2 \in Y$ and $\alpha \in R$ with $0 \leq \alpha \leq 1$, there exists $y_0 \in Y$ such that

$$F(x, y_0) \geq \alpha F(x, y_1) + (1-\alpha) F(x, y_2) \quad \text{for all } x \in X.$$

Similarly, F is *convexlike* in its second variable if for any $y_1, y_2 \in Y$ and $\alpha \in R$ with $0 \leq \alpha \leq 1$, there exists $y_0 \in Y$ such that

$$F(x, y_0) \leq \alpha F(x, y_1) + (1-\alpha) F(x, y_2) \quad \text{for all } x \in X.$$

Theorem 2.1 *Let X be a compact convex subset of a topological vector space E, let Y be a nonempty set, and let F be a real valued function on $X \times Y$ satisfying the following conditions:*

(1) *For each $y \in Y$, the function $x \mapsto F(x, y)$ is lower semicontinuous and convex;*

(2) *F is concavelike in its second variable.*

Then the following holds:

$$\sup_{y \in Y} \min_{x \in X} F(x, y) = \min_{x \in X} \sup_{y \in Y} F(x, y).$$

Proof. Let $c = \sup_{y \in Y} \min_{x \in X} F(x, y)$ and let $\{y_1, y_2, \ldots, y_n\}$ be a finite set of Y. Then, for any n nonnegative numbers $\alpha_1, \alpha_2, \ldots, \alpha_n$ with $\sum_{i=1}^{n} \alpha_i = 1$, by (2) there exists $y_0 \in Y$ such that

$$\sum_{i=1}^{n} \alpha_i F(x, y_i) \leq F(x, y_0) \quad \text{for all } x \in X.$$

From the definition of c, there exists $x_0 \in X$ with $F(x_0, y_0) \leq c$. That is, for any n nonnegative numbers $\alpha_1, \alpha_2, \ldots, \alpha_n$ with $\sum_{i=1}^{n} \alpha_i = 1$, there exists $x_0 \in X$ such that

$$\sum_{i=1}^{n} \alpha_i F(x_0, y_i) \leq F(x_0, y_0) \leq c.$$

By Fan's system theorem, we have

$$\bigcap_{i=1}^{n} \{x : F(x, y_i) \leq c\} \neq \phi.$$

Since X is compact, we have

$$\bigcap_{y \in Y} \{x : F(x, y) \leq c\} \neq \phi.$$

This implies

$$\min_{x \in X} \sup_{y \in Y} F(x, y) \leq c = \sup_{y \in Y} \min_{x \in X} F(x, y).$$

Since the reverse inequality is obvious, we complete the proof. □

By using Theorem 2.1, we can prove a generalization [22] of Fan's system theorem.

Theorem 2.2 *Let X be a nonempty set and let $\{f_1, f_2, \ldots, f_n\}$ be real valued functions on X. Let $I = \{1, 2, \ldots, n\}$ and define a real valued function F on $X \times I$ by*

$$F(x, i) = f_i(x) \quad \textit{for all } (x, i) \in X \times I.$$

Let $c \in R$ and suppose that F is convexlike in its first variable and for any n nonnegative numbers $\alpha_1, \alpha_2, \ldots, \alpha_n$ with $\sum_{i=1}^{n} \alpha_i = 1$, there exists $x_0 \in X$ such that

$$\sum_{i=1}^{n} \alpha_i f_i(x_0) \leq c.$$

Then the following holds:

$$\inf_{x \in X} \max_{i} f_i(x) \leq c.$$

Proof. Let $Y = \{\alpha = (\alpha_1, \alpha_2, \ldots, \alpha_n) : \alpha_i \geq 0, \sum_{i=1}^{n} \alpha_i = 1\}$ and define

$$f(x, \alpha) = \sum_{i=1}^{n} \alpha_i f_i(x) \quad \text{for all } x \in X \text{ and } \alpha = (\alpha_1, \alpha_2, \ldots, \alpha_n) \in Y.$$

Then, f is convexlike in its first variable. In fact, for any $x, y \in X$ and $a, b \in R$ with $a, b \geq 0$ and $a + b = 1$, there exists $z \in X$ such that

$$\begin{aligned} af(x, \alpha) + bf(y, \alpha) &= \sum_{i=1}^{n} \alpha_i \{af_i(x) + bf_i(y)\} \\ &\geq \sum_{i=1}^{n} \alpha_i f_i(z) \\ &= f(z, \alpha) \quad \text{for all } \alpha \in Y. \end{aligned}$$

So, by Theorem 2.1, we have

$$\inf_{x \in X} \max_{\alpha \in Y} f(x, \alpha) = \max_{\alpha \in Y} \inf_{x \in X} f(x, \alpha) \leq c.$$

Therefore from $f(x, \alpha) = \sum_{i=1}^{n} \alpha_i f_i(x)$, we have

$$\inf_{x \in X} \max_{i} f_i(x) \leq \inf_{x \in X} \max_{\alpha \in Y} f(x, \alpha) \leq c. \qquad \square$$

As a direct consequence of Theorem 2.2, we obtain the following:

Theorem 2.3 *Let X be a convex subset of a linear space E and let $\{f_1, f_2, \ldots, f_n\}$ be real valued convex functions on X. Suppose that for n nonnegative numbers $\alpha_1, \alpha_2, \ldots, \alpha_n$ with $\sum_{i=1}^{n} \alpha_i = 1$, there exists $x_0 \in X$ such that*

$$\sum_{i=1}^{n} \alpha_i f_i(x_0) \leq c.$$

Then the following holds:

$$\inf_{x \in X} \max_{i} f_i(x) \leq c.$$

Using Theorem 2.2, we prove a minimax theorem obtained by Simons [17]. Let X be a nonempty set and let $F(X)$ be the set of all nonempty finite subsets of X.

Theorem 2.4 (Simons' minimax theorem) *Let X, Y be nonempty sets and let f, g be real valued functions on $X \times Y$ satisfying the following conditions:*

(1) *$f(x, y) \leq g(x, y)$ for all $(x, y) \in X \times Y$;*
(2) *f is convexlike in its second variable;*
(3) *g is concavelike in its first variable.*

Then the following hold:

(i) *For $X_0 \in F(X)$ and $Y_0 \in F(Y)$,*

$$\inf_{y \in Y} \max_{x \in X_0} f(x, y) \leq \sup_{x \in X} \min_{y \in Y_0} g(x, y);$$

(ii) *if $\sup_{x \in X} \inf_{y \in Y} g(x, y) > -\infty$, then for $X_0 \in F(X)$,*

$$\inf_{y \in Y} \max_{x \in X_0} f(x, y) \leq \sup_{x \in X} \inf_{y \in Y} g(x, y);$$

(iii) *if Y is compact and f is lower semicontinuous in its second variable, then*

$$\min_{y \in Y} \sup_{x \in X} f(x, y) \leq \sup_{x \in X} \inf_{y \in Y} g(x, y).$$

Proof. (i) Let $c = \sup_{x \in X} \min_{y \in Y_0} g(x, y)$ and $X_0 = \{x_1, x_2, \ldots, x_n\}$. Then, for any n nonnegative numbers $\alpha_1, \alpha_2, \ldots, \alpha_n$ with $\sum_{i=1}^{n} \alpha_i = 1$, there exists $x_0 \in X$ such that

$$\sum_{i=1}^{n} \alpha_i f(x_i, y) \leq \sum_{i=1}^{n} \alpha_i g(x_i, y) \leq g(x_0, y) \quad \text{for all } y \in Y.$$

Since, from the definition of c, there exists $y_1 \in Y$ such that $g(x_0, y_1) \leq c$, we obtain

$$\sum_{i=1}^{n} \alpha_i f(x_i, y_1) \leq g(x_0, y_1) \leq c.$$

By Theorem 2.2,

$$\inf_{y \in Y} \max_{x \in X_0} f(x, y) \leq c = \sup_{x \in X} \min_{y \in Y_0} g(x, y).$$

(ii) Let $c = \sup_{x \in X} \inf_{y \in Y} g(x, y)$ and $\varepsilon > 0$. For any finite set $X_0 = \{x_1, x_2, \ldots, x_n\}$ and n nonnegative numbers $\alpha_1, \alpha_2, \ldots, \alpha_n$ with $\sum_{i=1}^{n} \alpha_i = 1$, there exists $y_1 \in Y$ such that

$$\sum_{i=1}^{n} \alpha_i f(x_i, y_1) \leq c + \varepsilon,$$

as in the proof of (i). So, by Theorem 2.2, we obtain

$$\inf_{y\in Y}\max_{x\in X_0} f(x,y) \le c = \sup_{x\in X}\inf_{y\in Y} g(x,y).$$

(iii) Let $X_0 = \{x_1, x_2, \ldots, x_n\}$. Then, $\max_{x\in X_0} f(x,y)$ is lower semicontinuous. Since Y is compact, there exists $y_0 \in Y$ such that $\max_{x\in X_0} f(x,y_0) \le c$, where $c = \sup_{x\in X}\inf_{y\in Y} g(x,y)$. So, we have

$$\bigcap_{x\in X}\{y \in Y : f(x,y) \le c\} \ne \phi$$

and hence

$$\min_{y\in Y}\sup_{x\in X} f(x,y) \le c = \sup_{x\in X}\inf_{y\in Y} g(x,y). \quad \square$$

Next, we extend Fan's system theorem to lower semicontinuous convex functions with values in $(-\infty,\infty]$.

Theorem 2.5 **([16])** *Let X be a nonempty compact convex subset of a topological vector space. Let $\{f_1, f_2, \ldots, f_n\}$ be a finite family of lower semicontinuous convex functions on X with values in $(-\infty,\infty]$. Then the following are equivalent:*

(1) *The system of inequalities*

$$f_i(x) \le 0 \quad \textit{for all } i \in \{1, 2, \ldots, n\}$$

is consistent on X;

(2) *for any n nonnegative numbers $\alpha_1, \alpha_2, \ldots, \alpha_n$ with $\sum_{i=1}^{n} \alpha_i = 1$, there exists a $y \in X$ such that*

$$\sum_{i=1}^{n} \alpha_i f_i(y) \le 0.$$

To prove Theorem 2.5, we need the following lemma.

Lemma 2.6 *Let X and Y be topological spaces. Let β be a continuous function of X into $[0,\infty)$ and let f be a lower semicontinuous function of Y into $(-\infty,\infty]$. If $\beta \cdot f$ is defined by*

$$(\beta \cdot f)(x,y) = \beta(x)f(y) \quad \textit{for all } (x,y) \in X \times Y,$$

then $\beta \cdot f$ is lower semicontinuous on $X \times Y$.

Sketch of the proof of Theorem 2.5. It is clear that (1) implies (2). To show that (2) implies (1), put

$$A = \bigcap_{i=1}^{n} \{x \in X : f_i(x) < \infty\}$$

and $Y = \{\alpha = (\alpha_1, \alpha_2, \ldots, \alpha_n) : \alpha_i \geq 0, \sum_{i=1}^{n} \alpha_i = 1\}$. If F is defined by

$$F(x, \alpha) = \sum_{i=1}^{n} \alpha_i f_i(x) \quad \text{for all } x \in A \text{ and } \alpha \in Y,$$

then by Theorem 2.1, we have

$$\inf_{x \in A} \max_{\alpha \in Y} F(x, \alpha) = \max_{\alpha \in Y} \inf_{x \in A} F(x, \alpha).$$

To prove Theorem 2.5, it is sufficient to prove that for any $\alpha \in Y$, there exists $x_0 \in A$ such that $F(x_0, \alpha) \leq 0$. This inequality is proved by the compactness of X and Lemma 2.6; see, in details, [16]. So we obtain

$$\inf_{x \in X} \max_{\alpha \in T} F(x, \alpha) \leq \inf_{x \in A} \max_{\alpha \in Y} F(x, \alpha) = \max_{\alpha \in Y} \inf_{x \in A} F(x, \alpha) \leq 0,$$

where $T = \{(1, 0, \ldots, 0), (0, 1, 0, \ldots, 0), \ldots, (0, \ldots, 0, 1)\}$. Therefore, there exists an $x_0 \in X$ such that

$$f_i(x_0) \leq 0 \quad \text{for all } i = \{1, 2, \ldots, n\}. \quad \square$$

Finally in this section, we extend Fan's system theorem to functions with vector-values. Let C be a closed convex cone of a locally convex topological vector space F and define its *polar* in F^* to be the set

$$C^* = \{\phi \in F^* : \phi(y) \geq 0 \quad \text{for all } y \in C\}.$$

Theorem 2.7 *Let X be a compact convex subset of a topological vector space E and let C be a closed convex cone of a locally convex topological vector space F. Let G be a mapping of X into F such that for any $\phi \in C^*$, $\phi G : X \to \mathbf{R}$ is lower semicontinuous and convex. Then the following are equivalent:*

(1) *There exists $x \in X$ such that $G(x) \leq 0$;*
(2) *for any $\phi \in C^*$, there exists $y \in X$ such that $\phi G(y) \leq 0$.*

Proof. (1) $\Rightarrow$ (2) is obvious. We show that (2) implies (1). Define a two variable function F on $X \times C^*$ to be

$$F(x, \phi) = \phi G(x) \quad \text{for all } x \in X \text{ and } \phi \in C^*.$$

Then we have

$$\sup_{\phi \in C^*} \min_{x \in X} F(x, \phi) \leq 0.$$

By Theorem 2.1, we also know

$$\min_{x \in X} \sup_{\phi \in C^*} F(x, \phi) = \sup_{\phi \in C^*} \min_{x \in X} F(x, \phi).$$

So, we have $\min_{x \in X} \sup_{\phi \in C^*} \phi G(x) = \min_{x \in X} \sup_{\phi \in C^*} F(x, \phi) \leq 0$. Hence there exists $x_0 \in X$ such that $\sup_{\phi \in C^*} \phi G(x_0) \leq 0$. This implies that $G(x_0) \leq 0$. This completes the proof. □

3. Basic Results in Functional Analysis

In this section, we first prove the Mazur-Orlicz theorem by using Fan's system theorem (Theorem A) and Markov-Kakutani's fixed point theorem [2]. Let E be a linear space and let p be a real valued function on E. Then p is said to be *sublinear* if $p(x + y) \leq p(x) + p(y)$ for all $x, y \in E$ and let $p(\lambda x) = \lambda p(x)$ for all $\lambda \geq 0$ and $x \in E$. If E is a linear space, we denote by E' the dual space of E which is the set of all linear functionals from E into $\mathbf{R}$. The following lemma is proved by using Markov Kakutani's fixed point theorem; see [6].

Lemma 3.1 *If p is a sublinear functional on a linear space E and $x_0 \in E$, then there exists an $f \in E'$ such that $f(x) \leq p(x)$ for all $x \in E$ and $f(x_0) = p(x_0)$.*

Using Lemma 3.1 and Fan's system theorem, we can prove the following [13]:

Theorem 3.2 (the Mazur-Orlicz theorem) *Let p be a sublinear functional on a linear space E, let $\{x_\nu : \nu \in I\}$ be a family of elements of E and let $\{\alpha_\nu : \nu \in I\}$ be a corresponding family of real numbers. Then, the following are equivalent:*

(1) *There exists $f \in E'$ such that*

$$f(x) \leq p(x) \quad \textit{for all } x \in E$$

and

$$\alpha_\nu \leq f(x_\nu) \quad \textit{for all } \nu \in I;$$

(2) *for any* n *nonnegative numbers* $\delta_1, \delta_2, \ldots, \delta_n$ *with* $\sum_{i=1}^{n} \delta_i = 1$ *and* $\nu_1, \nu_2, \ldots, \nu_n \in I$, *the inequality*

$$\sum_{i=1}^{n} \delta_i \alpha_{\nu_i} \leq p(\sum_{i=1}^{n} \delta_i x_{\nu_i})$$

holds.

Proof. It is clear that (1) implies (2). We show that (2) implies (1). Let F be the product space R^E. Then F is a topological vector space. Let X_0 be the compact convex subset

$$\prod_{x \in E} [-p(-x), p(x)]$$

of $F = R^E$ and let $B = \{g \in E' : g(x) \leq p(x) \text{ for all } x \in E\}$. Then B is nonempty, by Lemma 3.1. Since X_0 is compact, B is compact and convex. For each $\nu \in I$, we define a real valued function F_ν on B by

$$F_\nu(f) = \alpha_\nu - f(x_\nu) \quad \text{for all } f \in B,$$

then F_ν is continuous and affine. If for each n nonnegative numbers $\delta_1, \delta_2, \ldots$ $\ldots, \delta_n$ with $\sum_{i=1}^{n} \delta_i = 1$ and $\nu_1, \nu_2, \ldots, \nu_n \in I$, the inequality

$$\min_{f \in B} \sum_{i=1}^{n} \delta_i F_{\nu_i}(f) \leq 0$$

holds, the conclusion is true by Fan's system theorem and compactness of B. But this inequality is easily obtained as follows by using Lemma 3.1;

$$\begin{aligned} \sum_{i=1}^{n} \delta_i \alpha_{\nu_i} &\leq p(\sum_{i=1}^{n} \delta_i x_{\nu_i}) \\ &\leq \max_{f \in B} f(\sum_{i=1}^{n} \delta_i x_{\nu_i}) \\ &= \max_{f \in B} \sum_{i=1}^{n} \delta_i f(x_{\nu_i}). \quad \Box \end{aligned}$$

As a direct consequence of Theorem 3.2, we have the following:

Theorem 3.3 **([4])** *Let* $\{x_\nu : \nu \in I\}$ *be a family of elements of a normed linear space* E *and let* $\{\alpha_\nu : \nu \in I\}$ *be a corresponding family of real numbers. Then, for any* $\rho \geq 0$, *the following are equivalent:*

(1) *There exists a continuous linear functional f on E with $\|f\| \leq \rho$ such that*

$$f(x_\nu) \geq \alpha_\nu \quad \text{for all } \nu \in I;$$

(2) *for any n nonnegative numbers $\delta_1, \delta_2, \ldots, \delta_n$ with $\sum_{i=1}^{n} \delta_i = 1$ and $\nu_1, \nu_2, \ldots, \nu_n \in I$, the inequality*

$$\sum_{i=1}^{n} \delta_i \alpha_{\nu_i} \leq \rho \| \sum_{i=1}^{n} \delta_i x_{\nu_i} \|$$

holds.

Let p be a sublinear functional on a linear space E and let T be a nonempty subset of E. Then, T is said to be *p-midconvex* if for any $u, v \in T$, there is an $x \in T$ such that $p(x - \frac{u+v}{2}) \leq 0$. Using the Mazur-Orlicz theorem, we have the following [7]:

Theorem 3.4 (König's theorem) *Let p be a sublinear functional on a linear space E and let T be a nonempty $p-$midconvex subset of E. Then the following holds:*

$$\inf_{x \in T} p(x) = \max_{f \leq p} \inf_{x \in T} f(x).$$

Proof. If $\inf_{x \in T} p(x) = -\infty$, the conclusion is trivial. So we may assume that $\inf_{x \in T} p(x) > -\infty$. Let $\{x_\nu : \nu \in I\} = T$ and let $\alpha_\nu = \inf_{x \in T} p(x)$ for all $\nu \in I$. Then, from hypothesis, for any $n \in N$, $\nu_1, \nu_1, \ldots, \nu_n \in I$ and nonnegative numbers $\delta_1, \delta_2, \ldots, \delta_n$ with $\sum_{i=1}^{n} \delta_i = 1$, there is a $y_0 \in T$ such that

$$p(y_0 - \sum_{i=1}^{n} \delta_i x_{\nu_i}) \leq 0.$$

Hence we have

$$\begin{aligned} p(\sum_{i=1}^{n} \delta_i x_{\nu_i}) &\geq p(y_0 - \sum_{i=1}^{n} \delta_i x_{\nu_i}) + p(\sum_{i=1}^{n} \delta_i x_{\nu_i}) \\ &\geq p(y_0) \geq \inf_{x \in T} p(x). \end{aligned}$$

By Theorem 3.2, there is an $f \in E'$ such that $f(x) \leq p(x)$ for all $x \in E$ and

$$f(y) \geq \inf_{x \in T} p(x) \quad \text{for all } y \in T.$$

Then, from compactness of $\prod_{x \in E} [-p(-x), p(x)]$, we have

$$\max_{f \leq p} \inf_{x \in T} f(x) \geq \inf_{x \in T} p(x).$$

It is obvious that

$$\max_{f \leq p} \inf_{x \in T} f(x) \leq \inf_{x \in T} p(x). \quad \square$$

Using König's theorem, we can prove the following separation theorem and best approximation theorem. Let p be a sublinear functional on a linear space E. Then for two subsets A and B of E, the *distance* $p(A, B)$ between A and B is given by

$$\inf\{p(x - y) : x \in A, y \in B\}.$$

Theorem 3.5 ([6], [16]) *Let p be a sublinear functional in a linear space E. If C and D are nonempty subsets of E such that $C - D$ is $p-$midconvex and $p(C, D) > -\infty$, then there is an $f \in E'$ such that*

$$\inf\{f(x) : x \in C\} = p(C, D) + \sup\{f(y) : y \in D\}$$

and

$$f(x) \leq p(x) \quad \textit{for all } x \in E.$$

Proof. Let $T = C - D$. Then, since T is $p-$midconvex, by König's theorem, there is an $f \in E'$ such that

$$f(x) \leq p(x) \quad \text{for all } x \in E$$

and

$$f(y - z) \geq p(C, D) \quad \text{for all } y \in C \text{ and } z \in D. \tag{$*$}$$

From (*), we obtain

$$f(y) \geq p(C, D) + f(z) \quad \text{for all } y \in C \text{ and } z \in D$$

and hence

$$\inf_{y \in C} f(y) \geq p(C, D) + \sup_{z \in D} f(z).$$

Then, we have

$$\begin{aligned} p(C, D) &\leq \inf_{y \in C} f(y) - \sup_{z \in D} f(z) \\ &\leq \inf\{f(y - z) : \ y \in C, \ z \in D\} \\ &\leq \inf\{p(y - z) : \ y \in C, \ z \in D\} \\ &= p(C, D). \end{aligned}$$

This implies

$$\inf\{f(y) : y \in C\} = p(C, D) + \sup\{f(z) : z \in D\}. \quad \square$$

Theorem 3.6 *Let p be a sublinear functional on a linear space E and let K be a nonempty $p-$midconvex subset of E. Let $x \in E$ and $x_0 \in K$. Then the following are equivalent:*

(1) $p(x_0 - x) = \inf\{p(z-x) : z \in K\}$;

(2) *there exists an $f \in E'$ such that*

$$f(y) \leq p(y) \quad \textit{for all } y \in E$$

and

$$\inf_{y \in K} f(y-x) = p(x_0 - x);$$

(3) *there exists an $f \in E'$ such that*

$$\inf\{f(y) : y \in K\} = f(x_0), \ f(x_0 - x) = p(x_0 - x)$$

and

$$f(y) \leq p(y) \quad \textit{for all } y \in E.$$

Proof. (1) $\Rightarrow$ (2): Let $T = \{y - x : y \in K\}$. Then, by König's theorem, there is an $f \in E'$ such that

$$f(y) \leq p(y) \quad \text{for all } y \in E$$

and

$$\inf_{y \in K} f(y-x) = p(x_0 - x).$$

(2) $\Rightarrow$ (3): By (2), we have $f(x_0 - x) \leq p(x_0 - x)$ and $f(x_0 - x) \geq p(x_0 - x)$. Hence, $f(x_0 - x) = p(x_0 - x)$. We also have

$$\begin{aligned} f(y - x_0) &= f(y-x) + f(x - x_0) \\ &\geq p(x_0 - x) - p(x_0 - x) = 0 \end{aligned}$$

for all $y \in K$. This implies $\inf_{y \in K} f(y) = f(x_0)$.

(3) $\Rightarrow$ (1): For any $z \in K$,

$$p(x_0 - x) = f(x_0 - x) \leq f(z-x) \leq p(z-x).$$

Hence we have

$$p(x_0 - x) = \inf\{p(z-x) : z \in K\}. \quad \square$$

4. Applications

In this section, using Fan's system theorem (Theorem 3.3) concerning linear inequalities, we first discuss cores of games. Let P be a nonempty set and let Σ be a field of subsets of P. A nonnegative real valued function v defined on Σ is called a *game* if $v(\phi) = 0$. The space $B(P, \Sigma)$ consists of all uniform limits of finite linear combinations of characteristic functions of sets in Σ. The norm in $B(P, \Sigma)$ is given by the formula

$$\|f\| = \sup_{s \in P} |f(s)|.$$

The space $ba(P, \Sigma)$ consists of all bounded additive scalar functions defined on Σ. The norm $\|\mu\|$ is the total variation of μ on P, that is,

$$\|\mu\| = \sup \sum_{i=1}^{n} |\mu(E_i)|,$$

where the supremum is taken over all finite sequences $\{E_i\}$ of disjoint sets in Σ. It is well known that there is an isometric isomorphism between $B(P, \Sigma)^*$ and $ba(P, \Sigma)$. An *outcome* of a game v is an element $\mu \in ba(P, \Sigma)$ such that

$$v(P) = \mu(P) \quad \text{and} \quad v(S) \leq \mu(S) \quad \text{for all} \quad S \in \Sigma.$$

The set of all outcomes of a game v is called the *core* of the game v and denoted by $\mathcal{C}(v)$. Let v be a game defined on a field Σ of an arbitrary set P and let T be a mapping of P into itself such that for all $S \in \Sigma$, $T^{-1}S \in \Sigma$. We define the *T-core* of the game to be the set

$$\mathcal{C}_T(v) = \{\mu \in \mathcal{C}(v) : \mu T^{-1} = \mu\}.$$

Theorem 4.1 ([16]) *Let v be a game defined on a field Σ of an arbitrary set P and let T be a mapping of P into itself such that for all $S \in \Sigma$, $T^{-1}S \in \Sigma$. Then the following are equivalent:*

(1) $\mathcal{C}_T(v) \neq \phi$;

(2) *for any $\lambda_1, \ldots, \lambda_n \geq 0$, $A_1, \ldots, A_n \in \Sigma$, $\eta_1, \ldots, \eta_m \in \mathbf{R}$, and $S_1, \ldots$ $\ldots, S_m \in \Sigma$ with $\sum_{i=1}^{n} \lambda_i 1_{A_i} + \sum_{j=1}^{m} \eta_j (1_{S_j} - 1_{T^{-1}S_j}) \leq 1_P$, the inequality*

$$\sum_{i=1}^{n} \lambda_i v(A_i) \leq v(P) \qquad (*)$$

holds;

(3) *for any $\lambda_1, \ldots, \lambda_n \geq 0$, $A_1, \ldots, A_n \in \Sigma$, $\eta_1, \ldots, \eta_m \in \mathbf{R}$ and $S_1, \ldots$ $\ldots, S_m \in \Sigma$, the inequality*

$$\sum_{i=1}^{n} \lambda_i v(A_i) \leq v(P) \Big\| \sum_{i=1}^{n} \lambda_i 1_{A_i} + \sum_{j=1}^{m} \eta_j (1_{S_j} - 1_{T^{-1}S_j}) \Big\| \qquad (**)$$

holds.

Proof of (3) ⇒ (1). From Theorem 3.3, there is a $\mu^* \in B(P,\Sigma)^*$ such that

$$\|\mu^*\| \leq v(P),$$
$$\mu^*(1_A) \geq v(A) \quad \text{for all} \quad A \in \Sigma,$$
$$\mu^*(1_S - 1_{T^{-1}S}) \geq 0 \quad \text{for all} \quad S \in \Sigma,$$

and

$$\mu^*(1_{T^{-1}S} - 1_S) \geq 0 \quad \text{for all} \quad S \in \Sigma,$$

if and only if for any $\delta = (\delta_1, \ldots, \delta_n) \in T_n$, the inequality

$$\sum_i \delta_i v(A_i) + \sum_j \delta_j \cdot 0 + \sum_k \delta_k \cdot 0$$

$$\leq v(P) \| \sum_i \delta_i 1_{A_i} + \sum_j \delta_j (1_{s_j} - 1_{T^{-1}S_j}) + \sum_k \delta_k (1_{T^{-1}S_k} - 1_{S_k}) \|$$

holds, where $\sum_i \delta_i + \sum_j \delta_j + \sum_k \delta_k = 1$ and $i + j + k = n$. Therefore (3) implies (1). □

As a direct consequence of Theorem 4.1, we have the following:

Corollary 4.2 *Let v be a game defined on a field Σ of an arbitrary set P. Then the following are equivalent:*

(1) *$\mathcal{C}(v)$ is nonempty;*

(2) *for any $n \in N$, $\lambda_1, \ldots, \lambda_n \geq 0$, and $A_1, \ldots, A_n \in \Sigma$ with $\sum_{i=1}^{n} \lambda_i 1_{A_i} \leq 1_P$, the inequality*

$$\sum_{i=1}^{n} \lambda_i v(A_i) \leq v(P)$$

holds;

(3) *for any $n \in N$, $\lambda_1, \ldots, \lambda_n \geq 0$, and $A_1, \ldots, A_n \in \Sigma$, the inequality*

$$\sum_{i=1}^{n} \lambda_i v(A_i) \leq v(P) \| \sum_{i=1}^{n} \lambda_i 1_{A_i} \|$$

holds.

A game v is called *exact* if for any $A_0 \in \Sigma$, there is a $\mu \in \mathcal{C}(v)$ such that $\mu(A_0) = v(A_0)$.

Theorem 4.3 ([16]) *Let v be a game defined on a field Σ of an arbitrary set P and let T be a mapping of P into itself such that for all $S \in \Sigma$, $T^{-1}S \in \Sigma$. The following are equivalent:*

(1) *For any* $A_0 \in \Sigma$, *there is a* $\mu \in C_T(v)$ *such that* $\mu(A_0) = v(A_0)$;

(2) *for any* $\lambda_0, \dots, \lambda_n \geq 0$, $A_0, \dots, A_n \in \Sigma$, $\eta_1, \dots, \eta_m \in R$ *and* $S_1, \dots$ $\dots, S_m \in \Sigma$ *with* $\sum_{i=1}^{n} \lambda_i 1_{A_i} - \lambda_0 1_{A_0} + \sum_{j=1}^{m} \eta_j (1_{S_j} - 1_{T^{-1}S_j}) \leq 1_P$, *the inequality*

$$\sum_{i=1}^{n} \lambda_i v(A_i) - \lambda_0 v(A_0) \leq v(P)$$

hodls;

(3) *for any* $\lambda_0, \dots, \lambda_n \geq 0$, $A_0, \dots, A_n \in \Sigma$, $\eta_1, \dots, \eta_m \in R$ *and* $S_1, \dots$ $\dots, S_m \in \Sigma$, *the inequality*

$$\sum_{i=1}^{n} \lambda_i v(A_i) - \lambda_0 v(A_0) \leq v(P) \| \sum_{i=1}^{n} \lambda_i 1_{A_i} - \lambda_0 1_{A_0} + \sum_{j=1}^{m} \eta_j (1_{S_j} - 1_{T^{-1}S_j}) \|$$

holds.

Proof of (3) ⇒ (1). Let $A_0 \in \Sigma$. Then, by (3) and Theorem 3.3, there is a $\mu^* \in B(P, \Sigma)^*$ such that

$$\begin{aligned} &\|\mu^*\| \leq v(P), \\ &\mu^*(1_A) \geq v(A) \quad \text{for all } A \in \Sigma, \\ &\mu^*(1_S - 1_{T^{-1}S}) \geq 0 \quad \text{for all } S \in \Sigma, \\ &\mu^*(1_{T^{-1}S} - 1_S) \geq 0 \quad \text{for all } S \in \Sigma, \end{aligned}$$

and

$$\mu^*(-1_{A_0}) \geq -v(A_0).$$

Therefore there is a $\mu \in C_T(v)$ such that $\mu(A_0) = v(A_0)$. □

Finally, using Fan's system theorem (Theorem A), we prove a fixed point theorem for noncommutative semigroups of nonexpansive mappings in Banach spaces. Before proving it, we give some definitions. A closed convex subset C of a Banach space E is said to have *normal structure* if for each closed convex bounded subset K of C which contains at least two points, there exists an element of K which is not a diametral point of K. It is well known that a closed convex subset of a uniformly convex Banach space has normal structure and a compact convex subset of a Banach space has normal structure. We also know the following:

Lemma 4.4 ([11]) *A closed convex subset* C *of a Banach space has normal structure if and only if it does not contain a sequence* $\{x_n\}$ *such that for* *some* $c > 0$,

$$\|x_n - x_m\| \le c \quad \textit{and} \quad \|x_{n+1} - \overline{x_n}\| \ge c - \frac{1}{n^2}$$

for all $n \ge 1$ *and* $m \ge 1$, *where* $\overline{x_n} = \frac{1}{n}\sum_{i=1}^{n} x_i$.

Let S be a semitopological semigroup, i.e., a semigroup with Hausdorff topology such that for each $s \in S$, the mappings $t \mapsto ts$ and $t \mapsto st$ of S into itself are continuous. Let $B(S)$ be the Banach space of all bounded real valued functions on S with supremum norm and let X be a subspace of $B(S)$ containing constants. Then, an element μ of X^* (the dual space of X) is called a *mean* on X if $\|\mu\| = \mu(1) = 1$. We know that $\mu \in X^*$ is a mean on X if and only if

$$\inf\{f(s) : s \in S\} \le \mu(f) \le \sup\{f(s) : s \in S\}$$

for every $f \in X$. A real valued function μ on X is called a *submean* on X if the following properties are satisfied :

(i) $\mu(f + g) \le \mu(f) + \mu(g)$ for every $f, g \in X$;
(ii) $\mu(\alpha f) = \alpha\mu(f)$ for every $f \in X$ and $\alpha \ge 0$;
(iii) for $f, g \in X$, $f \le g$ implies $\mu(f) \le \mu(g)$;
(iv) $\mu(c) = c$ for every constant function c.

Clearly every mean on X is a submean. The notion of submeans was first introduced by Mizoguchi and Takahashi [14]. See also [9, 25]. For a submean μ on X and $f \in X$, we sometimes use $\mu_t(f(t))$ instead of $\mu(f)$. For each $s \in S$ and $f \in B(S)$, we define elements $\ell_s f$ and $r_s f$ of $B(S)$ given by

$$(\ell_s f)(t) = f(st) \quad \text{and} \quad (r_s f) = f(ts)$$

for all $t \in S$. Let X be a subspace of $B(S)$ containing constants and invariant under ℓ_s, $s \in S$. Then a submean μ on X is said to be *left invariant* if $\mu(f) = \mu(\ell_s f)$ for all $f \in X$ and $s \in S$. A submean μ on X is also said to be *left subinvariant* if $\mu(f) \le \mu(\ell_s f)$ for all $f \in X$ and $s \in S$. Let S be a semitopological semigroup. Then S is called *left reversible* if any two closed right ideals of S have non-void intersection. If S is left reversible, $(S, \le)$ is a directed system when the binary relation "$\le$" on S is defined by $a \le b$ if and only if $\{a\} \cup \overline{Sa} \supset \{b\} \cup \overline{Sb}$, $a, b \in S$. Let $RUC(S)$ be the space of right uniformly continuous functions on S. We may comment on the relationship between "$RUC(S)$ has an invariant mean" and "S is left reversible". As well known, they do not imply each other in general. But if $RUC(S)$ has sufficiently many functions to separate closed sets, then "$RUC(S)$ has an invariant mean" would imply "S is left and right reversible".

Let S be a semitopological semigroup and let C be a nonempty subset of a Banach space E. Then a family $\mathcal{S} = \{T_s : s \in S\}$ of mappings of C into itself is called a *nonexpansive semigroup* on C if it satisfies the following:

(i) $T_{st}x = T_sT_tx$ for all $s, t \in S$ and $x \in C$;
(ii) for each $x \in C$, the mapping $s \mapsto T_sx$ is continuous;
(iii) for each $s \in S$, T_s is a nonexpansive mapping of C into itself, i.e.,

$$\|T_sx - T_sy\| \leq \|x - y\|$$

for all $x, y \in C$.

Recently, Lau and Takahashi [10] generalized Lim's result [12] and Takahashi and Jeong's result [26]. See also [18, 20, 21, 23].

Theorem 4.5 ([10]) *Let S be a semitopological semigroup, let C be a nonempty weakly compact convex subset of a Banach space E which has normal structure and let $\mathcal{S} = \{T_s : s \in S\}$ be a nonexpansive semigroup on C. Suppose $RUC(S)$ has a left subinvariant submean. Then $\mathcal{S}$ has a common fixed point in C.*

Sketch of the proof. For any $x, y \in C$, a real valued function h on S defined by

$$h(t) = \|T_tx - y\| \quad \text{for all } t \in S$$

is in $RUC(S)$. Define

$$U = \{K \subset C : K \neq \phi, \text{closed, convex, and } T_s - \text{invariant}\}.$$

Then by Zorn's lemma, we obtain a minimal element K of U. Suppose $\delta(K) > 0$, where $\delta(K)$ is the diameter of K and let μ be a left subinvariant submean. Then, for any $x \in K$,

$$A_x = \{z \in K : \mu_t\|T_tx - z\| = \min_{y \in K} \mu_t\|T_tx - y\|\}$$

is nonempty, closed, convex, and T_s-invariant. So, we have $A_x = K$ by minimality of K. Let $\varepsilon > 0$ and $\delta = \{x_1, x_2, \ldots, x_n\} \subset K$. Consider a system $\{h_i\}_{i=1}^n$ of convex weakly lower semicontinuous functions on K defined by

$$h_i(x) = \|x - x_i\| \quad \text{for all } x \in K \text{ and } i = 1, 2, \ldots, n.$$

Using Fan's system theorem (Theorem A), we obtain

$$D_\delta = \{z \in K : h_i(z) \leq \rho_0 + \varepsilon, \; i = 1, 2, \ldots, n\} \neq \phi,$$

where $\rho_0 = \min_{y \in K} \mu_t\|T_tx - y\|$. In fact, since $A_x = K$, we have $\mu_t\|T_tx - x_i\| = \rho_0$ for all $i = 1, 2, \ldots, n$. So, for any $\{\lambda_i\}_{i=1}^n$ with $\lambda_i \geq 0$ and $\sum_{i=1}^n \lambda_i = 1$, we have

$$\begin{aligned}\mu_t(\sum_{i=1}^{n}\lambda_i\|T_tx - x_i\|) &\leq \sum_{i=1}^{n}\lambda_i\mu_t\|T_tx - x_i\|\\ &= \rho_0 < \rho_0 + \varepsilon.\end{aligned}$$

Hence, there exists $t_0 \in S$ such that $\sum_{i=1}^{n}\lambda_i\|T_{t_0}x - x_i\| \leq \rho_0 + \varepsilon$, i.e., $\sum_{i=1}^{n}\lambda_i h_i(T_{t_0}x) \leq \rho_0 + \varepsilon$. Using Fan's system theorem, there exists $z \in K$ such that $h(z) \leq \rho_0 + \varepsilon$ for all $i = 1, 2, \ldots, n$. This implies that the set D_δ is nonempty. Hence by weak compactness of K, the set $D = \{z \in C : \|x - z\| \leq \rho_0 + \varepsilon \text{ for all } x \in K\}$ is also nonempty. So, we have

$$\begin{aligned}\rho_0 &= \min_{y\in K}\mu_t\|T_tx - y\|\\ &\leq \min_{y\in K}\sup_t\|T_tx - y\|\\ &\leq \min_{y\in K}\sup_{z\in K}\|z - y\| \leq \rho_0 + \varepsilon.\end{aligned}$$

Since $\varepsilon > 0$ is arbitrary, we have $\rho_0 = \inf_{y\in K}\sup_{z\in K}\|z - y\|$. Since $A_x = K$ for all $x \in K$, we have

$$\rho_0 = \mu_t\|T_tx - y\| \quad \text{for all } x, y \in K.$$

So, let

$$A_0 = \{z \in K : \sup_t\|T_tx - z\| \leq \rho_0 \quad \text{for all } x \in K\}.$$

Then A_0 is nonempty. Fix $z_0 \in A_0$. Since $\rho_0 = \mu_t\|T_tz_0 - z_0\|$, there exists $s_1 \in S$ such that $\|T_{s_1}z_0 - z_0\| \geq \rho_0 - 1$. Let $x_1 = z_0$ and $x_2 = T_{s_1}z_0$ and $\overline{x_2} = \frac{1}{2}z_0 + \frac{1}{2}T_{s_1}z_0$. Since $\rho_0 = \mu_t\|T_tz_0 - \overline{x_2}\| \leq \mu_t\|T_{s_1t}z_0 - \overline{x_2}\|$, there exists $s_2 \in S$ such that $\|T_{s_1s_2}z_0 - \overline{x_2}\| \geq \rho_0 - \frac{1}{2^2}$. So, let $x_3 = T_{s_1s_2}z_0$ and $\overline{x_3} = \frac{1}{3}x_1 + \frac{1}{3}x_2 + \frac{1}{3}x_3$. By mathematical induction, let $x_4 = T_{s_1s_2s_3}z_0$, $x_5 = T_{s_1s_2s_3s_4}z_0, \ldots$. Then we have a sequence $\{x_n\}$ in K such that

$$\|x_n - x_m\| \leq \rho_0 \quad \text{and} \quad \|x_{n+1} - \overline{x_n}\| \geq \rho_0 - \frac{1}{n^2}.$$

Using Lemma 4.4, K does not have normal structure. This is a contradiction. □

Theorem 4.5 answers affirmatively a problem posed during the Conference on Fixed Point Theory and Applications held at CIRM, Marseille-Luminy, 1989 (see [8, p. 307, Problem 5]).

References

1. F. E. Browder, *The fixed point theory of multi-valued mappings in topological vector spaces,* Math. Ann., **177** (1968), 283-301.
2. N. Dunford and J. T. Schwartz, *Linear Operators I,* Interscience, New York, 1958.
3. K. Fan, *Minimax theorems,* Proc. Nat. Acad. Sci. U.S.A. **39** (1953), 42-47.
4. K. Fan, *On systems of linear inequalities,* in Linear Inequalities and Related Systems (H. W. Kuhn and A. W. Tucker Eds.), Annals of Mathematics Studies, Vol. 38, Princeton Univ. Press, Princeton, NJ, 1956, 99-156.
5. K. Fan, *Existence theorems and extreme solutions for inequalities concerning convex functions or linear transformations,* Math. Z. **68** (1957), 205-217.
6. N. Hirano, H. Komiya and W. Takahashi, *A generalization of the Hahn-Banach theorem,* J. Math. Anal. Appl. **88** (1982), 333-340.
7. H. König, *On certain applications of the Hahn-Banach and minimax theorems,* Arch. Math. **21** (1970), 583-591.
8. A. T. Lau, *Amenability and fixed point property for semigroup of non-expansive mappings,* in Fixed Point Theory and Applications (M. A. Théra and J. B. Baillon Eds.), Pitman Research Notes in Mathematics Series #252, 1991, 303-313.
9. A. T. Lau and W. Takahashi, *Invariant means and fixed point properties for non-expansive representations of topological semigroups,* Topol. Methods Nonlinear Anal. **5** (1995), 39-57.
10. A. T. Lau and W. Takahashi, *Invariant submeans and semigroups of nonexpansive mappings on Banach spaces with normal structure,* J. Functional Analysis **142** (1996), 79-88.
11. T. C. Lim, *A fixed point theorem for families of nonexpansive mappings,* Pacific J. Math. **53** (1974), 484-493.
12. T. C. Lim, *Characterization of normal structure,* Proc. Amer. Math. Soc. **43** (1974), 313-319.
13. S. Mazur and W. Orlicz, *Sur les espaces métriques linéares (II),* Studia Math. **13** (1953), 137-179.
14. N. Mizoguchi and W. Takahashi, *On the existence of fixed points and ergodic retractions for Lipschitzian semigroups in Hilbert spaces,* Nonlinear Analysis **14** (1990), 69-80.
15. K. Sakamaki and W. Takahashi, *Systems of convex inequalities and their applications,* J. Math. Anal. Appl. **70** (1979), 445-459.
16. N. Shioji and W. Takahashi, *Fan's theorem concerning systems of convex inequalities and its applications,* J. Math. Anal. Appl. **135** (1988), 383-398.
17. S. Simons, *Minimax and variational inequalities. Are they of fixed point or Hahn-Banach type?* Game Theory and Mathematical Economics, North Holland Publishing Company, 1981, 379-387.
18. W. Takahashi, *Fixed point theorem for amenable semigroups of non-expansive mappings,* Kodai Math. Sem. Rep. **21** (1969), 383-386.
19. W. Takahashi, *Nonlinear variational inequalities and fixed point theorems,* J. Math. Soc. Japan **28** (1976), 168-181.
20. W. Takahashi, *A nonlinear ergodic theorem for an amenable semigroup of nonexpansive mappings in a Hilbert space,* Proc. Amer. Math. Soc. **81** (1981), 253-256.
21. W. Takahashi, *Fixed point theorems for families of nonexpansive mappings on unbounded sets,* J. Math. Soc. Japan **36** (1984), 543-553.
22. W. Takahashi, *Fixed point, minimax and Hahn-Banach theorems,* Proceedings of Symposia in Pure Mathematics, Amer. Math. Soc. **45** (1986), 417-427.
23. W. Takahashi, *A nonlinear ergodic theorem for a reversible semigroup of nonexpansive mappings in a Hilbert space,* Proc. Amer. Math. Soc. **96** (1986), 55-58.
24. W. Takahashi, *Nonlinear Functional Analysis (Japanese),* Kindai-kagakusha, Japan, 1988.
25. W. Takahashi, *Fixed point theorem and nonlinear ergodic theorem for nonexpansive

semigroups without convexity, Can. J. Math. **44** (1992), 880-887.
26. W. Takahashi and D. H. Jeong, *Fixed Point theorem for nonexpansive semigroups on Banach space,* Proc. Amer. Math. Soc. **122** (1994), 1175-1179.

AN ALGORITHM FOR THE MULTI-ACCESS CHANNEL PROBLEM

PENG-JUNG WAN
Computer Science Department
University of Minnesota
Minneapolis, MN 55455, U.S.A.
e-mail: wan@cs.umn.edu

DING-ZU DU
Computer Science Department
University of Minnesota
Minneapolis, MN 55455, U.S.A.
e-mail: dzd@cs.umn.edu.

This work is partially supported by the National Science Foundation under grant CCR-9208913

AND

PANOS M. PARDALOS
Center for Applied Optimization
University of Florida
Gainesville, FL 32611, U.S.A.
e-mail: pardalos@ufl.edu.

This work is partially supported by National Science Foundation grants DMI-9622200 and BIR-9505919

Abstract. Consider a multi-access communication channel which is shared by multiple users. Only one user can transmit a message at any time, otherwise conflicts resulting in noise will arise. The solution to this problem is to schedule transmission of messages so that conflicts are at a minimum. Previously, the best algorithm for this problem was a tree-based algorithm given by Capetanakis [1]. In this paper we give an algorithm which is asymptotically better than that of Capetanakis.

Keywords: multi-access channel, min-max, conflict, noise, group testing.

B. Ricceri and S. Simons (eds.), Minimax Theory and Applications, 261–270.

1. Introduction

We consider a communication network in which many users share a single multi-access communication channel [1, 2, 6, 7]. A user is said to be *active* if it has a message to transmit. For convenience, assume that each message is of unit length and can be transmitted in one time slot. Transmission is accomplished by broadcasting over the channel. Any user, active or not, can receive from the channel. At any given time, if more than one active user broadcasts, then the messages will conflict with each other and result in noise. The objective is to devise an algorithm which schedules transmissions of active users so that the number of conflicts are at a minimum.

A direct transmission algorithm queries sets of users, requesting the active users among the set to transmit. If the channel remains quiet, then the queried set contains no active users. If the channel broadcasts a message then the queried set contains exactly one active user which has transmitted successfully. If the channel is noisy, then the queried set contains 2 or more active users. In this case further queries are required to separate the transmissions of these active users. Note that the type of feedback is ternary: $0, 1, 2^+$; where 2^+ means at least 2.

The desirable algorithms should minimize the number of queries performed. Let $\mathcal{A}(n, d)$ denote the set of users of size n containing d active users. For an algorithm α and any set $s \in \mathcal{A}(n, d)$, let $N_\alpha(s)$ be the number of queries performed on s. a set s of size n. Denote

$$\begin{aligned} M_\alpha(d|n) &= \max_{s \in \mathcal{A}(n,d)} N_\alpha(s), \\ M(d|n) &= \min_\alpha M_\alpha(d|n). \end{aligned}$$

Then $M(d, n)$ is the *min-max* number of queries required to schedule the transmissions of d active users from total n users [3]. In general, the exact number of $M(d|n)$ is very hard to find and some works have been done to find some good heuristics [1, 2, 6, 7]. In this paper we propose an algorithm G for which $M_G(d|n) \leq \frac{3}{4} d \log \frac{n}{d} + \frac{7}{4} d + 1$ (all logarithms in this paper are to the base 2). Asymptotically this algorithm outperforms the previously best algorithm due to Capetanakis [1]. The reader may refer to [5] for more information.

2. The Algorithm

In the following we assume that the function *feedback(X)* queries a set X of users and returns $0, 1, 2^+$ as the set contains $0, 1$, or at least 2 active users. Note that if *feedback(X)* returns 2^+ when $|X| = 2$ then exactly two active users are identified. *Algorithm G* makes use of two procedures, *Big-Test* and *Dig*, which are given following it.

Algorithm *G:*

```
Input S;
while |S| > 2 do
    f := feedback(S) ;
    if f ≤ 1 then
        S := ∅;
    else
        k := 0 ;
        repeat {jumping process}
            if k = 0 then
                X := 2 items of S;
            else
                X := min{2^k + 2^(k+1), |S|} items of S;
            f := feedback(X);
            if f ≤ 1 then
                S := S − X;
                if S ≠ ∅ then
                    k := k + 2;
        until (f = 2+) or (S = ∅).
        if S ≠ ∅ then
            if |X| = 2 then
                S = S − X;
                /* Let the two users in X send the messages in the next two slots */;
            else
                Big-Test(X);
 end-while.
 if S ≠ ∅ then
     f := feedback(S);
     if f = 2+ then
         /* Let the two users in X send the messages in the next two slots */;
     S := ∅;
end-algorithm.
```

The procedure *Big-Test* is given as follows:

Procedure *Big-Test(X):*

if $|X| > 2^k$ then

```
        Y := 2^k items of X ;
        f := feedback(Y);
        if f = 2+ then
            X := Y;
        if f ≤ 1 then
            X := X − Y;
            S := S − Y;
        if f = 1 then
            if |X| = 1 then
                /* X must be active and let it send its message in the next
slot */
                S := ∅;
                end-procedure.
            else
                f := feedback(X);
                if f = 1 then
                    S := S − X;
                    end-procedure.
                else
                    if |X| = 2 then
                        /* Let them send their messages in the next two slots
*/
                        S := ∅;
                        end-procedure.
    Dig(X);
end-procedure.
```

The procedure *Dig* is given below:

```
Procedure Dig(X):
    m := 0;
    repeat
        Y := ⌊|X|/2⌋ elements of X ;
        f := feedback(Y) ;
        if f ≤ 1 then
            S := S − Y;
            X := X − Y;
            m := m + f;
        else
            if |Y| = 2 then
                m := m + 2
                /* Let the two users in Y send their messages in the next two
```

slots. */ **else**

$X := Y$

until $m \geq 2$.

end-procedure.

Note that the jumping process (lines 8-17 of Algorithm G) terminates when a set, X, of size at most $2^k + 2^{k+1}$ containing at least 2 active users is found or when S is exhausted. If $|X| > 2^k$ then the procedure Big-Test reduces the set with an unknown number of active users to either 2^k or $|X| - 2^k$. The *Dig* procedure performs binary search until at least 2 active users (total) have been found.

3. Analysis

To simplify the analysis define $f(d,n) = \frac{3}{4}d\log\frac{n}{d} + \frac{7}{4}d$. For convenience we assume that the value of $d\log\frac{n}{d}$ is 0 at $d = 0$. We begin with a lemma which gives certain useful properties of the function $f(d,n)$.

Lemma 3.1

1. *For* $d_1, d_2 \geq 0$ *and* $n_1, n_2 > 0$,

$$f(d_1,n_1) + f(d_2,n_2) \leq f(d_1+d_2, n_1+n_2).$$

2. $f(d,n)$ *is strictly increasing with respect to both* n *and* d.

Proof.

1. This is a direct corollary of lemma 3.2 of Du and Park [4].
2. It's obvious that f is strictly increasing with respect to n. Since

$$\begin{aligned} f'_d(d,n) &= \frac{3}{4}\log\frac{n}{d} - \frac{3}{4\ln 2} + \frac{7}{4} \\ &\geq \frac{7 - 3\log e}{4} \\ &> 0, \end{aligned}$$

f is also strictly increasing with respect to d. □

We now show the cost of the algorithm.

Theorem 3.2 $M_G(d|n) \leq f(d,n) + 1$.

Proof. The proof is given by induction on d. Note that for $d = 0$ and $d = 1$ the algorithm requires one call to *feedback()* (line 3) to analyze the entire set. For $d \geq 2$ assume that the jumping process (lines 8-17) terminates with $k = i$. If $i = 0$, then the total number of calls to *feedback()* is at most

$$\begin{aligned} & 1 + M_G(d-2|n-2) \\ < \; & f(2,2) + f(d-2,n-2) + 1 \\ \leq \; & f(d,n) + 1 \end{aligned}$$

We can assume $i \geq 2$. Then before the jumping process terminates, $\frac{i}{2}+2$ calls to *feedback()* have been performed. Moreover $2^i - 2$ users have successfully transmitted any messages they attempted to transmit. If the jumping process terminates because $S = \emptyset$, then $n \geq 2^i - 1$ and the total number of calls to *feedback()* is $\frac{i}{2}+2$. Since

$$\begin{aligned}
& f(d,n)+1 \\
\geq\ & f(2,2^i-1)+1 \\
=\ & \frac{3}{2}\log\frac{2^i-1}{2}+\frac{7}{2}+1 \\
\geq\ & \frac{3}{2}(i-2)+\frac{7}{2}+1 \\
=\ & \frac{3}{2}i+\frac{1}{2}+1 \\
=\ & \frac{i}{2}+i+\frac{1}{2}+1 \\
>\ & \frac{i}{2}+2,
\end{aligned}$$

we can further assume that the jumping process terminates because $f = 2^+$. Consider the set X when the jumping process terminates. If $|X| = 2$, then $n = 2^i$, and the total number of calls to *feedback()* is $\frac{i}{2}+2$. By using the same proof as stated above, the theorem also holds for this case. So now we can suppose the procedure *Big-Test(X)* is called in line 23. There are 4 cases in *Big-Test(X)*:

Case 1: $2 < |X| \leq 2^i$.

In this case , at most $i-1$ calls to *feedback()* in *Dig(X)* to identify at least 2 active users. Thus there are at most $\frac{3}{2}i+1$ calls to *feedback()* and at least 2^i users accounted for. Let n' be the remaining unaccounted for users and d' be the remaining active users. Note that $n' \leq n - 2^i$ and $d' \leq d-2$. then, by the induction hypothesis and appealing to lemma 3.1, the total number of calls to *feedback()* is at most

$$\begin{aligned}
& \frac{3}{2}i+1+M_G(d'|n') \\
<\ & f\left(2,2^i\right)+M_G(d'|n') \\
\leq\ & f\left(2,2^i\right)+M_G\left(d-2|n-2^i\right) \\
\leq\ & f\left(2,2^i\right)+f\left(d-2,n-2^i\right)+1 \\
\leq\ & f(d,n)+1.
\end{aligned}$$

Case 2: $|X| > 2^i$ and the first call to *feedback()* in procedure *Big-Test(X)* finds a set Y of size 2^i containing at least 2 active users.

The procedure *Dig(X)* then finds at least 2 active users in at most $i-1$ calls to *feedback()*. There are at most $\frac{3}{2}i+3$ calls to *feedback()* and at least 2^i users accounted for. Again, let n' and d' be the total remaining unaccounted for users and unaccounted for active users, respectively. We again have $n' \leq n-2^i$ and $d' \leq d-2$. In this case, the total number of calls to *feedback()* is at most

$$\begin{aligned} & \frac{3}{2}i+3+M_G\left(d'|n'\right) \\ = \; & f\left(2,2^i\right)+M_G\left(d'|n'\right)+1 \\ \leq \; & f\left(2,2^i\right)+f\left(d-2,n-2^i\right)+1 \\ \leq \; & f\left(d,n\right)+1. \end{aligned}$$

Case 3: $|X| > 2^i$ and the first call to *feedback()* in procedure *Big-Test(X)* finds a set Y of size 2^i.

Case 3.1: The remaining set $X-Y$ has only one user.

In this case, $n = 2^{i+1}-1$, and the total number of calls to *feedback()* is $\frac{i}{2}+3$. Since

$$\begin{aligned} f(d,n)+1 & \geq f(2,2^{i+1}-1)+1 \\ & = \frac{3}{2}\log\frac{2^{i+1}-1}{2}+\frac{7}{2}+1 \\ & \geq \frac{3}{2}(i-1)+\frac{9}{2} \\ & = \frac{3}{2}i+3 \\ & > \frac{i}{2}+3, \end{aligned}$$

the theorem holds for this case.

Case 3.2: The remaining set $X-Y$ has more than one user but only one active user.

In this case, $n \geq 2^{i+1}$ and the total number of calls to *feedback()* is $\frac{i}{2}+4$. Since

$$\begin{aligned} f(d,n)+1 & \geq f(2,2^{i+1})+1 \\ & = \frac{3}{2}i+\frac{9}{2} \\ & > \frac{i}{2}+4, \end{aligned}$$

the theorem holds for this case.

Case 3.3: The remaining set $X - Y$ has only two active users.

In this case, $n = 2^{i+1}$ and the total number of calls to *feedback()* is $\frac{i}{2} + 4$. The proof is the same as above.

Case 3.4: The remaining set $X - Y$ has more than two users and 2^{+} active users.

The procedure *Dig(X)* then finds at least 2 active users in at most $(i+1) - 1 = i$ calls to *feedback()*. Thus there are at most $\frac{3}{2}i + 4$ calls to *feedback()* , in which at least 3 active users are identified and at least 2^{i+1} users accounted for. Again, let n' be the number of remaining users and d' the number of active users among them, then $n' \leq n - 2^{i+1}$ and $d' \leq d - 3$. Since for $i \geq 2$

$$\begin{aligned}
& f(3, 2^{i+1}) - \left(\frac{3}{2}i + 4\right) \\
= \; & \frac{3}{4}i + \frac{7}{2} - \frac{9}{4}\log 3 \\
\geq \; & \frac{3}{2} + \frac{7}{2} - \frac{9}{4}\log 3 \\
= \; & 5 - \frac{9}{4}\log 3 \\
> \; & 5 - \frac{9}{4} \cdot 2 \\
= \; & \frac{1}{2} \\
> \; & 0,
\end{aligned}$$

the total number of calls to *feedback()* is at most

$$\begin{aligned}
& \frac{3}{2}i + 4 + M_G(d'|n') \\
\leq \; & f\left(3, 2^{i+1}\right) + M_G(d'|n') \\
\leq \; & f\left(2, 2^{i+1}\right) + f\left(d - 3, n - 2^{i+1}\right) + 1 \\
\leq \; & f(d, n) + 1.
\end{aligned}$$

Case 4: $|X| > 2^i$ and the first call to *feedback()* in procedure *Big-Test(X)* finds a set Y of size 2^i.

The procedure *Dig(X)* then finds at least 2 active users in at most $(i+1) - 1 = i$ calls to *feedback()*. There are at most $\frac{3}{2}i + 3$ calls to *feedback()* and at least 2^{i+1} users accounted for. Again, let n' be the number of remaining users and d' the number of active users among them, then $n' \leq n - 2^{i+1}$ and $d' \leq d - 2$. Since

$$f(2, 2^{i+1}) - \left(\frac{3}{2}i + 3\right) = \frac{1}{2} > 0,$$

the total number of calls to *feedback()* is at most

$$\begin{aligned} & \frac{3}{2}i + 3 + M_G\left(d'|n'\right) \\ \leq \; & f\left(2, 2^{i+1}\right) + M_G\left(d'|n'\right) \\ \leq \; & f\left(2, 2^{i+1}\right) + f\left(d-2, n-2^{i+1}\right) + 1 \\ \leq \; & f\left(d, n\right) + 1. \Box \end{aligned}$$

For Algorithm G, for every fixed d, as $n \to \infty$

$$M_G\left(d|n\right) \approx \frac{3}{4} d \log n.$$

Capetanakis [1] gave a tree algorithm, C, for the multi-access channel problem for $n = 2^t$ users. For this algorithm

$$M_C\left(d|n\right) = 2\lfloor \frac{d}{2} \rfloor \left(\log n - \left(\lceil \log d \rceil - 1\right)\right) + 2^{\lceil \log d \rceil} - 1.$$

Thus for every fixed d, as $n \to \infty$

$$M_C\left(d|n\right) \approx d \log n.$$

So algorithm G is asymptotically better than algorithm C.

4. Conclusion

In this paper, we have studied the multi-access channel problem. Due to the fact that the number of active users is unknown, the optimal min-max solution is very difficult to compute. The previous best known algorithm takes around $d \log n$ queries asymptotically to schedule the transmission of d active users among n total users. In this paper, we have a given algorithm which takes $\frac{3}{4} d \log n$ queries asymptotically to schedule the transmission of d active users among n total users. Therefore, it could in all likelihood save about twenty-five percent queries from the previous multi-access control algorithm.

References

1. J. I. Capetanakis, *Tree algorithms for packet broadcast channels,* IEEE Trans. Inform. Theory **25** (1979), 505-515.
2. X. M. Chang and F. K. Hwang, *The minimax number of calls for finite population multi-access channels,* in Computer Networking and Performance Evaluation, Ed: T. Hasegawa, H. Takagi and Y. Takahasi, Elsevier, Amsterdam, 1986, 381-388.

3. D.-Z. Du and P. M. Pardalos (Eds.), *Minimax and Applications,* Kluwer Academic Publishers, 1995.
4. D.-Z. Du and H. Park, *On Competitive Group Testing,* SIAM J. Computing **23**, 5, (1994) 1019-1025.
5. D.-Z. Du and F.K. Hwang, *Combinatorial Group Testing and Its Applications,* World Scientific, Singapore, 1993.
6. J. Komlos and A. G. Greenberg, *An asymptotically fast nonadaptive algorithm for conflict resolution in multiaccess channels,* IEEE Trans. Inform. Theory **31** (1985), 302-306.
7. B. S. Tsybakov and V. A. Mikhailov, *Free synchronous packet access in a broadcast channel with feedback,* Probl. Inform. Transm. **14** (1978), 259-280.

Author Index

Nonconvex Optimization and Its Applications

1. D.-Z. Du and J. Sun (eds.): *Advances in Optimization and Approximation*. 1994. ISBN 0-7923-2785-3
2. R. Horst and P.M. Pardalos (eds.): *Handbook of Global Optimization*. 1995 ISBN 0-7923-3120-6
3. R. Horst, P.M. Pardalos and N.V. Thoai: *Introduction to Global Optimization* 1995 ISBN 0-7923-3556-2; Pb 0-7923-3557-0
4. D.-Z. Du and P.M. Pardalos (eds.): *Minimax and Applications*. 1995 ISBN 0-7923-3615-1
5. P.M. Pardalos, Y. Siskos and C. Zopounidis (eds.): *Advances in Multicriteria Analysis*. 1995 ISBN 0-7923-3671-2
6. J.D. Pintér: *Global Optimization in Action*. Continuous and Lipschitz Optimization: Algorithms, Implementations and Applications. 1996 ISBN 0-7923-3757-3
7. C.A. Floudas and P.M. Pardalos (eds.): *State of the Art in Global Optimization*. Computational Methods and Applications. 1996 ISBN 0-7923-3838-3
8. J.L. Higle and S. Sen: *Stochastic Decomposition*. A Statistical Method for Large Scale Stochastic Linear Programming. 1996 ISBN 0-7923-3840-5
9. I.E. Grossmann (ed.): *Global Optimization in Engineering Design*. 1996 ISBN 0-7923-3881-2
10. V.F. Dem'yanov, G.E. Stavroulakis, L.N. Polyakova and P.D. Panagiotopoulos: *Quasidifferentiability and Nonsmooth Modelling in Mechanics, Engineering and Economics*. 1996 ISBN 0-7923-4093-0
11. B. Mirkin: *Mathematical Classification and Clustering*. 1996 ISBN 0-7923-4159-7
12. B. Roy: *Multicriteria Methodology for Decision Aiding*. 1996 ISBN 0-7923-4166-X
13. R.B. Kearfott: *Rigorous Global Search: Continuous Problems*. 1996 ISBN 0-7923-4238-0
14. P. Kouvelis and G. Yu: *Robust Discrete Optimization and Its Applications*. 1997 ISBN 0-7923-4291-7
15. H. Konno, P.T. Thach and H. Tuy: *Optimization on Low Rank Nonconvex Structures*. 1997 ISBN 0-7923-4308-5
16. M. Hajdu: *Network Scheduling Techniques for Construction Project Management*. 1997 ISBN 0-7923-4309-3
17. J. Mockus, W. Eddy, A. Mockus, L. Mockus and G. Reklaitis: *Bayesian Heuristic Approach to Discrete and Global Optimization*. Algorithms, Visualization, Software, and Applications. 1997 ISBN 0-7923-4327-1
18. I.M. Bomze, T. Csendes, R. Horst and P.M. Pardalos (eds.): *Developments in Global Optimization*. 1997 ISBN 0-7923-4351-4
19. T. Rapcsák: Smooth Nonlinear Optimization in R^n. 1997 ISBN 0-7923-4680-7
20. A. Migdalas, P.M. Pardalos and P. Värbrand (eds.): *Multilevel Optimization: Algorithms and Applications*. 1998 ISBN 0-7923-4693-9
21. E.S. Mistakidis and G.E. Stavroulakis: *Nonconvex Optimization in Mechanics*. Algorithms, Heuristics and Engineering Applications by the F.E.M. 1998 ISBN 0-7923-4812-5

Nonconvex Optimization and Its Applications

22. H. Tuy: *Convex Analysis and Global Optimization.* 1998 ISBN 0-7923-4818-4
23. D. Cieslik: *Steiner Minimal Trees.* 1998 ISBN 0-7923-4983-0
24. N.Z. Shor: *Nondifferentiable Optimization and Polynomial Problems.* 1998 ISBN 0-7923-4997-0
25. R. Reemtsen and J.-J. Rückmann (eds.): *Semi-Infinite Programming.* 1998 ISBN 0-7923-5054-5
26. B. Ricceri and S. Simons (eds.): *Minimax Theory and Applications.* 1998 ISBN 0-7923-5064-2

KLUWER ACADEMIC PUBLISHERS – DORDRECHT / BOSTON / LONDON

9 780792 350644

GPSR Compliance

The European Union's (EU) General Product Safety Regulation (GPSR) is a set of rules that requires consumer products to be safe and our obligations to ensure this.

If you have any concerns about our products, you can contact us on ProductSafety@springernature.com

In case Publisher is established outside the EU, the EU authorized representative is:

Springer Nature Customer Service Center GmbH
Europaplatz 3
69115 Heidelberg, Germany

Batch number: 10584469

Printed by Printforce, the Netherlands